Wan Chan
Oct '76
Berkeley, Ca

Studies in Dynamical Systems

During the last twenty years there has been a progressive increase in the complexity and degree of interconnection of systems of all kinds. The reasons are clear: recent progress in communication, data processing, and control have made possible a much greater degree of coordination between the parts of a system than was ever possible before. Digital computers have played a central part in this evolution, and are likely to play a still greater part in the future.

In the course of this development the traditional engineering disciplines have tended to polarize. At one extreme is the detailed consideration of devices, drawing heavily on applied physics and chemistry. At the other extreme is the consideration of aggregates of devices and their interaction and interrelation. The distinction is not always easy to maintain, because what is called a system in one context may become an element when still larger systems are considered. Nevertheless the two kinds of approach are sufficiently clear and each has its special techniques.

It is the aim of this series to give an account of the theory and practice involved in designing dynamical engineering systems. By this we imply an interest in deterministic and stochastic behaviour, control, optimization, and synthesis. Our effort has been to assemble a series which is oriented towards problem solving in these areas, but also presents a sound conceptual framework with an eye towards generalization.

The level of presentation will depend to some extent on the topic, but will generally cover the range from introductory undergraduate to first-year postgraduate courses. The success of the series must depend ultimately on its acceptance both by students and by practising engineers.

Roger W. Brockett
Harvard University
H. H. Rosenbrock
University of Manchester Institute of Science and Technology

Studies in Dynamical Systems

Editors

Roger W. Brockett, *Harvard University*

H. H. Rosenbrock, *University of Manchester Institute of Science and Technology*

H. H. ROSENBROCK and C. STOREY
Mathematics of Dynamical Systems

J. L. WILLEMS
Stability Theory of Dynamical Systems

Further titles in preparation

Mathematics of Dynamical Systems

H. H. Rosenbrock
University of Manchester Institute of Science and Technology

C. Storey
Loughborough University of Technology

Wiley Interscience Division
JOHN WILEY & SONS, INC., NEW YORK

SBN: 471 73650 3

Library of Congress Catalog Number: 78–113603

Made and printed in Great Britain by William Clowes and Sons, Limited
London and Beccles

And the ends of all these Inquiries they intend to be the Pleasure *of Contemplative minds, but above all, the* ease and dispatch *of the labours of mens hands.*

Robert Hooke, MICROGRAPHIA

Contents

6 Differential Equations

7 Complex Analysis

8 Laplace Transform and *z*-Transform

Preface

The scope of the series on dynamical systems, of which this is the first volume to appear, has already been defined in the Foreword. In this volume we have brought together the chief mathematical results which underlie and support this area of engineering. Because they come from widely diverse parts of mathematics, these results have not previously been available in a single volume. In bringing them together the opportunity has arisen of imposing a more coherent form, adapted particularly to the applications in view.

It seems to us important that when mathematics is presented to engineers it should be done in a way which sharpens their appreciation of sound mathematical argument. For this reason we have proved all the important results in a way which we believe will be acceptable to mathematicians. At the same time, the engineer's motivation springs from his desire to solve engineering problems. We have accordingly stressed the gap which the engineer has to bridge beween the behaviour of real systems and the axioms from which the mathematician starts his development. We have also in some places preferred the concrete development of a particular topic to a more abstract and general approach with only the single application in mind. The student can always advance to greater abstraction and generality at a later stage, if he feels the desire to do so.

To achieve these aims the scope of the results which are presented has had to be restricted. So we have given no general treatment, for example, of conditions for inverting the order of integration, and no proof of the formula for the inverse Laplace transform. On the other hand we have proved results, such as the uniqueness of the transform, which are essential in applications and are usually omitted. Probability theory is not dealt with here but will be introduced in the volume in which its applications are considered. In the same way the calculus of variations will be introduced in the volume on optimal control.

Our greatest departure from current practice is in the use we make of the Riemann–Stieltjes integral to give a sound yet simple treatment of convolution, transfer functions, and sampling. The treatment of matrices and their standard forms under transformations is more complete than usual. Free use has been made of the ideas of modern mathematics in order to give a framework for later and deeper study.

The level of presentation may appear higher at first sight than is in fact the case. The topics included do not reach any great degree of mathematical difficulty, though proofs often demand some expenditure of effort and concentration. If details of the proofs are omitted, or read only in outline, the book will take on a much simpler aspect. The details can then be studied on a second reading.

To achieve greater brevity we have regarded the examples and exercises as part of the text, and have referred to them as necessary. The exercises are that and no more: they are not intended to test the student but merely to give practice in filling in detail. The problems, on the other hand, are sometimes more demanding.

Our thanks are due to a number of colleagues for discussions and suggestions, and in particular to Mr. D. Bell who checked the manuscript in detail.

H.H.R.
C.S.

1 Elements

1 Sets

A *set* is a collection of arbitrary elements and can be specified by enumerating all the elements (when this is feasible) or by giving a rule which tells if an element is contained in the set or not. A set is *finite* or *infinite* according to whether it has a finite or infinite number of elements. The concept of element is regarded as being self evident. Synonyms for set are *collection*, *aggregate*, *family*, *class*, etc., although some of these are given special significance by some authors. For example, 'aggregate' is sometimes reserved for sets in which there is no ordering of the elements, and 'class' for sets having elements which are themselves sets.

Examples of finite sets are: the positive integers less than 100; the population of London; and all possible permutations of the numbers 1, 2, 3, 4. Some infinite sets are: the totality of odd numbers; the continuous functions of the real variable x for $0 \leqslant x \leqslant 1$; the square $n \times n$ matrices having as elements the real numbers; and the set of all circles with radii 2^n, $n = 0, 1, 2, \ldots$.

Usually, upper case italic letters will be used for sets, that is $A, B, C, \ldots$, and lower case italic letters $a, b, c, \ldots$ for elements. To signify that an element a is in a set A the notation $a \in A$ is used: the notation $a \notin A$ signifies that a is not a member of A. The set consisting of the finite number of elements, a, b, c will be written $\{a, b, c\}$ and the set of points x satisfying $0 \leqslant x \leqslant 1$ as $\{x \mid 0 \leqslant x \leqslant 1\}$, that is, the rule for specifying the elements of the set is given to the right of the bar. Obviously the set consisting of the positive integers $1, 2, 3, \ldots, 10$ could be written as

$$\{1, 2, 3, 4, 5, 6, 7, 8, 9, 10\}$$

or as

$$\{a \mid a \text{ integer and } 0 < a < 11\}$$

Clearly, a part of a set is itself a set; it is called a *subset* of the original set. Thus A is a subset of B (written $A \subseteq B$ or $B \supseteq A$) if every member of A is also a member of B. Notice that this does not preclude the possibility that each member of B is also a member of A, in which case A and B are identical, or $A = B$. If each element of A is a member of B, but B has elements not in A, then A is called a *proper* subset of B and we write $A \subset B$ or $B \supset A$.

1.1 Operations on sets

The *union* or *sum* of two sets A and B is defined as the set consisting of elements either in A or in B and is denoted by $A \cup B$. The *intersection* of A and B is the set of elements in both A and B and is denoted by $A \cap B$. If A and B have no elements in common then their intersection $A \cap B$ is a set without any elements. This set, called the *vacuous* (or *null*) set, is denoted by $\varnothing$; it is obviously a subset of all other sets. When $A \cap B = \varnothing$ then A and B are said to be *disjoint*.

If $A \subset B$ then the set of elements of B which are not also in A is called the *difference* between B and A and is written $B - A$. All sets are regarded as subsets of a set called the *universal set* U and the difference between U and one of its subsets A is called the complement of A with respect to U.

The ideas underlying the above definitions can be illustrated by the use of plane geometric figures, called Venn diagrams. Thus in Fig. 1(a) the total shaded area is $A \cup B$ and the double shaded area is $A \cap B$. In Fig. 1(b) the sets A and B

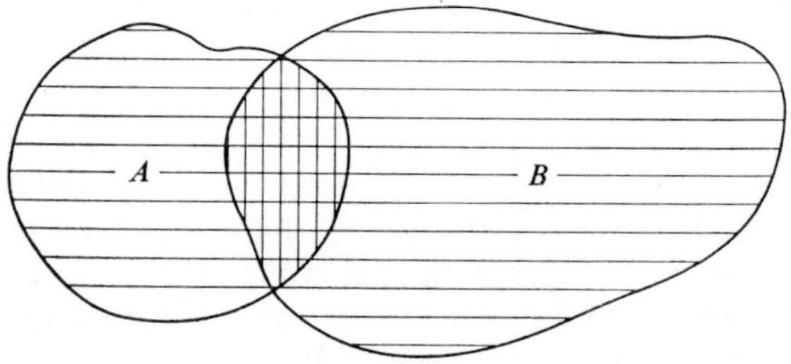

(a)

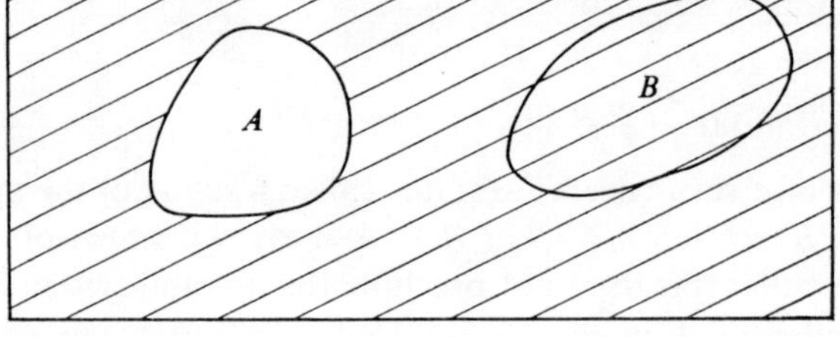

(b)

FIG. 1

are disjoint so that $A \cap B = \varnothing$ and the complement of A with respect to U (the rectangle) is shown shaded.

The concepts of union and intersection have a natural extension to more than two sets. The union $\cup A_i$ of the sets A_i, where $i = 1, 2, \ldots,$ n and n can be finite or infinite, is the set of elements which lie in at least one of the A_i. Similarly, the intersection $\cap A_i$ is the set of elements which lie in all of the A_i.

There are close connections between set-theoretic union and intersection on the one hand and the sum and product of ordinary algebra on the other. Thus

$$A \cup B = B \cup A \tag{1.1}$$

$$A \cup (B \cup C) = (A \cup B) \cup C \tag{1.2}$$

so that union is commutative and associative. For more details of these links the reader is referred to Apostol (1957) and Birkhoff and MacLane (1965).

The *cartesian product* (or sometimes just *product set*) of two sets A and B is the set of all ordered pairs (a, b) with the first entry an element of A and the second an element of B; the cartesian product of A and B is written $A \times B$. Two elements (a_1, b_1) and (a_2, b_2) in $A \times B$ are equal if and only if $a_1 = a_2$ and $b_1 = b_2$. This idea can be extended to the cartesian product $A \times B \times C \times \ldots$ of more than two sets by the consideration of ordered associations of elements $(a, b, c, \ldots)$ with $a \in A$, $b \in B$, $c \in C$, and so on. The study of sets is continued in Chapter 4.

2 Groups

The sets discussed in the previous section were examples of very general systems. Rather less general systems are obtained by considering sets together with operations between their elements. If with any pair of elements a, b of a set A there can be associated a third unique element $c \in A$ then A is said to be *closed under a binary operation*. Generally, the result of applying a binary operation to elements a and b is written ab so that $ab = c$. It must be remembered that ab does not necessarily mean the product $a \times b$, although ordinary multiplication is a valid binary operation for many sets, just as is ordinary addition. When confusion may arise a different notation can be used, such as $a \circ b = c$. More generally a set A is said to be closed under an n-ary operation (or rule of composition) when it is possible to associate with any subset of n elements $a, b, c, \ldots$ in A another unique element $f \in A$.

Example 2.1 Let S be the set of even numbers $\{2, 4, 6, \ldots\}$: then S is closed under the binary composition of ordinary arithmetic multiplication.

Example 2.2 The finite set $S = \{2, 4, 6, 8, 10\}$ is not closed under addition since $4 + 8 = 12 \notin S$.

A set A together with a single binary operation defined for its elements is called a *groupoid* with respect to this operation, provided that A is closed under the operation. It is obviously possible to carry out the binary operation repeatedly (just as powers can be formed in ordinary numerical multiplication), so that if $ab = c$ and $cd = e$ we have $(ab)d = e$. In a groupoid, however, there is no insistence that the law of composition is *associative* so that $(ab)d$ may not equal $a(bd)$. Similarly, in a groupoid the *commutative law* $ab = ba$ does not necessarily hold.

Example 2.3 The infinite set $S = \{2, 4, 6, \ldots\}$ under multiplication is a groupoid, and multiplication can be carried out repeatedly. Thus,

$$(2 \times 2) \times 2 = 4 \times 2 = 8 \in S$$
$$(2 \times 4) \times 10 = 8 \times 10 = 80 \in S$$

It is also easily verified that multiplication in S is associative [for example, $(2 \times 4) \times 6 = 2 \times (4 \times 6)$] and commutative (for example, $2 \times 4 = 4 \times 2$).

Example 2.4 Consider the finite set $\{a, b, c\}$ with multiplication defined by the table below

	a	*b*	*c*
a	*b*	*b*	*b*
b	*c*	*c*	*c*
c	*a*	*a*	*a*

This is a groupoid because it is closed under multiplication, but it is neither associative nor commutative. For example,

$$a(bc) = ac = b$$
$$(ab)c = bc = c$$

and

$$ab = b \neq ba = c$$

A groupoid in which the law of operation is associative is called a *semigroup*. If also the operation is commutative the semigroup is said to be *commutative* or *abelian*. An element i_l of a groupoid A is a *left identity* if $i_l b = b$ for all $b \in A$. Similarly, i_r is a *right identity* if $bi_r = b$ for all $b \in A$. It is easy to see that if a groupoid has a left identity and a right identity then the two are equal and unique so that there is a single 'two-sided' *identity*.

Example 2.5 In Example 2.1, S is an abelian semigroup which has neither a left nor a right identity. On the other hand, the set $\{0, 2, 4, \ldots\}$ with addition as a binary composition is an abelian semigroup with the element 0 as its identity.

Suppose that A is a groupoid with an identity i (by which we understand a two-sided identity). Then if an element $x_l \in A$ can be found such that $x_l b = i$ for some $b \in A$, x_l is called a *left inverse* of b. In the same way, a solution x_r of $bx_r = i$ is called a *right inverse* of b. If a left inverse x_l is also a right inverse of an element b, that is, if $x_l b = bx_l = i$, then x_l is called a 'two-sided' inverse or just an *inverse* and written b^{-1}.

Suppose now that S is a semigroup and an element b of S has a left inverse x_l and a right inverse x_r. It is easily shown using the associative law that $x_l = x_r = x$, say, and is the only inverse of b. If S has an identity i then the elements of S which have inverses are called the *units* of S. Not every element of a semigroup need be a unit, of course.

Example 2.6 The set $\{0, \pm 1, \pm 2, \ldots\}$ forms an abelian semigroup under addition, with every element a unit because

$$0 = n + (-n) = (-n) + n$$

We are now in a position to define a *group*. A group is a semigroup with identity in which every element is a unit. Thus a group is a set S closed under a binary operation for which the following properties hold:

$(ab)c = a(bc)$ for $a, b, c \in S$.
There is an element $i \in S$ such that $xi = ix = x$ for all $x \in S$.
For every element $x \in S$ there is an element $x^{-1} \in S$ such that $x^{-1}x = xx^{-1} = i$.

If also the binary operation is commutative the group is said to be *commutative* or *abelian.*

It is possible to define a group by a number of different sets of rules. One other such set is given below and the reader should convince himself that the two definitions are equivalent.

A set of elements with a binary operation is a group if the following conditions hold:

$a(bc) = (ab)c, \quad a, b, c \in S.$
For all $a, b \in S$ there exist elements $p, q \in S$ such that $ap = b$ and $qa = b$.

The number of elements in the group set S is called the *order* of the group: if S has a finite number of elements the group is called a finite group; otherwise it is said to be infinite.

Example 2.7 Consider the set S consisting of the three elements a, b, c with a binary operation defined by the multiplication table below:

	a	b	c
a	a	b	c
b	b	c	a
c	c	a	b

It is easily verified by inspection that S is closed under the binary operation and that the associative law holds. The element a is the identity since $ab = ba = b$, $ac = ca = c$. Each element is a unit, for $aa = a$, $bc = a$, $cb = a$. Notice also that this is a commutative finite group.

Example 2.8 Consider the set of all possible permutations of the three numbers 1, 2, 3. These are listed below:

$$a = \begin{pmatrix} 1 & 2 & 3 \\ 1 & 2 & 3 \end{pmatrix} \quad b = \begin{pmatrix} 1 & 2 & 3 \\ 2 & 1 & 3 \end{pmatrix} \quad c = \begin{pmatrix} 1 & 2 & 3 \\ 2 & 3 & 1 \end{pmatrix}$$

$$d = \begin{pmatrix} 1 & 2 & 3 \\ 1 & 3 & 2 \end{pmatrix} \quad e = \begin{pmatrix} 1 & 2 & 3 \\ 3 & 1 & 2 \end{pmatrix} \quad f = \begin{pmatrix} 1 & 2 & 3 \\ 3 & 2 & 1 \end{pmatrix}$$

where in each case the permutation is written beneath the original numbers. If two permutations are carried out consecutively the result is called the *product* of the two permutations. Since the product of two permutations is itself a unique permutation of the set it follows that product is a valid binary operation. To illustrate this, carry out b and then c: this gives

$$1 \rightarrow 2 \rightarrow 3$$
$$2 \rightarrow 1 \rightarrow 2$$
$$3 \rightarrow 3 \rightarrow 1$$

so that $cb = f$. In this way, it is possible to build up the full multiplication table:

	a	b	c	d	e	f
a	a	b	c	d	e	f
b	b	a	d	c	f	e
c	c	f	e	b	a	d
d	d	e	f	a	b	c
e	e	d	a	f	c	b
f	f	c	b	e	d	a

It is easy now to verify that the result is a group. That the associative law holds can be seen directly from the table or, in fact, from the definition of product. The permutation a which leaves the numbers unchanged is the identity and from the table it is apparent that each element has an inverse. Notice also that the alternative definition of a group is easy to verify in this case because each row and each column has all six permutations in it.

In contrast to Example 2.7, however, the commutative law does not hold for this group. For example, $bc = d \neq cb = f$, so that this is a finite non-commutative group. It is possible to extend these ideas to the $n!$ possible permutations of the n numbers $1, 2, \ldots, n$, the resulting group being known as the nth degree *symmetric group*.

Example 2.9 The set of all positive and negative integers is a group with respect to ordinary addition. The associative law $a + (b + c) = (a + b) + c$ is evident, the identity is obviously 0, and the inverse of an element a is $-a$. In general, where the binary operation of a group is addition it is known as an *additive group*, the identity is called the *zero*, and the inverse of an element is called its *negative*. Notice that the set of positive and negative integers under multiplication form a semigroup but not a group, because the only two units are ± 1.

Example 2.10 If zero is excluded, the set of real numbers with ordinary multiplication as the law of composition forms a group. In this case the identity is 1 and the inverse of any element a is its reciprocal $1/a$. The associative law $a(bc) = (ab)c$ is again obvious.

When each element of a set A is associated with a unique element of a set B and each element of B is associated with a unique element of A, there is said to be a *one-to-one correspondence* between the elements of A and those of B. More will be said about such correspondences in Section 8. When the two sets are the same the one-to-one correspondence is said to be a *one-to-one transformation* of the set into itself.

Example 2.11 Consider the set of all one-to-one transformations $\{t_1, t_2, \ldots\}$ of a set A into itself, with the 'product' of two transformations (that is, the result of carrying them out consecutively) as a binary composition. That the associative law holds is seen from the following. Let a be any element of A; then

$$[(t_i t_j)t_k]a = (t_i t_j)(t_k a) = t_i[t_j(t_k a)]$$

and

$$[t_i(t_j t_k)]a = t_i[(t_j t_k)a] = t_i[t_j(t_k a)]$$

showing that the effect of $t_i(t_j t_k)$ on any element is the same as that of $(t_i t_j)t_k$.

The transformation which leaves all elements unchanged is obviously an identity. Since the transformations are one-to-one, each of the t_i has as an inverse the transformation t_i^{-1} which associates each element with the one from which it arose, because $t_i^{-1}t_i$ is the identity.

Thus the set of all transformations of a set into itself is a group known as a *transformation group*. Such groups are closely connected with geometry (see, for example, Benson, 1966). The symmetric groups of Example 2.8 are, of course, transformation groups.

EXERCISE 2.1 Prove that in any finite group there must be an integer $n > 0$ for which $a^n = e$, where a is any element of the group and e is the identity.†

2.1 Group theory

A brief introduction is now given to some of the concepts of group theory. For fuller details the reader is referred to Jacobson (1951) and Birkhoff and MacLane (1965).

Two groups T and S are said to be *isomorphic* to each other if it is possible to set up a one-to-one correspondence between their elements in such a way that if $a, b \in S$ correspond to $c, d \in T$ then ab corresponds to cd. If T and S are isomorphic we write $T \sim S$.

Isomorphism between groups is a relation of a type known as an *equivalence relation*. If a quite general set A with elements $a, b, c, \ldots$ and a relationship R are given and it is possible to say whether the relationship R holds or does not hold for any two elements of A, then R is said to be an equivalence relation if it has the following properties:

$a\,R\,a$ for any $a \in A$.
If $a\,R\,b$ then $b\,R\,a$ for any $a, b \in A$.
If $a\,R\,b$ and $b\,R\,c$ then $a\,R\,c$ for any $a, b, c \in A$.

These are called the *reflexive*, *symmetric*, and *transitive* properties respectively.

Two groups that are isomorphic can be considered as identical from an abstract standpoint even though their elements and law of composition are quite different. An interesting fact that we note without proof is that any group is isomorphic to a transformation group (Jacobson, 1951). If the condition that the correspondence which determines an isomorphism should be one-to-one is relaxed so that it becomes many-to-one, the relationship is known as a *homomorphism*.

Example 2.12 Two numbers a and b are said to be *congruent modulo m* if they differ from each other by an integral multiple of m. This relationship is denoted by $a = b(\text{mod } m)$. It is obvious that $a = a(\text{mod } m)$, Also if $a = b(\text{mod } m)$ then $a - b = rm$ for some integer r, so that $b - a = (-r)m$, which implies $b = a(\text{mod } m)$. Again, if $a = b(\text{mod } m)$ and $b = c(\text{mod } m)$ then $a - b = rm$ and $b - c = sm$, so that $a - c = (r + s)m$, implying $a = c(\text{mod } m)$. Thus the relationship 'congruence modulo m' is an equivalence relationship.

EXERCISE 2.2 Prove that the (positive, negative, and zero) integers modulo 3 form an additive group, homomorphic to the additive group of integers.

† The symbol e is often used for the group identity when the group operation is unspecified. If the group operation is addition, the identity is almost always written 0, while when the group operation is multiplication the identity is often written 1, or I if the group elements are matrices.

Suppose that S is a semigroup with an identity e and that G is the subset consisting of the units of S. Let a and b be any two elements of G: then a^{-1} and b^{-1} exist and are in G since they themselves are units (their inverses are a and b). The identity $e \in G$ since it is a unit, that is, $ee = e$. Also the subset G is closed under the law of composition of the semigroup since any product ab is a unit with inverse $b^{-1}a^{-1}$, because $abb^{-1}a^{-1} = e = b^{-1}a^{-1}ab$. The subset of units of a semigroup with identity therefore forms a group under the binary operation of the semigroup and with the semigroup identity as the group identity. Such a group G is called the *group of units* of S and is an important source of groups. As a very simple example we mention the group of units $+1$, -1 of the semigroup of integers under multiplication.

We are now naturally led to inquire what are the conditions for a non-vacuous subset of a group to be itself a group? One such set of conditions is as follows: a set $H \subseteq G$ is a group under the binary operation in G if and only if it is closed under this operation and the inverse h^{-1} of any element $h \in H$ is also in H. That H is a group is clear, because if $h \in H$ then $h^{-1} \in H$ and by the closure property $hh^{-1} = e \in H$. Thus H contains the identity of G and the other group conditions hold by hypothesis. On the other hand if H is a group it is closed by definition, the identity e' of H must also be the identity e of G (because $e'e' = e' = ee'$) and so the unique inverse of any element $h \in H$ is also in H. An equivalent set of conditions for a non-vacuous set H to be a subgroup of G is that H is a subgroup if and only if, for any pair of elements h_1, h_2 in H, $h_1h_2^{-1} \in H$.

EXERCISE 2.3 Verify the truth of the last statement.

The identity e of a group G, and G itself, are obviously subgroups of G according to the definitions; subgroups of G other than these are known as *proper subgroups*. Notice that if the order of the group is finite any subset of G closed under the binary operation of G is a subgroup. This follows from the fact (Exercise 2.1) that in a finite group there must be some power a^n of any element a such that $aa^{n-1} = a^n = e$, and therefore $a^{n-1} = a^{-1}$, showing that any (non-vacuous) closed subset of a finite group contains the inverse of each of its elements.

3 Rings†

So far we have specialized the concept of a set by introducing a binary operation between its elements. A further step in this direction is to consider a set together with more than one law of composition between its elements. In particular, a structure consisting of a set R with two binary operations which we call multiplication and addition (although the operations need not, of course, be

† The definition of a ring can vary according to different authors. Some authors do not require multiplication to be associative and others insist that every ring has an identity.

multiplication or addition in the usual sense) is known as a *ring* if the following conditions hold:

R is a commutative group under addition.
R is a semigroup under multiplication.
Multiplication is distributive with respect to addition, that is, given any three elements $a, b, c \in R$, then $a(b + c) = ab + ac$ and $(b + c)a = ba + ca$.

If the operation of multiplication is commutative then R is said to be a *commutative ring* and the two distributive laws in the third condition above become identical. If multiplication has an identity, which we denote by 1, then R is said to be a *ring with identity* and the identity must be unique (R under multiplication is a semigroup). That a ring need not necessarily have a multiplicative identity is easily seen by considering the set of even numbers.

An interesting result which can be proved directly from the definition of a ring is that the product of any two elements is zero when one of them is zero; the converse of this is not necessarily true, however. Any element $a \in R$, for which the equation $ab = 0$ (where 0 is the identity of the additive group) has a solution $b \neq 0$, is called a *left divisor of zero*: a *right divisor of zero* is defined similarly. Obviously 0 is itself both a left and right divisor of zero. Rings with no divisors of zero other than 0 and for which the multiplication is commutative are called *integral domains*. For example, if the elements of any ring are numbers (integers, rational, real, or complex) and the operations are ordinary addition and multiplication, the ring is an integral domain. It follows directly from our definition that an integral domain is a ring in which the non-zero elements are closed under multiplication so that these elements form a sub-semigroup of the multiplicative semigroup. Notice that in an integral domain the 'cancellation rule' holds; that is, if $ab = ac$ and $a \neq 0$ then $b = c$. This is easily seen because if $a(b - c) = 0$ and $a \neq 0$ then $b - c = 0$.

Since the set R of a ring need not necessarily form a group under the multiplicative operation, the elements of R need not have inverses with respect to multiplication. There may not even be a multiplicative identity. If, however, we consider a ring with identity† then any element $a \in R$ which has an inverse $a^{-1} \in R$ is said to be a *regular* element of R. Obviously, the set of regular elements of R forms the subgroup of units of the multiplicative semigroup. We now define a *division ring* as a ring with identity with all of its non-zero elements regular. Thus the set of real numbers under multiplication and addition is a division ring; but the set of integers is not because the group of units is just ± 1 in this case.

Example 3.1 If multiplication and addition have their ordinary meanings, the set $\{0, \pm 2, \pm 4, \ldots\}$ is a ring. The additive identity is 0 and there are obviously

† It is assumed that the multiplicative identity 1 is not the same as the additive identity 0, otherwise R must be the single element 0: compare the set of integers modulo the integer 1.

no divisors of zero. Multiplication is commutative so that the ring is an integral domain. There is no multiplicative identity and hence there are no regular elements.

Example 3.2 The set of real numbers $\mathscr{R}$ forms a ring under addition and multiplication. The additive identity is 0 and the multiplicative identity is 1. Every element (except 0) obviously has an inverse and so $\mathscr{R}$ is a division ring.

Example 3.3 The set of all integers forms a ring. The additive identity is 0 and the multiplicative identity is 1. Multiplication is commutative and there are no divisors of zero other than zero; hence it is an integral domain and the cancellation rule holds. The equation $a.b = b.a = 1$ has no solutions other than $a = b = \pm 1$; hence the only regular elements in the ring are ± 1.

If a and b are two elements in an integral domain I then a is said to *divide* b if another element $c \in I$ can be found such that $ac = ca = b$. This relationship is also written $c = b/a$. A regular element of I will divide any element $b \in I$ since it is possible to take $c = a^{-1}b = ba^{-1}$. (Of course, a regular element is just an element that will divide the multiplicative identity.)

On the other hand a non-zero element p of I is called *prime* if:

p is not regular;
when a divides p, then a is regular or is of the form bp, where b is regular.

Example 3.4 In Example 3.3 the prime elements are just the 'prime numbers' of arithmetic: for example the only numbers which will divide 13 are ± 1 or ± 13. Also, 13 is certainly not regular.

EXERCISE 3.1 Prove that regular elements of an integral domain are divisible only by regular elements.

4 Fields

A *field* is a commutative division ring. For convenience, this definition will be restated in terms of elements, and a number of properties of fields will be derived.

A field F is a set of elements together with two binary operations which are called addition and multiplication for which the following conditions hold:

1. For any pair of elements $a, b \in F$ there is a unique sum $a + b$ in F and a unique product ab in F.
2. For any three elements $a, b, c \in F$

$$\left.\begin{aligned} a + (b + c) &= (a + b) + c \\ a(bc) &= (ab)c \end{aligned}\right\} \text{associative laws}$$

$$\left.\begin{aligned} a + b &= b + a \\ ab &= ba \end{aligned}\right\}\text{ commutative laws}$$

$$a(b + c) = ab + ac \qquad \text{distributive law}$$

3. There exists an element $0 \in F$ such that $a + 0 = 0 + a = a$ for any $a \in F$, and an element $-a \in F$ such that $a + (-a) = -a + a = 0$ for any $a \in F$: $a + (-a)$ is written $a - a$.
4. There exists an element $1 \in F\ (1 \neq 0)$ such that $1a = a1$ for any $a \in F$, and an element $a^{-1} \in F$ such that $aa^{-1} = a^{-1}a = 1$, provided $a \neq 0$.

An important property of a division ring, and therefore of a field, which is not necessarily true for an integral domain is that the equations $ax = b$ and $ya = b$, with $a \neq 0$, have the unique solutions $x = a^{-1}b$, $y = ba^{-1}$ (compare the alternative definition of a group given previously). In a field, because of the commutative rule for multiplication, $x = y$. The unique solution x of the equation $ax = b$ is then called the quotient of b by a. Thus division is always possible in a field. Notice also that a field has no divisors of zero.

Exercise 4.1 Verify that the usual rules for division of numbers are satisfied (Birkhoff and MacLane, 1965).

Exercise 4.2 Show that the integers modulo 3 (Exercise 2.2) form a field.

Example 4.1 The set of real numbers $\mathscr{R}$ has already been used as an example of a ring and the reader will easily verify that $\mathscr{R}$ is a field. (It has been assumed, of course, that the real numbers exist and can be combined by the usual rules of arithmetic.)

The real numbers have the further property that there exists a relationship ($\leqslant$) between pairs of elements of $\mathscr{R}$ which satisfies the following axioms:†

1. If $a, b, c \in \mathscr{R}$, then $a \leqslant b$ and $b \leqslant c \Rightarrow a \leqslant c$.
2. If $a \leqslant b$ and $b \leqslant a$ then $a = b$.
3. Either $a \leqslant b$ or $b \leqslant a$.
4. $a \leqslant b \Rightarrow a + c \leqslant b + c$.
5. If $0 \leqslant a$ and $0 \leqslant b$ then $0 \leqslant ab$.
6. If $0 < a$ and $0 \leqslant b$ there exists an integer n such that $b \leqslant na$, where $0 < a$ has been used to indicate $0 \leqslant a$ and $0 \neq a$.
7. The intersection of any sequence of closed intervals (Chapter 4, Section 3) $\{[a_n, b_n]\}$ with $a_n \leqslant a_{n+1}$ and $b_{n+1} \leqslant b_n$ for all n, is non-vacuous.

The *absolute value* or *modulus* of a real number a is defined as $(a^2)^{1/2} \geqslant 0$ and is written $|a|$.

† The symbol $\Rightarrow$ means that what precedes implies what follows.

Example 4.2 (Complex numbers) Consider the set $\mathscr{C}$ of ordered pairs (x, y) of real numbers with binary operations $+$ and $\times$ defined as follows:

$$(x, y) + (r, s) = (x + r, y + s) \tag{4.1}$$

$$(x, y) \times (r, s) = (xr - ys, xs + yr) \tag{4.2}$$

Clearly, $\mathscr{C}$ together with $+$ and $\times$ is a field, with $(0, 0)$ the identity of the additive group and $(1, 0)$ the identity of the multiplicative group; in accordance with previous notation the element $(0, 0)$ of $\mathscr{C}$ may be written alternatively as 0, and $(1, 0)$ may be written as 1. The inverse of any nonzero element $z = (x, y)$ is given by

$$\left(\frac{x}{x^2 + y^2}, \frac{-y}{x^2 + y^2}\right)$$

since

$$(x, y)\left(\frac{x}{x^2 + y^2}, \frac{-y}{x^2 + y^2}\right) = \left(\frac{x^2}{x^2 + y^2} + \frac{y^2}{x^2 + y^2}, \frac{-xy}{x^2 + y^2} + \frac{xy}{x^2 + y^2}\right)$$
$$= (1, 0)$$

Any element $z \in \mathscr{C}$ is called a *complex number* and $\mathscr{C}$ is called the *field of complex numbers*.

The element $(0, 1)$ is of particular importance and is denoted by i. From equations (4.1) and (4.2) it follows that any element $(x, y) = z \in \mathscr{C}$ can be written

$$(1, 0)x + (0, 1)y = 1x + iy$$

where in the last expression $1, i \in \mathscr{C}$, $x, y \in \mathscr{R}$. The expression $1x + iy$ is usually abbreviated to $x + iy$. When z is written in this form x is called its *real part* (Re $\{z\}$) and y its *imaginary part* (Im $\{z\}$). When y is zero (x, y) is real and when x is zero (x, y) is said to be *purely imaginary*. The element i has the property $i \times i = i^2 = (0, 1)(0, 1) = (-1, 0) = -1$.

If $z = (x, y) = x + iy \in \mathscr{C}$ then the complex number $\bar{z} = (x, -y) = x - iy \in \mathscr{C}$ is called the *conjugate* of z. It is quickly verified that $\bar{\bar{z}} = z$ and that

$$z \times \bar{z} = z\bar{z} = (x + iy)(x - iy) = x^2 + y^2$$

The real number $\sqrt{(z\bar{z})} = \sqrt{(x^2 + y^2)} \geqslant 0$ is defined to be the *absolute value* or *modulus* of z and coincides with the definition of absolute value of a real number when z is real; it is also written $|z|$. Complex numbers are studied in greater detail in Chapter 7.

A subset of F is said to be a *subfield* of F if it is a field under the operations of addition and multiplication in F. A similar definition applies to a *subdomain* of an integral domain or a *subring* of a ring. Thus a subset of F is a subfield of F if it is closed under addition and multiplication and contains the additive inverse and multiplicative inverse of each of its elements. If I is a subring of a ring R and for any element $a \in I$, $ar \in I(ra \in I)$ for every element $r \in R$, then I is said to be a *right-hand* (*left-hand*) *ideal.* In a commutative ring, right- and left-hand ideals are the same.

EXERCISE 4.3 Prove that a commutative ring R with identity e is a field if and only if it has no ideals other than R itself and the zero of R. [For $r \in R$, re does not generate 0, so must generate R.]

Two rings (or integral domains, or fields) R_1 and R_2 are *homomorphic* if each element of R_1 can be made to correspond to a unique element of R_2 in such a way that the operations of addition and multiplication are preserved. (The correspondence may be many-to-one.) That is, if $a, b \in R_1$ correspond to $c, d \in R_2$ then $a + b$ corresponds to $c + d$ and ab to cd. Here we have used the same notation for the operations in the two rings but this should not lead to confusion as long as the reader remembers that they by no means need to be the same. If the correspondence is one-to-one then the rings are *isomorphic*, written $R_1 \sim R_2$.

EXERCISE 4.4 Prove that a homomorphism between two fields is either an isomorphism or a correspondence between one field and the zero element of the other.

Example 4.3 Consider the one-to-one mapping of $\mathscr{R}$ into $\mathscr{C}$ defined by $x \to (x, 0)$ (compare Section 8). The set of elements forms a subfield of $\mathscr{C}$. Also, if $x \to (x, 0)$, $y \to (y, 0)$ then from (4.1) and (4.2) above

$$x + y \to (x + y, 0) = (x, 0) + (y, 0)$$

and

$$xy \to (xy, 0) = (x, 0) \times (y, 0)$$

showing that addition and multiplication are preserved under the mapping. Thus the subfield of elements of $\mathscr{C}$ of the form $(x, 0)$ is isomorphic to $\mathscr{R}$.

5 Polynomial rings

The reader is no doubt familiar with algebraic expressions of the form

$$a_0 + a_1x + a_2x^2 + \cdots + a_nx^n$$

or

$$\sum_{k=0}^{n} a_kx^k \tag{5.1}$$

which are known as *polynomials* in the 'unknown' or 'indeterminate' x, the coefficients a_k being real or complex numbers. We give now a brief treatment of polynomials with coefficients in a general ring.

Let R be a ring with identity 1, and R_1 a subring of R containing 1. If x is an element of R with the property that $xa = ax$ (x is said to *commute* with a) for every $a \in R_1$, then it is possible to form the set $R_1[x]$ consisting of the elements $\sum_{k=0}^{n} a_k x^k$ with $a_k \in R_1$ for $k = 1, 2, \ldots, n$ and $a_n \neq 0$. Each element of the form $\sum_{k=0}^{n} a_k x^k$ is called a polynomial in x with coefficients in R_1. The *degree* of a polynomial is defined as n and a_n is called the *leading coefficient*. A single non-zero element of R_1 is obviously a polynomial of degree 0 and the zero of R_1 is also admitted as a polynomial (with degree undefined). When the leading coefficient of a polynomial is 1 the polynomial is said to be *monic*.

Using the operations of addition and multiplication defined in R we see that if $\sum_{k=0}^{n} a_k x^k$ and $\sum_{k=0}^{m} b_k x^k$ are any two polynomials (assume $m \geqslant n$, say) then their sum is

$$\sum_{k=0}^{n} (a_k + b_k)x^k + \sum_{k=n+1}^{m} b_k x^k \tag{5.2}$$

and their product is

$$\sum_{k=0}^{n+m} c_k x^k \tag{5.3}$$

with

$$c_k = \sum_{r+s=k} a_r b_s \tag{5.4}$$

The negative (additive inverse) of the polynomial $\sum_{k=1}^{n} a_k x^k$ is $\sum_{k=1}^{n} (-a_k)x^k$. Thus we have shown that $R_1[x]$ is a subring of R: it is called the ring of polynomials in x over R_1, and the association of x with R_1 is known as adjoining x to R_1. It is obvious that if R_1 is commutative then so is $R_1[x]$.

If now the condition is imposed on the element x that the equation $\sum_{k=0}^{n} a_k x^k = 0$ implies $a_k = 0, k = 1, 2, \ldots, n$, then x is said to be an *indeterminate* over R_1 or to be *transcendental* with respect to R_1. Notice that if x is transcendental with respect to R_1 it cannot be an element b of R_1, because then it would satisfy $x - b = 0$. If x is not transcendental with respect to R_1 it is said to be *algebraic* with respect to R_1. An important property of a polynomial ring $R_1[x]$ where x is an indeterminate over R_1 is that the representation of any polynomial is unique. For if $\sum_{k=1}^{n} a_k x^k = \sum_{k=1}^{m} b_k x^k$, with $m \geqslant n$, then $\sum_{k=1}^{n} (a_k - b_k)x^k - \sum_{k=n+1}^{m} b_k x^k = 0$ so that $a_k = b_k$, $k = 1, 2, \ldots, n$, $b_k = 0$, $k = n + 1, n + 2, \ldots, m$. We notice that if R_1 is any ring there is always a ring R which has R_1 as a subring and contains an element x which is transcendental over R_1. (For a proof see Mostow (1963) or Jacobson (1951).)

In the case where the ring R_1 is an integral domain I then so is $I[x]$. To show this it is only necessary to prove that $I[x]$ has no non-zero divisors of zero, since we have already seen that $I[x]$ is commutative. If $a(x) = \sum_{k=1}^{n} a_k x^k$ with degree n and $b(x) = \sum_{k=1}^{m} b_k x^k$ with degree m are any two polynomials in $I[x]$, then their product $a(x)b(x) = \sum_{k=1}^{m+n} c_k x^k$, with $c_{m+n} = a_n b_m \neq 0$ since I is an integral domain. It follows that $a(x)b(x)$ is a non-zero polynomial of degree $n + m$ which shows that $I[x]$ has no divisors of zero. If, however, R_1 is not an integral domain then the leading coefficient $a_n b_m$ of the product $a(x)b(x)$ may be zero and the best we can say is that $\deg a(x)b(x) \leqslant \deg a(x) + \deg b(x)$. Should a_n or b_n be a regular element of R_1 the equality holds, since if say a_n is regular then a_n^{-1} exists and $a_n^{-1}(a_n b_m) = b_m \neq 0$ by hypothesis. It also follows directly from the definition of a polynomial that the degree of the sum of two polynomials of degrees n and m does not exceed the greater of n, m.

Let $I[x]$ be a polynomial ring in the indeterminate x over the integral domain I. If a polynomial $a(x)$ is regular in $I[x]$ then it must have an inverse $a^{-1}(x)$ such that $a(x)a^{-1}(x) = 1$, the identity in I. Since, however, the degree of 1 (considered as a polynomial) is 0, it follows from the previous paragraph that $\deg a(x) = \deg a^{-1}(x) = 0$. Thus the regular elements of $I[x]$ are the regular elements of I and, in particular, if I is a field F then the regular elements of $F[x]$ are the non-zero elements of F. As a direct consequence we see that the ring of polynomials $F[x]$ over a field F cannot itself be a field; it is an integral domain, however, by a result given above since every field F is an integral domain.

If a and b are any two elements in a field F and $b \neq 0$ then the element ab^{-1} exists and is the quotient of a by b. It can be shown that if I is any integral domain then there is a field F which contains I as a subring and every element of F can be expressed as a quotient of two elements of I. (See, for example, Jacobson, 1951, p. 87.) The field F is called the *quotient* (or *fraction*) field of I and is unique in the sense that any two quotient fields of the same integral domain must be isomorphic. This process is known as embedding the domain I in the field F. It is clear that the field of rational numbers is just the quotient field of the integers.

Now recall that the ring of polynomials $F[x]$ over a field F is not itself a field but an integral domain. It is possible, however, to form the quotient field of $F[x]$; this consists of elements which are written $a(x)/b(x)$, where $a(x)$ and $b(x)$ are polynomials, and is called the field of *rational forms* in the indeterminate (or variable) x over the base field F. The field of rational forms is denoted by $F(x)$ and is of particular importance in the theory of dynamical systems. When x is an element of the field F then $a(x)/b(x)$ is also an element of F, provided $b(x) \neq 0$.

6 Factorization of polynomials

Consider the domain of polynomials $F[x]$ over the field F. If x is replaced by $f \in F$ in the polynomial $a(x) = \sum_{k=1}^{n} a_k x^k$ then $a(f) = \sum_{k=1}^{n} a_k f^k$ is an element of F and is known as the *value* of $a(x)$ at $x = f$. If it happens that $a(f) = 0$ then f is said to be a *root* of the polynomial $a(x)$.

We have already seen that the regular (invertible) elements of $F[x]$ are the non-zero elements of F. The prime elements of $F[x]$ are called *irreducible*. The *greatest common divisor* (g.c.d.) of a set of elements $p_1, p_2, \ldots, p_n$ of $F[x]$ which are not all zero is an element d of $F[x]$ which divides $p_1, p_2, \ldots, p_n$ and is such that any element of $F[x]$ which divides $p_1, p_2, \ldots, p_n$ also divides d. It is easily seen that the greatest common divisor of a set of elements is unique up to multiplication by a non-zero element of F. A set of polynomials $p_1, p_2, \ldots, p_n$ having a greatest common divisor of degree zero are said to be *relatively prime*.

Theorem 6.1 Let $p_1(x)$ and $p_2(x)$ be any two non-zero elements of $F[x]$. Then there are two unique elements $g(x)$ and $r(x)$ in $F[x]$ such that

$$p_1(x) = p_2(x)g(x) + r(x) \tag{6.1}$$

with $r(x) = 0$ or $\deg r(x) < \deg p_2(x)$. The process for obtaining $g(x)$ and $r(x)$ (and for proving the theorem) is an inductive one known as the *division algorithm* and is, of course, related to ordinary long division.

Proof Assume that the result is true for $\deg p_1(x) \leqslant n$. Write $p_1(x) = \sum_{k=0}^{n+1} a_k x^k$ with $a_{n+1} \neq 0$, and let $p_2(x) = \sum_{k=0}^{m} b_k x^k$ with $b_m \neq 0$. Two possibilities arise: either $n + 1 < m$ or $n + 1 \geqslant m$. In the first case set $g(x) = 0$ and $r(x) = p_1(x)$: the result is then true for $\deg p_1(x) = n + 1$. In the second case it is obvious that the degree of

$$p_1(x) - \frac{a_{n+1}}{b_m} x^{n+1-m} p_2(x) \tag{6.2}$$

must be less than or equal to n and hence, by hypothesis, (6.2) is of the form $p_2(x)g(x) + r_1(x)$ with $r_1(x) = 0$ or $\deg r_1(x) < \deg p_2(x)$. Thus

$$\begin{aligned} p_1(x) &= \frac{a_{n+1}}{b_m} x^{n+1-m} p_2(x) + p_2(x)g(x) + r_1(x) \\ &= \left[\frac{a_{n+1}}{b_m} x^{n+1-m} + g(x)\right] p_2(x) + r_1(x) \end{aligned}$$

showing that the result holds for $\deg p_1(x) = n + 1$. The result is trivially true for $\deg p_1(x) = 0$ and hence is true for all values of n by induction.

Example 6.1 Take

$$p_1(x) = 3x^4 + 2x^3 + 5x^2 + x + 4$$
$$p_2(x) = 3x^2 - x + 3$$

Then

$$g(x) = x^2 + x + 1$$

and

$$r(x) = -x + 1$$

EXERCISE 6.1 The existence of a suitable $g(x)$ and $r(x)$ in the division algorithm has just been proved; show that they are unique.

The polynomial $g(x)$ is called the *quotient* of $p_1(x)$ by $p_2(x)$, and $r(x)$ is the *remainder*. If $r(x) = 0$ then $p_2(x)$ obviously divides $p_1(x)$. If the degree of a divisor of a polynomial is less than that of the polynomial but greater than 0, it is said to be a *proper divisor*; using this terminology an irreducible polynomial is a polynomial with no proper divisors.

It follows directly from the division algorithm, on taking $p_2(x) = x - f$ with f any element of F, that any polynomial $p_1(x) \in F[x]$ can be written

$$p_1(x) = (x - f)g(x) + r(x) \tag{6.3}$$

This equation must also hold when x is replaced by f, so that $p_1(f) = r(f)$, showing that $x - f$ will divide $p_1(x)$ if and only if $p_1(f) = 0$, that is if f is a root of $p_1(x)$. This is known as the *remainder theorem*.

Theorem 6.2 The greatest common divisor $g(x)$ of any two non-zero polynomials $p_1(x)$ and $p_2(x)$ in $F[x]$ can be expressed in the form

$$g(x) = r(x)p_1(x) + s(x)p_2(x) \tag{6.4}$$

where $r(x)$ and $s(x)$ are polynomials in $F[x]$.

Proof Applying the division algorithm to $p_2(x)$ and $p_1(x)$ gives (dropping the x for convenience)

$$p_2 = q_1p_1 + r_1$$

where $\deg r_1 < \deg p_1$. If $r_1 \neq 0$ the division algorithm can be applied to p_1 and r_1 to give

$$p_1 = q_2r_1 + r_2$$

with $\deg r_2 < \deg r_1$, and so on, until

$$r_{k-2} = q_kr_{k-1} + r_k$$

with $\deg r_1 > \deg r_2 > \deg r_3 > \cdots > \deg r_{k-1}$ and $r_k = 0$. Notice that this last result will be obtained after applying the division algorithm at most ($\deg p_2$) times.

From the above equations it is obvious that r_{k-1} divides r_{k-2} and hence it divides r_{k-3} and so on, showing that it divides p_1 and p_2. On the other hand if any polynomial p divides both p_1 and p_2 it follows from the equations (working downward from the first) that $p(x)$ divides $r_1, r_2, \ldots, r_{k-1}$. Thus r_{k-1} is the g.c.d. of p_0 and p_1.

The result now follows from writing

$$\begin{aligned} r_1 &= p_2 - q_1 p_1 \\ r_2 &= p_1 - q_2 r_1 \\ &= -q_2 p_2 + (1 + q_1 q_2) p_1 \\ \cdots \end{aligned}$$

whence (by an obvious induction)

$$r_{k-1} = rp_1 + sp_2$$

where r and s are polynomials.

Example 6.2 Find the g.c.d. of $x^3 + 2x^2 + 2x + 1$ and $x^2 - 1$.

$$\begin{aligned} x^3 + 2x^2 + 2x + 1 &= (x + 2)(x^2 - 1) + (3x + 3) \\ x^2 - 1 &= 3(x + 1).\tfrac{1}{3}(x - 1) + 0 \end{aligned}$$

showing that $(x + 1)$ is the required g.c.d. It is also easily seen that

$$(x + 1) = -\tfrac{1}{3}(x + 2)(x^2 - 1) + \tfrac{1}{3}(x^3 + 2x^2 + 2x + 1).$$

As a direct consequence of Theorem 6.2 it follows that if $p_1(x)$ and $p_2(x)$ are any two non-zero polynomials in $F[x]$ and $g(x)$ is an irreducible polynomial in $F[x]$ which divides the product $p_1(x)p_2(x)$, then $g(x)$ must divide $p_1(x)$ or $p_2(x)$. For either $g(x)$ divides $p_1(x)$, or the g.c.d. of $g(x)$ and $p_1(x)$ is 1; in which case $1 . p_2(x) = [g(x)r(x) + p_1(x)s(x)]p_2(x)$ and $g(x)$ divides both terms on the right-hand side.

This section is concluded with a theorem regarding the possibility of representing a polynomial $p(x)$ in $F[x]$ as the product of prime factors.

Theorem 6.3 Any polynomial in $F[x]$ can be expressed as a product of monic irreducible polynomials and such a factorization is unique (except, perhaps, for the ordering of the factors).

Proof Let $p(x)$ be any polynomial in $F[x]$. Assume first that $\deg p(x) = 1$ and that

$$p(x) = p_1(x)p_2(x)$$

with $\deg p_1(x) = r$ and $\deg p_2(x) = s$. Then

$$1 = r + s$$

so that $r = 0$ or $s = 0$. It follows that either $p_1(x)$ or $p_2(x)$ is just an element of F and hence $p(x)$ is irreducible.

Now assume that the result is true for polynomials of degree less than n and use induction on n. If $p(x)$ is irreducible the result obviously holds, so it is taken to be reducible. Let $\deg p(x) = n$ and

$$p(x) = p_1(x)p_2(x)$$

with $\deg p_1(x) = r > 0$, $\deg p_2(x) = s > 0$. Then

$$n = r + s$$

so that $n > r$ and $n > s$ and by the induction hypothesis $p_1(x)$ and $p_2(x)$ can be expressed as products of irreducible factors. It follows directly that $p(x)$ can be expressed in this manner. The factors can be made monic by multiplying by an appropriate element of F.

The uniqueness of the factorization has still to be demonstrated and this again involves induction on n. Assume $\deg p(x) = n$ and that the result is true for polynomials of degree less than n. The result is obvious when $n = 1$. Now (dropping the x for convenience) let

$$p = f_1 p_1 p_2 \dots p_r = f_2 q_1 q_2 \dots q_s$$

where $f_1, f_2 \in F$ and the p_i and q_j are monic irreducible polynomials.

As a consequence of Theorem 6.2, p_1 is a factor of q_1 or of $f_2 q_2 \dots q_r$. If it is a factor of $f_2 q_2 \dots q_r$ then it is a factor of q_2 or of $f_2 q_3 \dots q_r$ and by continuing in this way it is clear that p_1 is a factor of one of the q_j (p_i cannot divide f_2 since f_2 is of zero degree). Thus by making any necessary reordering

$$q_1 = g_1 p_1$$

Because q_1 is irreducible, g_1 is of zero degree (that is, $g_1 \in F$) and because q_1 and p_1 are monic, $g_1 = 1$.

It follows that $f_1 p_1 p_2 \dots p_r = f_2 p_1 q_2 \dots q_s$ and hence (by the cancellation law)

$$f_1 p_2 \dots p_r = f_2 q_2 \dots q_s$$

By the induction hypotheses it is now seen that

$$r - 1 = s - 1$$

and, by reordering where necessary, that

$$q_i = g_i p_i \quad i = 2, 3, \dots, r$$

with $g_i \in F$: because the q_i and p_i are monic, the $g_i = 1$ and $f_1 = f_2$. The proof of the theorem is now complete.

In the factored form of a polynomial $p(x)$, repeated factors can be grouped together to give

$$p(x) = f p_1(x)^{k_1} p_2(x)^{k_2} \dots p_u(x)^{k_u} \tag{6.5}$$

where the $p_i(x)$ are now distinct and monic and $f \in F$.

It is possible to show that any polynomial over the complex field $\mathscr{C}$ has at least one root. This fact is known as the *fundamental theorem of algebra* and cannot be demonstrated by purely algebraic techniques. (See, however, Chapter 7.) With this result and the remainder theorem, together with induction on the degree of $p(x)$, the reader should have no difficulty in proving the following special case of Theorem 6.3.

EXERCISE 6.2 Show that any polynomial $p(x) \in \mathscr{C}[x]$ can be expressed by the product

$$c(x - c_1)(x - c_2)\ldots(x - c_n) \tag{6.6}$$

where $c, c_1, \ldots, c_n \in \mathscr{C}$ and $n(>0)$ is the degree of $p(x)$. (Some of the factors may be repeated.)

EXERCISE 6.3 Let F be the field of integers modulo 3. Show that $p(x) = x^3 + 2x + 1$ is irreducible over F. [If $p(x)$ is reducible, one of its factors has degree 1; use the remainder theorem.]

7 Matrix rings

Let R be any ring and consider the square array of elements of R,

$$A = \begin{pmatrix} a_{11} & a_{12} & \ldots & a_{1n} \\ a_{21} & a_{22} & \ldots & a_{2n} \\ \vdots & \vdots & & \vdots \\ a_{n1} & a_{n2} & \ldots & a_{nn} \end{pmatrix}$$

Such an array is called a *square matrix of order n*. Matrices will be discussed in more detail in the next two chapters but it is of interest to show here that square matrices with certain prescribed rules for multiplication and addition form a ring.

Let B be a second square matrix of order n with b_{ij} the element at the intersection of the ith row and jth column. Then we define $C = A + B$ as the matrix having as its ijth element $c_{ij} = a_{ij} + b_{ij}$, $i, j, = 1, 2, \ldots, n$, where $+$ in the last equation is the operation of addition in R. Clearly, the set R^n of square matrices of order n with elements in R forms a commutative group under the rule for addition just defined. The additive identity is the matrix with all its elements equal to the zero of R; this is called the zero matrix. The additive inverse of any matrix A is the matrix having as its ijth element $-a_{ij}$.

The product of two matrices A and B is written AB and is defined to be the matrix C with ijth element

$$c_{ij} = \sum_{k=1}^{n} a_{ik}b_{kj}, \quad i, j, = 1, 2, \ldots, n$$

where the multiplication is that defined in R. We notice first that multiplication is associative. This follows directly from the definition: the ijth element of $A(BC)$ is $\sum_{k,l=1}^{n} a_{ik}(b_{kl}c_{lj})$ and that of $(AB)C$ is $\sum_{k,l=1}^{n} (a_{ik}b_{kl})c_{lj}$, but these are equal because multiplication in the ring R is associative. Similarly, it can be shown that the distribution laws for matrices

$$A(B + C) = AB + AC$$

and

$$(B + C)A = BA + CA$$

follow from the distributive rules in R.

This shows that R^n is a ring: it is called the *ring of $n \times n$ matrices over the ring R.* If the ring R has a multiplicative identity then so has R^n; it is the matrix

$$I = \begin{pmatrix} 1 & 0 & \dots & 0 \\ 0 & 1 & \dots & 0 \\ \vdots & \vdots & & \vdots \\ 0 & 0 & \dots & 1 \end{pmatrix}$$

that is, the matrix having 0 as its ijth element when $i \neq j$, and 1 when $i = j$. Notice that the multiplicative identity in R^n has been given the symbol I to avoid confusion with the multiplicative identity of R. When it is desired to emphasize the order of the $n \times n$ matrix I it is written I_n.

It is clear from the rule for multiplication of matrices that R^n need not be commutative and can have divisors of zero even if R itself is commutative, or an integral domain or field. The following examples taken from the ring of 2×2 matrices over the field of real numbers will illustrate these facts.

Example 7.1

$$A = \begin{pmatrix} 0 & r \\ r & 0 \end{pmatrix} \qquad B = \begin{pmatrix} 0 & -r \\ r & 0 \end{pmatrix}$$

gives

$$AB = \begin{pmatrix} r^2 & 0 \\ 0 & -r^2 \end{pmatrix} \neq BA = \begin{pmatrix} -r^2 & 0 \\ 0 & r^2 \end{pmatrix}$$

and

$$A = \begin{pmatrix} 0 & r \\ 0 & 0 \end{pmatrix} \qquad B = \begin{pmatrix} 0 & s \\ 0 & 0 \end{pmatrix}$$

gives $AB = 0$, where r and s are any real numbers.

More will be said regarding the regular elements of the matrix ring R^n after determinants have been defined in Chapter 2.

8 Mappings and functions

The concepts of many-to-one and one-to-one correspondences have already been dealt with. This chapter is concluded with a more careful look at these and related ideas.

Let S and T be two quite arbitrary sets; then a rule (recipe, relationship) f which associates with each element $s \in S$ a unique element $t \in T$ is called a *mapping* of S into T. We take the terms *function* and *many-to-one correspondence* to be synonymous with mapping. Although they are explicitly ruled out from the definition of functions, many-to-many correspondences can easily be included by defining as functions the single-valued branches of the many-to-many correspondence. Thus, although $y = \sqrt{x}$ is not a function according to the definition, we can obtain two functions by taking the positive and negative roots separately.

Another way of regarding a function is to think of it as the set of ordered pairs (s, t), the first entry being from S and the second from T. This will be a subset of $S \times T$. In this way it is seen that the set of ordered pairs (s, t) defines a function if, for every $s \in S$, there is a pair (s, t) and if when the first entry is the same in any two pairs then so is the second. The element $t \in T$ corresponding to an element $s \in S$ is called the *image* of s and is written† $f(s)$. Furthermore if $S_1 \subseteq S$ then $\bigcup_{s \in S_1} f(s)$ is called the *image set* of S_1 and written $f(S_1)$. The set $\{s \in S | f(s) = t\}$ is called the *inverse image* of t, and is often written $f^{-1}(t)$; notice that this does not imply that there is a function f^{-1}. The inverse image $f^{-1}(T_1)$ is defined similarly, where $T_1 \subseteq T$.

The set S is called the *domain* of the function f and the set $f(S)$ its *range*. In the case where $f(S) = T$ (that is, every element of T is the image of some element of S) then f is said to be *surjective* or to be a mapping of S *onto* T. A mapping of a set S onto itself is called a transformation as was noted previously (Example 2.11). A function f of S onto T is said to be *one-to-one* or *bijective* (sometimes *univalent*) if each point of T is the image of only one point of S, so that $f(s_1) = f(s_2)$ implies $s_1 = s_2$. (This is the one-to-one correspondence between sets defined in Section 2.) A mapping f of S into T is said to be *injective* if $f(s_1) = f(s_2)$ implies $s_1 = s_2$. Notice that a mapping which is both injective and surjective is one-to-one or bijective. In terms of ordered pairs, a function f is one-to-one if for every $t \in T$ there is a pair (s, t), and $(s_1, t_1) = (s_2, t_1)$ implies $s_1 = s_2$.

One-to-one functions are of particular importance because they always possess an inverse function which is itself one-to-one. This is easily seen by defining the inverse f^{-1} of a function f as the function which associates with any element $t \in T$ the unique element $s \in S$ for which $f(s) = t$. Then

$$f^{-1}(t) = f^{-1}f(s) = s$$

† The function f and the image $f(s)$ are quite distinct, but it is sometimes convenient to refer loosely to a typical image as the 'function'. An example has already been given: the 'function $y = \sqrt{x} \geqslant 0$' is more accurately 'the function y which associates with each $x \in \mathscr{R}$ the image $\sqrt{x} \geqslant 0$'. The notation $f(s) \equiv 0$ is used in such cases to mean $f = 0$ on S.

Obviously, from the definition, f^{-1} is one-to-one of T onto S so that the range and domain of f are the domain and range of f^{-1}. In what follows we shall avoid the notations $f^{-1}(t)$ and $f^{-1}(T_1)$ for the inverse image unless an inverse function f^{-1} exists.

The *composition* or *product* of two mappings is defined in the following way. Let f_1 be a mapping of S into T, and f_2 a mapping of T into R. Then if $s \in S$, with $f_1(s) = t \in T$ and $f_2(t) = r \in R$, the product $f_2 f_1$ is the mapping such that $f_2 f_1(s) = r$; this means that $f_2 f_1(s) = f_2\{f_1(s)\}$.

Functions of more than one variable can now be introduced. Suppose we have two sets S_1 and S_2 and a rule which associates with every element of the product set $S_1 \times S_2$ a unique element of a third set T. This is called a function of two variables in S_1 and S_2 into T, and is a mapping of $S_1 \times S_2$ into T. In the special case $S_1 = S_2 = T$ we have the binary compositions defined previously. The extension of these ideas to more than two variables is obvious.

9 Rings of functions

Let S and R be an arbitrary set and an arbitrary ring respectively and let F be the set of all functions $f: S \to R$. Two functions $f, g \in F$ are said to be equal if $f(s) = g(s)$ for all $s \in S$. If addition and multiplication of functions are defined by

$$(f + g)(s) = f(s) + g(s) \tag{9.1}$$

$$(fg)(s) = f(s)\, g(s) \tag{9.2}$$

for all $s \in S$, then it is easily verified that the axioms for a ring in Section 3 are satisfied. Thus F together with addition and multiplication as defined in (9.1) and (9.2) forms a ring which is called a *ring of functions*.

Now consider the particular case when $S = R$ and R is an integral domain I. The set of functions of the form

$$f: s \to a_0 + a_1 s + \cdots + a_n s^n \tag{9.3}$$

where s and the coefficients $a_0, a_1, \ldots, a_n$ are in I is known as the set of *polynomial functions* over I. Notice that in (9.3) $s \in I$, whereas in the polynomial form $a_0 + a_1 x + \cdots + a_n x^n$, x is allowed to belong to a larger ring containing I. Each polynomial form $a_0 + a_1 x + \cdots + a_n x^n$ determines a unique polynomial function (namely the function obtained by restricting x to I) so that there is evidently a homomorphism between $I[x]$ and the corresponding set of polynomial functions. That this homomorphism is not an isomorphism is shown by the following counter-examples.

Example 9.1 Let I be the field of integers modulo 3. Then the two distinct polynomials $x + x^4$ and $x^2 + x^3$ define the same polynomial function

$$\{f: 0 \to 0, 1 \to 2, 2 \to 0\}$$

Example 9.2 Let I be the finite integral domain $\{a_1, a_2, a_3\}$. Then

$$(x - a_1)(x - a_2)(x - a_3) \neq 0$$

if x is indeterminate over I. The corresponding polynomial function

$$f: s \to (s - a_1)(s - a_2)(s - a_3)$$

on the other hand, is zero for every $s \in I$. Thus at least two polynomial forms give rise to the zero function.

It is an important fact that if I has an infinite number of elements then no single polynomial can have among its roots all the elements of I. From this arises the following theorem.

Theorem 9.1 If I is infinite then the homomorphism between $I[x]$ and the corresponding set of polynomial functions is an isomorphism.

To prove the theorem the following lemma is required.

Lemma A polynomial form of degree n in $I[x]$ has at most n distinct roots.

Proof of lemma Suppose that $a_1, a_2, \ldots, a_m \in I$ are distinct roots of the polynomial $f(x) \in I[x]$. Then (by the remainder theorem)

$$f(x) = (x - a_1)\, g(x) \tag{9.4}$$

Since a_2 is a root of $f(x)$

$$f(a_2) = (a_2 - a_1)\, g(a_2) = 0$$

showing that $g(a_2) = 0$ (by the cancellation rule, since $a_2 \neq a_1$) and hence

$$g(x) = (x - a_2)\, h(x)$$

From (9.4) it follows that

$$f(x) = (x - a_1)(x - a_2)\, h(x) \tag{9.5}$$

and on repeating this process

$$f(x) = (x - a_1)(x - a_2) \cdots (x - a_m)\, r(x) \tag{9.6}$$

Thus the degree of $f(x)$ must be greater than or equal to the number of its distinct roots, proving the lemma.

Proof of theorem Let $f(x)$ and $g(x)$ be any two polynomials in $I[x]$ and let the polynomial functions f and g be equal on I. The difference $\phi(x) = f(x) - g(x)$ is a polynomial; assume, contrary to what is to be proved, that $\phi(x)$ is not the zero polynomial but has degree $n \geqslant 0$. Then by the lemma $\phi(x)$ has at most n

distinct roots, and because I is infinite there is an $s \in I$ for which $\phi(s) \neq 0$. The contradiction establishes the theorem.

Rational forms $a(x)/b(x)$ in $F(x)$ give rise to *rational functions* $a(s)/b(s)$ when s is restricted to be an element of F. It follows from what has been proved above that if F is an infinite field the rational forms and rational functions are isomorphic. Note that the rational function $a(s)/b(s)$ is defined only where $b(s) \neq 0$.

10 Denumerable sets

Finite and infinite sets were defined in Section 1. The definition of a finite set means, of course, that a one-to-one correspondence can be set up between the elements of the set and a subset $\{1, 2, \ldots, n\}$ of the set of positive integers. An infinite set is said to be *denumerable* (*enumerable, countable*) if a one-to-one correspondence can be established between its elements and the set of positive integers $\{1, 2, \ldots\}$.

EXERCISE 10.1 Prove that the union of a denumerable collection of denumerable sets is itself denumerable. [Arrange the elements of the sets in a suitable rectangular array and count along diagonals.]

EXERCISE 10.2 Prove that the rational numbers are denumerable. [Consider the set of sets which have the form $\{1/n, 2/n, \ldots\}$ and use the previous Exercise, omitting elements which have already been counted.]

Further reading

It has been assumed that the reader is familiar with the various number systems (integers, rationals, reals, etc.). Good introductory accounts of these systems can be found in Courant and John (1965) and Graves (1956). For a more detailed treatment Burrill (1967) is recommended.

Most of the work in this chapter is covered in the texts referred to in the remaining chapters of this book. In particular Apostol (1957), Jacobson (1951), and Birkhoff and MacLane (1965) will give relevant supplementary material.

References

APOSTOL, T., 1957, *Mathematical analysis*, Addison-Wesley, Reading, Mass.

BENSON, R. V., 1966, *Euclidean geometry and convexity*, McGraw-Hill, New York.

BIRKHOFF, G. and S. MACLANE, 1965, *A survey of modern algebra*, Macmillan, New York.

BURRILL, C., 1967, *Foundations of real numbers*, McGraw-Hill, New York.

COURANT, R. and F. JOHN, 1965, *Introduction to calculus and analysis* (Vol. 1), John Wiley (Interscience), New York.

GRAVES, L. M., 1956, *The theory of functions of real variables*, McGraw-Hill, New York.

JACOBSON, N., 1951, *Lectures in abstract algebra, Vol. 1—Basic concepts*, Van Nostrand, Princeton, N.J.
MOSTOW, G. D., J. H. SAMPSON, and J. P. MEYER, 1963, *Fundamental structures of algebra*, McGraw-Hill, New York.

Problems

1 Show by counter-examples that subtraction of sets does not necessarily obey the usual subtraction rules of algebra.

2 Show that an arbitrary set S with the binary composition defined by $ab = b$, $a, b \in S$, forms a semigroup. What can be said regarding the identities of this semigroup?

3 Consider a semigroup S with a right identity s_r and such that the equation $xa = s_r$, $a \in S$, has a unique solution for any $x \in S$ (i.e., every element of S has a left inverse with respect to s_r). Is the resulting system necessarily a group?

4 If z is an element of a field F show that $z^n + z^{-n}$ is a polynomial in $z + z^{-1}$.

5 Find the greatest common divisor of $x^4 + x^3 - x - 1$ and $x^4 - 1$ and express it in the form of equation (6.4).

6 Break down the polynomial $x^3 + 2x^2 - x + 1$ into its irreducible factors over (a) the real field, (b) the complex field.

7 Show that the set of all 2×2 matrices of the form

$$\begin{pmatrix} a & b \\ -b & a \end{pmatrix}$$

with $a, b \in \mathscr{R}$, is isomorphic to the field of complex numbers.

8 Prove that the set of all numbers $a + \sqrt{2}\, b$, where a and b are rational, forms a ring under ordinary addition and multiplication.

9 Prove that the set Q of all matrices of the form

$$\begin{pmatrix} a & b \\ -\bar{b} & \bar{a} \end{pmatrix}$$

with $a, b \in \mathscr{C}$ forms a division ring. (This ring is called the ring of *quaternions*.)

10 Prove that the set of all $p \times q$ matrices, with entries elements of an additive group, forms a group with respect to matrix addition.

2 Matrices and determinants

1 Matrices

In Chapter 1 square matrices over a ring were defined and their addition and multiplication were dealt with. In this chapter matrices that are not necessarily square are considered.

The rectangular array

$$A = \begin{pmatrix} a_{11} & a_{12} & \cdots & a_{1n} \\ a_{21} & a_{22} & \cdots & a_{2n} \\ \vdots & \vdots & & \vdots \\ a_{m1} & a_{m2} & \cdots & a_{mn} \end{pmatrix}$$

where a_{ij}, $i = 1, 2, \ldots, m$, $j = 1, 2, \ldots, n$ are elements of a ring R, is said to be a *rectangular* $m \times n$ *matrix over* R. Thus m is the number of rows and n the number of columns in the array and when $m = n$ it is a square matrix of order n as defined in Chapter 1. In a square matrix of order n, the diagonal containing $a_{11}, a_{22}, \ldots, a_{nn}$ is called the *principal diagonal.*

Addition of two rectangular matrices A and B is possible only when they each have the same number of rows and of columns (such matrices are said to be *conformable for addition*) and is defined in the same way as for square matrices. Thus if A and B are $m \times n$ matrices their sum, C, is the matrix with ijth entry $c_{ij} = a_{ij} + b_{ij}$ for $i = 1, 2, \ldots, m$, $j = 1, 2, \ldots, n$. We will often write the matrix with ijth entry a_{ij} as (a_{ij}) and with this notation $(c_{ij}) = (a_{ij}) + (b_{ij}) = (a_{ij} + b_{ij})$.

Similarly, if A is an $m \times n$ matrix and B is an $r \times s$ matrix, the product AB is defined only when $n = r$ (that is, the number of columns of A is equal to the number of rows of B) and is then given by $C = AB$ where $c_{ij} = \sum_{k=1}^{m} a_{ik}b_{kj}$: the matrices A and B are said to be *conformable for the product AB.* Notice with rectangular matrices that not only is the multiplication generally non-commutative, but BA may not even exist; in fact, AB and BA both exist only when $m = s$ and $n = r$.

Example 1.1

$$\begin{pmatrix} 2 & 1 \\ 1 & 1 \end{pmatrix}\begin{pmatrix} 3 & 1 & 1 \\ 1 & 2 & 1 \end{pmatrix} = \begin{pmatrix} 7 & 4 & 3 \\ 4 & 3 & 2 \end{pmatrix}$$

but

$$\begin{pmatrix} 3 & 2 & 1 \\ 1 & 2 & 1 \end{pmatrix}\begin{pmatrix} 2 & 1 \\ 1 & 1 \end{pmatrix}$$

is not defined. On the other hand,

$$\begin{pmatrix} 1 & 2 \\ 1 & 1 \\ 3 & 1 \end{pmatrix}\begin{pmatrix} 3 & 1 & 1 \\ 1 & 2 & 1 \end{pmatrix} = \begin{pmatrix} 5 & 5 & 3 \\ 4 & 3 & 2 \\ 10 & 5 & 4 \end{pmatrix}$$

and

$$\begin{pmatrix} 3 & 1 & 1 \\ 1 & 2 & 1 \end{pmatrix}\begin{pmatrix} 1 & 2 \\ 1 & 1 \\ 3 & 1 \end{pmatrix} = \begin{pmatrix} 7 & 8 \\ 6 & 5 \end{pmatrix}$$

Clearly, the product AB of an $n \times m$ matrix A with an $m \times r$ matrix B is an $n \times r$ matrix. For both AB and BA to exist $n = r$ so that one product is $n \times n$ and the other $m \times m$. The two products can only be of the same size when the factors are square, showing that commutative multiplication is only possible for square matrices and, as shown in Chapter 1, is not necessarily found even then.

In the mathematics of dynamical systems we are mainly concerned with matrices over a field (or perhaps an integral domain) and so in what follows the elements of a matrix will always be regarded as in a field F unless anything is said to the contrary; elements of F will be called *scalars* and multiplication of a matrix $A = (a_{ij})$ by a scalar $a \in F$ is defined by $aA = (aa_{ij}) = (a_{ij}a) = Aa$.

With any $m \times n$ matrix $A = (a_{ij})$ we can associate the $n \times m$ matrix obtained by interchanging the rows and columns of A. This new matrix is called the *transpose* of A and denoted by A^{T}; thus $A^{\mathrm{T}} = (a_{ji})$. Other notations for the transpose are A' and A^{t}. The transpose of the transpose of a matrix is obviously the matrix itself: $(A^{\mathrm{T}})^{\mathrm{T}} = A$.

EXERCISE 1.1 Verify that $(aA)^{\mathrm{T}} = aA^{\mathrm{T}}$ and $(aA + bB)^{\mathrm{T}} = aA^{\mathrm{T}} + bB^{\mathrm{T}}$, where a and b are scalars, and consequently that the transpose of any sum of products, of a matrix with a scalar, is equal to the sum of the corresponding scalar multiples of the transposed matrices.

A matrix which is equal to its own transpose is said to be a *symmetric matrix* and obviously must be square. A matrix which is equal to the negative of its transpose is called *skew-symmetric* (*anti-symmetric*) and again must be square. It is an important fact that any square matrix A over a field F in which $1 + 1 \neq 0$ can be written as the sum of a symmetric and a skew-symmetric matrix in the following way:

$$A = (a_{ij}) = B + C$$

where

$$B = 2^{-1}(A + A^{\mathrm{T}}) = 2^{-1}(a_{ij} + a_{ji})$$

and

$$C = 2^{-1}(A - A^{\mathrm{T}}) = 2^{-1}(a_{ij} - a_{ji})$$

Since $B^{\mathrm{T}} = 2^{-1}(A^{\mathrm{T}} + A) = B$, we see that B is symmetric and because $C^{\mathrm{T}} = 2^{-1}(A^{\mathrm{T}} - A) = -C$, C is skew-symmetric.

Example 1.2

$$\begin{pmatrix} 2 & 1 & 0 \\ 3 & 0 & 2 \\ 2 & 0 & 2 \end{pmatrix} = \begin{pmatrix} 2 & 2 & 1 \\ 2 & 0 & 1 \\ 1 & 1 & 2 \end{pmatrix} + \begin{pmatrix} 0 & -1 & -1 \\ 1 & 0 & 1 \\ 1 & -1 & 0 \end{pmatrix}$$

The first term on the right is symmetric and the second skew-symmetric.

Example 1.3 The decomposition of A into a symmetric and skew-symmetric part fails when F is the field of integers modulo 2. For then $2B = (a_{ij} + a_{ji})$, but $2 = 0 \bmod 2$, and so 2 does not have an inverse in F.

1.1 Partitioning of matrices

Consider the third-order matrix

$$A = \left(\begin{array}{cc:c} a_{11} & a_{12} & a_{13} \\ a_{21} & a_{22} & a_{23} \\ \hdashline a_{31} & a_{32} & a_{33} \end{array}\right) \tag{1.1}$$

with broken lines inserted as shown: A is then said to be *partitioned* and can be written as a *block matrix*

$$A = \begin{pmatrix} A_{11} & A_{12} \\ A_{21} & A_{22} \end{pmatrix} \tag{1.2}$$

where A_{11} is 2×2, A_{12} is 2×1, A_{21} is 1×2 and A_{22} is 1×1. If a second 3×3 matrix, B, is similarly partitioned to give

$$B = \begin{pmatrix} B_{11} & B_{12} \\ B_{21} & B_{22} \end{pmatrix} \tag{1.3}$$

then it is easy to verify by direct calculation that

$$AB = \begin{pmatrix} A_{11}B_{11} + A_{12}B_{21} & A_{11}B_{12} + A_{12}B_{22} \\ A_{21}B_{11} + A_{22}B_{21} & A_{21}B_{12} + A_{22}B_{22} \end{pmatrix} \tag{1.4}$$

In general the two matrices $A(m \times n)$ and $B(n \times p)$ are said to be *conformably partitioned* if

$$A = \begin{pmatrix} A_{11} & A_{12} & \cdots & A_{1r} \\ A_{21} & A_{22} & \cdots & A_{2r} \\ \vdots & \vdots & & \vdots \\ A_{s1} & A_{s2} & \cdots & A_{sr} \end{pmatrix} \tag{1.5}$$

$$B = \begin{pmatrix} B_{11} & B_{12} & \cdots & B_{1t} \\ B_{21} & B_{22} & \cdots & B_{2t} \\ \vdots & \vdots & & \vdots \\ B_{r1} & B_{r2} & \cdots & B_{rt} \end{pmatrix} \tag{1.6}$$

where A_{ij} is $m_i \times n_j$, with $\sum_{i=1}^{s} m_i = m$, $\sum_{j=1}^{r} n_j = n$; B_{ij} is $n_i \times p_j$ with $\sum_{i=1}^{r} n_i = n$, $\sum_{j=1}^{t} p_j = p$.

EXERCISE 1.2 Prove that the multiplication of conformably partitioned matrices can be carried out by regarding the blocks as single elements. That is to say, if A and B are given by (1.5) and (1.6), their product $C = AB$ is (C_{ij}) where

$$C_{ij} = \sum_{k=1}^{r} A_{ik}B_{kj} \tag{1.7}$$

Of particular importance (Chapter 3, Section 6) are partitioned matrices of the form

$$C = \begin{pmatrix} C_1 & & & \\ & C_2 & & 0 \\ & & \ddots & \\ & 0 & & \\ & & & C_n \end{pmatrix} \tag{1.8}$$

where the C_i, $i = 1, 2, \ldots, n$, are square matrices along the principal diagona and all the other elements are zero. Such matrices are said to be *quasi-diagonal* or *block-diagonal.*

2 Determinants

An important function which associates a single scalar with the n^2 elements of a square matrix of order n is known as the *determinant* of the matrix and can be defined in a variety of ways. Two of these definitions are pertinent to subsequent development.

Consider the pair of simultaneous homogeneous algebraic equations

$$a_{11}x_1 + a_{12}x_2 = 0 \tag{2.1}$$

$$a_{21}x_1 + a_{22}x_2 = 0 \tag{2.2}$$

Multiplying (2.1) by a_{22} and (2.2) by a_{12} and subtracting the results gives

$$(a_{11}a_{22} - a_{12}a_{21})x_1 = 0 \tag{2.3}$$

Similarly, by eliminating x_1

$$(a_{11}a_{22} - a_{12}a_{21})x_2 = 0 \tag{2.4}$$

and it can be seen that, if a solution of (2.1) and (2.2) exists for which at least one of $x_1, x_2 \neq 0$, then

$$a_{11}a_{22} - a_{12}a_{21} = 0 \tag{2.5}$$

This last expression is defined as the *determinant* of the matrix A,

$$A = (a_{ij}) = \begin{pmatrix} a_{11} & a_{12} \\ a_{21} & a_{22} \end{pmatrix}$$

and we write it $|A|$ or $|(a_{ij})|$ or $\begin{vmatrix} a_{11} & a_{12} \\ a_{21} & a_{22} \end{vmatrix}$. The expressions det A or det (a_{ij}) will also be used to denote determinants. Notice that equations (2.1) and (2.2) can be written in the form

$$\begin{pmatrix} a_{11} & a_{12} \\ a_{21} & a_{22} \end{pmatrix}\begin{pmatrix} x_1 \\ x_2 \end{pmatrix} = \begin{pmatrix} 0 \\ 0 \end{pmatrix}$$

Now consider the three simultaneous equations

$$a_{11}x_1 + a_{12}x_2 + a_{13}x_3 = 0 \tag{2.6}$$

$$a_{21}x_1 + a_{22}x_2 + a_{23}x_3 = 0 \tag{2.7}$$

$$a_{31}x_1 + a_{32}x_2 + a_{33}x_3 = 0 \tag{2.8}$$

Multiplying (2.7) by a_{33} and (2.8) by a_{23} and subtracting the results gives

$$(a_{21}a_{33} - a_{23}a_{31})x_1 + (a_{22}a_{33} - a_{23}a_{32})x_2 = 0 \tag{2.9}$$

and eliminating x_2 from these same two equations gives

$$(a_{21}a_{32} - a_{22}a_{31})x_1 + (a_{23}a_{32} - a_{22}a_{33})x_3 = 0 \tag{2.10}$$

Now, on substituting from (2.9) and (2.10) into (2.6) there results

$$[a_{11}(a_{22}a_{33} - a_{23}a_{32}) - a_{12}(a_{21}a_{33} - a_{23}a_{31}) \\ + a_{13}(a_{21}a_{32} - a_{22}a_{31})]x_1 = 0 \tag{2.11}$$

Had x_1 and x_3, or x_1 and x_2, been eliminated from (2.6), (2.7), and (2.8) then equations with the same expression in the square brackets, multiplying x_2 and x_3 respectively, would have been obtained in place of (2.11). For the existence of a solution of (2.6), (2.7), and (2.8) in which not all of x_1, x_2, and x_3 are zero the expression multiplying x_1 in (2.11) must vanish. We define this expression as the determinant of the third-order matrix

$$\begin{pmatrix} a_{11} & a_{12} & a_{13} \\ a_{21} & a_{22} & a_{23} \\ a_{31} & a_{32} & a_{33} \end{pmatrix}$$

and again write it as the same array with straight lines instead of brackets. Of course equations (2.6), (2.7), and (2.8) could be written in matrix form just as (2.1) and (2.2) were.

On comparing equations (2.4) and (2.11) it is easily seen that the third-order determinant can be expressed in terms of three second-order determinants as follows:

$$\begin{vmatrix} a_{11} & a_{12} & a_{13} \\ a_{21} & a_{22} & a_{23} \\ a_{31} & a_{32} & a_{33} \end{vmatrix} = a_{11}\begin{vmatrix} a_{22} & a_{23} \\ a_{32} & a_{33} \end{vmatrix} - a_{12}\begin{vmatrix} a_{21} & a_{23} \\ a_{31} & a_{33} \end{vmatrix} + a_{13}\begin{vmatrix} a_{21} & a_{22} \\ a_{31} & a_{32} \end{vmatrix}$$

with each element of the first row multiplying the determinant of the matrix obtained by deleting the row and column containing that element from the original matrix. Notice also that the signs of the products alternate: the first is positive, the second negative, and the third positive again.

If this process is repeated for four equations and then five equations, and so on a definition of an nth-order determinant in terms of $(n - 1)$th-order determinants will be obtained. In fact,

$$|(a_{ij})| = \sum_{k=1}^{n} a_{1k}(-1)^{k+1}M_{1k} \tag{2.12}$$

where M_{1k} is the $(n - 1)$th-order determinant obtained by striking out the 1st row and kth column from the original matrix. The determinant M_{1k} is called the *minor* of the element a_{1k} and, in general, the minor of any element a_{ij} is the determinant of the $(n - 1)$th-order matrix obtained by striking out the ith row and th column from the original nth-order matrix (a_{ij}).

Now determinants will be discussed from a somewhat different viewpoint. From equation (2.11) the third-order determinant $|(a_{ij})|$ can be written in the expanded form:

$$|(a_{ij})| = a_{11}a_{22}a_{33} - a_{11}a_{23}a_{32} - a_{12}a_{21}a_{33} + a_{12}a_{23}a_{31} + a_{13}a_{21}a_{32} - a_{13}a_{22}a_{31} \quad (2.13)$$

An examination of this expression will show that it consists of positive and negative terms each of which is a product of three elements taken from different rows and columns of the matrix (a_{ij}). Thus each term is of the form $\pm a_{1i}a_{2j}a_{3k}$ where i, j, k is some permutation of the integers 1, 2, 3.

In the same way if we took the trouble to write out a determinant of 4th order in expanded form we should find that it consisted of $4!(= 24)$ terms of the form $\pm a_{1i}a_{2j}a_{3k}a_{4l}$ with i, j, k, l a permutation of 1234. Proceeding further, it becomes apparent that the nth-order determinant $|(a_{ij})|$ consists of $n!$ terms of the form $\pm a_{1i}a_{2j}\ldots a_{nr}$ with $i, j, \ldots, r$ a permutation of the integers $1, 2, 3, \ldots, n$. A careful examination of the signs of the terms in the general expression would lead to the result that a positive sign occurs when an even number of entries in the permutation $i, j, k, \ldots, r$ have their natural order reversed, and a negative sign occurs when the number of inversions of order is odd.

These ideas will now be formalized and an alternative definition of a determinant will be given.

2.1 Odd and even permutations

Denote the permutation $i, j, k, l, \ldots, r$ of the integers $1, 2, \ldots, n$ by p; then, as in Chapter 1, write

$$p = \begin{pmatrix} 1 & 2 & 3 & \ldots & n \\ i & j & k & \ldots & r \end{pmatrix}$$

The notation

$$p = \begin{pmatrix} 1 & 2 & \ldots & n \\ p(1) & p(2) & \ldots & p(n) \end{pmatrix}$$

is also used, where $p(i)$ is the image of i under the permutation p. A pair of integers i, j with $j > i$ suffers a reversal of order by the permutation p if $p(j) < p(i)$. If the number of reversals of order caused by p is odd, then p is called an *odd permutation*; if the number of reversals of order is even, then p is said to be *even*. We now define the *signum* or *sign* (abbreviated to sgn) of a permutation p as $(-1)^n$ where n is the number of reversals of order in p. Thus $\operatorname{sgn} p = +1$ if p is even and -1 if p is odd.

Example 2.1 Consider the permutation

$$p = \begin{pmatrix} 1 & 2 & 3 & 4 & 5 \\ 3 & 5 & 2 & 4 & 1 \end{pmatrix}$$

Working from left to right it is seen that (3 2) and (3 1) are reversals of order as are (5 2), (5 4), (5 1), (2 1), and (4 1), so that p is odd and $\operatorname{sgn} p = -1$.

There are a number of other definitions of odd and even permutations and of the sign of a permutation but they depend on a rather deeper treatment of the symmetric group of permutations (Steward, 1963; Marcus and Minc, 1965). The above definition is sufficient for our present purpose.

EXERCISE 2.1 Prove that the sign of the product of two permutations is the product of their signs and that exchanging any two elements in a permutation but leaving the others unchanged is an odd permutation. The results of this exercise will be useful later on (Section 2.2).

An alternative definition of a *determinant* can now be given. A determinant $|(a_{ij})|$ of order n is the function

$$|(a_{ij})| = \sum_{p} \operatorname{sgn} p . a_{1p(1)}a_{2p(2)} \ldots a_{np(n)} \tag{2.14}$$

where the sum is taken over all permutations of the integers 1, 2, 3, . . ., n. The reader should verify that this definition agrees with the previous one by writing out (2.14) for the third-order case, for example; a proof is given later (Section 2.3).

EXERCISE 2.2 Prove that an alternative expression to (2.14) for $|A|$ is

$$|A| = \sum_{j} \delta^{r_1 r_2 \ldots r_n}_{j_1 j_2 \ldots j_n} a_{r_1 j_1} a_{r_2 j_2} \ldots a_{r_n j_n} \tag{2.15}$$

where the symbol $\delta^{r_1 r_2 \ldots r_n}_{j_1 j_2 \ldots j_n}$ is $+1$ if $r_1, r_2, \ldots, r_n$ is an even permutation of $j_1, j_2, \ldots, j_n$, -1 if $r_1, r_2, \ldots, r_n$ is an odd permutation of $j_1, j_2, \ldots, j_n$, 0 if $r_1, r_2, \ldots, r_n$ is not a permutation of $j_1, j_2, \ldots, j_n$. The summation in (2.15) is to be taken over all permutations j of 1, 2, . . ., n into $j_1, j_2, \ldots, j_n$.

2.2 Some properties of determinants

Directly from the definition it follows that the determinant of a scalar multiple of a matrix of order n is the product of the nth power of the scalar and the determinant of the matrix, that is,

$$|kA| = |(ka_{ij})| = k^n \, |(a_{ij})|$$

If only a single row or column of A is multiplied by a scalar k then the determinant of the product is $k\,|A|$.

EXERCISE 2.3 Show that the determinant of the transpose of a matrix A is equal to the determinant of A, that is,

$$|A^{\mathrm{T}}| = |A|$$

This important fact means that any result established on determinants concerning the rows of the corresponding matrices also holds for the columns.

Theorem 2.1 Interchanging any two rows of a matrix does not change the the numerical value of its determinant, but reverses its sign.

Proof Suppose that rows r and s in (a_{ij}) are interchanged. Then the expression

$$\sum_p \operatorname{sgn} p \,.\, a_{1p(1)} \ldots a_{rp(r)} \ldots a_{sp(s)} \ldots a_{np(n)} \tag{2.16}$$

becomes

$$\sum_p \operatorname{sgn} p \,.\, a_{1p(1)} \ldots a_{sp(r)} \ldots a_{rp(s)} \ldots a_{np(n)} \tag{2.17}$$

The factors in each term are unaltered but the corresponding permutation has two entries interchanged with each other. This single exchange regarded as a separate permutation is odd and therefore the net effect, resulting from the product of this permutation and the original one, is a reversal of sign (by the two results regarding the signs of permutations in Exercise 2.1). Thus each term in the determinant has its sign changed but its numerical value unchanged, proving the result.

EXERCISE 2.4 Prove that the effect of replacing the 1st, 2nd, 3rd, ..., nth rows of (a_{ij}) by the $p(1)$th, $p(2)$th, ..., $p(n)$th rows of (a_{ij}) is to multiply the determinant by sgn p where p is the permutation

$$\begin{pmatrix} 1 & 2 & \ldots & n \\ p(1) & p(2) & \ldots & p(n) \end{pmatrix}$$

Theorem 2.2 Adding a scalar multiple of any row of a matrix to any other row does not change the value of its determinant.

Proof Let (a_{ij}) be the original matrix and (b_{ij}) the matrix obtained by adding c times row r in (a_{ij}) to row s in (a_{ij}). Then

$$b_{sj} = a_{sj} + c a_{rj} \qquad j = 1, 2, \ldots, n \tag{2.18}$$

$$b_{ij} = a_{ij} \qquad \begin{cases} i = 1, 2, \ldots, s-1, s+1, \ldots, n \\ j = 1, 2, \ldots, n \end{cases} \tag{2.19}$$

so that

$$\begin{aligned} |(b_{ij})| &= \sum_p \operatorname{sgn} p \,.\, b_{1p(1)} \ldots b_{rp(r)} \ldots b_{sp(s)} \ldots b_{np(n)} \\ &= \sum_p \operatorname{sgn} p \,.\, a_{1p(1)} \ldots a_{rp(r)} \ldots (a_{sp(s)} + c a_{rp(r)}) \ldots a_{np(n)} \\ &= \sum_p \operatorname{sgn} p \,.\, a_{1p(1)} \ldots a_{rp(r)} \ldots a_{sp(s)} \ldots a_{np(n)} \\ &\qquad + c \sum_p \operatorname{sgn} p \,.\, a_{1p(1)} \ldots a_{rp(r)} \ldots a_{rp(r)} \ldots a_{np(n)} \\ &= |(a_{ij})| \end{aligned} \tag{2.20}$$

since the expression multiplying c is the determinant of a matrix with two identical rows.

Another useful fact is that any matrix (a_{ij}) having a row each entry of which is zero has determinant zero. This follows directly from the definition of a determinant (2.14) since each term in $|(a_{ij})|$ must contain an element of each row of (a_{ij}). This and the previous result can be used to prove the earlier rcsult that a matrix with two equal rows has determinant zero. Subtracting one of the identical rows from the other does not affect the value of the determinant but produces a row of zeros, proving that the value of the determinant must be zero.

Theorem 2.3 The determinant of the product of two matrices is equal to the product of their determinants.

Proof Let

$$A = (a_{ij}), \qquad B = (b_{ij})$$

and

$$C = (c_{ij}) = (a_{ij})(b_{ij}) = \left(\sum_{k=1}^{n} a_{ik} b_{kj}\right)$$

Then

$$\begin{aligned} |(c_{ij})| &= \sum_p \operatorname{sgn} p \,.\, c_{1p(1)} c_{2p(2)} \ldots c_{np(n)} \\ &= \sum_p \operatorname{sgn} p \sum_{r_1=1}^{n} a_{1r_1} b_{r_1 p(1)} \sum_{r_2=1}^{n} a_{2r_2} b_{r_2 p(2)} \ldots \sum_{r_n=1}^{n} a_{nr_n} b_{r_n p(n)} \\ &= \sum_{r_1=1}^{n} \sum_{r_2=1}^{n} \cdots \sum_{r_n=1}^{n} a_{1r_1} a_{2r_2} \ldots a_{nr_n} \\ &\qquad \times \sum_p \operatorname{sgn} p \,.\, b_{r_1 p(1)} b_{r_2 p(2)} \ldots b_{r_n p(n)} \end{aligned} \tag{2.21}$$

Now consider the expression

$$\sum_p \operatorname{sgn} p \,.\, b_{r_1 p(1)} b_{r_2 p(2)} \ldots b_{r_n p(n)} \tag{2.22}$$

This is just the determinant of the matrix having as its 1st, 2nd, ..., nth rows the r_1th, r_2th, ..., r_nth rows of B, provided that $r_1, r_2, \ldots, r_n$ is some rearrangement of $1, 2, \ldots, n$ (otherwise it is zero). By Exercise 2.4 the expression (2.22) is $\operatorname{sgn} r . |B|$ where r is the permutation

$$\begin{pmatrix} 1 & 2 & \ldots & n \\ r_1 & r_2 & \ldots & r_n \end{pmatrix}$$

Using this in (2.21) gives

$$\begin{aligned} |C| &= |B| \sum_{r_1=1}^{n} \sum_{r_2=1}^{n} \ldots \sum_{r_n=1}^{n} \operatorname{sgn} r . a_{1r_1} a_{2r_2} \ldots a_{nr_n} \\ &= |B| \, |A| \end{aligned}$$

This completes the proof.

Example 2.2

$$AB = \begin{pmatrix} 3 & 1 & 1 \\ 0 & 1 & 0 \\ 1 & 0 & 1 \end{pmatrix} \begin{pmatrix} 1 & 1 & 0 \\ 0 & 1 & 1 \\ 1 & 0 & 0 \end{pmatrix} = \begin{pmatrix} 4 & 4 & 1 \\ 0 & 1 & 1 \\ 2 & 1 & 0 \end{pmatrix} = C$$

$$|A| = -\begin{vmatrix} 0 & 1 & 0 \\ 3 & 1 & 1 \\ 1 & 0 & 1 \end{vmatrix} = \begin{vmatrix} 3 & 1 \\ 1 & 1 \end{vmatrix} = 2$$

$$|B| = \begin{vmatrix} 1 & 0 & 0 \\ 1 & 1 & 0 \\ 0 & 1 & 1 \end{vmatrix} = \begin{vmatrix} 1 & 0 \\ 1 & 1 \end{vmatrix} = 1$$

$$|C| = -\begin{vmatrix} 0 & 1 & 1 \\ 4 & 4 & 1 \\ 2 & 1 & 0 \end{vmatrix} = \begin{vmatrix} 4 & 1 \\ 2 & 0 \end{vmatrix} - \begin{vmatrix} 4 & 4 \\ 2 & 1 \end{vmatrix} = 2$$

2.3 Expansions of a determinant

The minor M_{ij} of the element a_{ij} in the matrix (a_{ij}) has already been defined as the determinant of the array obtained by striking out the ith row and the jth column from (a_{ij}). By definition we have

$$M_{11} = \sum_{p'} \operatorname{sgn} p' . a_{2p'(2)} a_{3p'(3)} \ldots a_{np'(n)} \tag{2.23}$$

where the permutations p' are taken over $2, 3, \ldots, n$. Now consider the determinant $|(a_{ij})|$ itself,

$$|(a_{ij})| = \sum_{p} \operatorname{sgn} p \,.\, a_{1p(1)}a_{2p(2)}\cdots a_{np(n)} \tag{2.24}$$

Collecting all the terms in (2.24) that multiply a_{11} shows that their sum is equal to

$$\sum_{p''} \operatorname{sgn} p'' \,.\, a_{2p''(2)}a_{3p''(3)}\cdots a_{np''(n)} \tag{2.25}$$

where p'' is a permutation of $1, 2, \ldots, n$ which leaves the first entry 1 fixed. But every permutation p'' corresponds to the permutation p' obtained by omitting the 1 from the extreme left and since inversion of order in p'' cannot involve 1, $\operatorname{sgn} p'' = \operatorname{sgn} p'$. Thus the expression (2.25) is just M_{11} and

$$|(a_{ij})| = a_{11}M_{11} + \text{terms not involving } a_{11}$$

Now notice that any element a_{ij} of the matrix can be brought to the top left-hand corner (that is, the 11 position) by $i - 1$ exchanges of rows and $j - 1$ exchanges of columns without disturbing the relative positions of the other rows and columns, and so without changing the minor M_{ij}. By the earlier result on the effect of a single change of rows or columns, it is seen that the process of bringing a_{ij} to the top left-hand corner multiplies $|(a_{ij})|$ by $(-1)^{i-1+j-1} = (-1)^{i+j}$ and so the sum of the terms of $|(a_{ij})|$ which multiply a_{ij} is $(-1)^{i+j}M_{ij}$. Each element of any row of (a_{ij}) occurs only once in each term of $|(a_{ij})|$, so that we must have

$$|(a_{ij})| = \sum_{j=1}^{n} a_{ij}(-1)^{i+j}M_{ij} \tag{2.26}$$

A similar result in terms of the elements of column j,

$$|(a_{ij})| = \sum_{i=1}^{n} a_{ij}(-1)^{i+j}M_{ij} \tag{2.27}$$

follows in the same way.

It is usual to simplify the expressions in (2.26) and (2.27) by giving one symbol A_{ij} to the expression $(-1)^{i+j}M_{ij}$: A_{ij} is called the *cofactor* of the element a_{ij}. Then equation (2.26), for example, becomes simply

$$|(a_{ij})| = \sum_{j=1}^{n} a_{ij}A_{ij} \tag{2.28}$$

If $|(a_{ij})|$ is expanded in terms of the first row of (a_{ij}) by putting $i = 1$ in (2.26) the result is

$$|(a_{ij})| = \sum_{j=1}^{n} a_{1j}(-1)^{j+1}M_{1j} \tag{2.29}$$

which is the same as (2.12), so the equivalence of the two definitions is established.

The sum of the products of the elements of any row of a matrix with their cofactors is equal to the determinant. What will be the result of multiplying the elements of a row by the cofactors of the corresponding elements in a different row? Clearly, the result will be the expansion of the determinant of a matrix with two identical rows, whence

$$\sum_{k=1}^{n} a_{ik}A_{jk} = 0, \quad i \neq j \tag{2.30}$$

A similar result holds for columns

$$\sum_{k=1}^{n} a_{ki}A_{kj} = 0, \quad i \neq j \tag{2.31}$$

These results will be useful when the inversion of matrices is discussed later in this chapter.

The $(n-1)$th-order minors of an $n \times n$ matrix A were defined above. A natural extension is that the determinants of order $n - m$ obtained by striking out any m rows and m columns of A should be called the *$(n - m)$th-order minors* of A. Thus the $(n - m)$th-order minors of A are just the determinants of the $(n - m)$th-order submatrices of A. The determinant of order m of the submatrix formed by the elements of the rows and columns deleted to form some minor M of A is known as the *complementary minor* of M.

Example 2.3

In the determinant

$$|A| = \begin{vmatrix} a_{11} & a_{12} & a_{13} & a_{14} \\ a_{21} & a_{22} & a_{23} & a_{24} \\ a_{31} & a_{32} & a_{33} & a_{34} \\ a_{41} & a_{42} & a_{43} & a_{44} \end{vmatrix}$$

the minors

$$\begin{vmatrix} a_{11} & a_{13} \\ a_{31} & a_{33} \end{vmatrix}, \quad \begin{vmatrix} a_{22} & a_{24} \\ a_{42} & a_{44} \end{vmatrix}$$

are complementary.

The expansion of the determinant of an nth-order matrix in terms of its rows and columns (that is, alternatively, in terms of its $(n - 1)$th-order minors) has already been demonstrated. Now consider the expansion of $|A|$ in terms of minors of any order: the result is known as *Laplace's expansion of* $|A|$.

Number an arbitrary selection of $q < n$ rows of A by $r_1, r_2, \ldots, r_q$ and the remaining $n - q$ rows (taken in any order) by $r_{q+1}, r_{q+2}, \ldots, r_n$. Notice that $r_1, r_2, \ldots, r_n$ is a permutation of $1, 2, \ldots, n$. By Exercise 2.2, $|A|$ can be written

$$|A| = \sum_{j} \delta^{r_1 r_2 \ldots r_n}_{j_1 j_2 \ldots j_n} a_{r_1 j_1} a_{r_2 j_2} \ldots a_{r_n j_n} \tag{2.32}$$

Now let $i_1, i_2, \ldots, i_n$ represent a permutation of $1, 2, \ldots, n$ such that

$$\left.\begin{array}{l} i_1 < i_2 < \ldots < i_q \\ i_{q+1} < i_{q+2} < \ldots < i_n \end{array}\right\} \tag{2.33}$$

and consider the terms on the right-hand side of (2.32) in which $j_1, j_2, \ldots, j_q$ is a derangement of $i_1, i_2, \ldots, i_q$ and $j_{q+1}, j_{q+2}, \ldots, j_n$ is a derangement of $i_{q+1}, i_{q+2}, \ldots, i_n$. There are $q!\,(n-q)!$ such terms and for any one of them it is easily seen that

$$\delta^{r_1 r_2 \ldots r_n}_{j_1 j_2 \ldots j_n} = \delta^{r_1 r_2 \ldots r_n}_{i_1 i_2 \ldots i_n} \delta^{i_1 i_2 \ldots i_q i_{q+1} \ldots i_n}_{j_1 j_2 \ldots j_q j_{q+1} \ldots j_n} \tag{2.34}$$

$$= \delta^{r_1 r_2 \ldots r_n}_{i_1 i_2 \ldots i_n} \delta^{i_1 i_2 \ldots i_q}_{j_1 j_2 \ldots j_q} \delta^{i_{q+1} i_{q+2} \ldots i_n}_{j_{q+1} j_{q+2} \ldots j_n} \tag{2.35}$$

It follows that the total contribution to $|A|$ from these terms is obtained by summing with respect to all permutations $j_1, j_2, \ldots, j_q$ of $i_1, i_2, \ldots, i_q$ and all permutations $j_{q+1}, j_{q+2}, \ldots, j_n$ of $i_{q+1}, i_{q+2}, \ldots, i_n$ to give

$$\delta^{r_1 r_2 \ldots r_n}_{i_1 i_2 \ldots i_n} \sum_j \delta^{i_1 i_2 \ldots i_q}_{j_1 j_2 \ldots j_q} a_{r_1 j_1} a_{r_2 j_2} \ldots a_{r_q j_q}$$
$$\times \sum_j \delta^{i_{q+1} i_{q+2} \ldots i_n}_{j_{q+1} j_{q+2} \ldots j_n} a_{r_{q+1} j_{q+1}} a_{r_{q+2} j_{q+2}} \ldots a_{r_n j_n} \tag{2.36}$$

The first sum in (2.36) is just the minor formed from the r_1th, r_2th, ..., r_qth rows and j_1th, j_2th, ..., j_qth columns and the second sum the corresponding complementary minor (Exercise 2.2) and so (2.36) can be written

$$\delta^{r_1 r_2 \ldots r_n}_{i_1 i_2 \ldots i_n} \begin{vmatrix} a_{r_1 i_1} & \ldots & a_{r_1 i_q} \\ \vdots & & \vdots \\ a_{r_q i_1} & \ldots & a_{r_q i_q} \end{vmatrix} \begin{vmatrix} a_{r_{q+1} i_{q+1}} & \ldots & a_{r_{q+1} i_n} \\ \vdots & & \vdots \\ a_{r_n i_{q+1}} & \ldots & a_{r_n i_n} \end{vmatrix} \tag{2.37}$$

Finally $|A|$ is obtained by summing (2.37) with respect to all possible selections of $i_1 < i_2 < \cdots < i_q$ and $i_{q+1} < i_{q+2} < \cdots < i_n$.

Denoting this summation by $\sum_i$ gives

$$|A| = \sum_i \delta^{r_1 r_2 \ldots r_n}_{i_1 i_2 \ldots i_n} \begin{vmatrix} a_{r_1 i_1} & \ldots & a_{r_1 i_q} \\ \vdots & & \vdots \\ a_{r_q i_1} & \ldots & a_{r_q i_q} \end{vmatrix} \begin{vmatrix} a_{r_{q+1} i_{q+1}} & \ldots & a_{r_{q+1} i_n} \\ \vdots & & \vdots \\ a_{r_n i_{q+1}} & \ldots & a_{r_n i_n} \end{vmatrix} \tag{2.38}$$

which is the Laplace expansion of $|A|$.

The following interpretation in words of equation (2.38) is helpful. Select any q rows of A and form all possible qth-order minors of A from these rows, keeping the original order of the rows unchanged. Now form the sum of products of these minors with their complementary minors, multiplying each term in the sum by $\delta^{r_1 r_2 \ldots r_n}_{i_1 i_2 \ldots i_n}$, where $r_1, r_2, \ldots, r_q, i_1, i_2, \ldots, i_q$ represent the rows and columns of the minor and $r_{q+1}, r_{q+2}, \ldots, r_n, i_{q+1}, i_{q+2}, \ldots, i_n$ the rows and columns of the complementary minor. The result is $|A|$. It is equally possible, of course, to carry out a similar expansion in terms of a selected number of columns rather than rows.

Example 2.4 Consider the expansion of $|A|$ of Example 2.3 in terms of its 1st and 3rd rows. The relevant minors of order 2 are

$$\begin{vmatrix} a_{11} & a_{12} \\ a_{31} & a_{32} \end{vmatrix} \quad \begin{vmatrix} a_{11} & a_{13} \\ a_{31} & a_{33} \end{vmatrix} \quad \begin{vmatrix} a_{11} & a_{14} \\ a_{31} & a_{34} \end{vmatrix}$$

$$\begin{vmatrix} a_{12} & a_{13} \\ a_{32} & a_{33} \end{vmatrix} \quad \begin{vmatrix} a_{12} & a_{14} \\ a_{32} & a_{34} \end{vmatrix} \quad \begin{vmatrix} a_{13} & a_{14} \\ a_{33} & a_{34} \end{vmatrix}$$

and their respective complementary minors are

$$\begin{vmatrix} a_{23} & a_{24} \\ a_{43} & a_{44} \end{vmatrix} \quad \begin{vmatrix} a_{22} & a_{24} \\ a_{42} & a_{44} \end{vmatrix} \quad \begin{vmatrix} a_{22} & a_{23} \\ a_{42} & a_{43} \end{vmatrix}$$

$$\begin{vmatrix} a_{21} & a_{24} \\ a_{41} & a_{44} \end{vmatrix} \quad \begin{vmatrix} a_{21} & a_{23} \\ a_{41} & a_{43} \end{vmatrix} \quad \begin{vmatrix} a_{21} & a_{22} \\ a_{41} & a_{42} \end{vmatrix}$$

Now consider the sign of

$$\begin{vmatrix} a_{11} & a_{12} \\ a_{31} & a_{32} \end{vmatrix} \begin{vmatrix} a_{23} & a_{24} \\ a_{43} & a_{44} \end{vmatrix}$$

which is just

$$\delta^{1324}_{1234} = -1.$$

The sign of

$$\begin{vmatrix} a_{11} & a_{13} \\ a_{31} & a_{33} \end{vmatrix} \begin{vmatrix} a_{22} & a_{24} \\ a_{42} & a_{44} \end{vmatrix}$$

is

$$\delta^{1324}_{1324} = +1$$

and so on. The final expansion is

$$\begin{aligned} |A| = &- \begin{vmatrix} a_{11} & a_{12} \\ a_{31} & a_{32} \end{vmatrix} \begin{vmatrix} a_{23} & a_{24} \\ a_{43} & a_{44} \end{vmatrix} + \begin{vmatrix} a_{11} & a_{13} \\ a_{31} & a_{33} \end{vmatrix} \begin{vmatrix} a_{22} & a_{24} \\ a_{42} & a_{44} \end{vmatrix} \\ &- \begin{vmatrix} a_{11} & a_{14} \\ a_{31} & a_{34} \end{vmatrix} \begin{vmatrix} a_{22} & a_{23} \\ a_{42} & a_{43} \end{vmatrix} - \begin{vmatrix} a_{12} & a_{13} \\ a_{32} & a_{33} \end{vmatrix} \begin{vmatrix} a_{21} & a_{24} \\ a_{41} & a_{44} \end{vmatrix} \\ &+ \begin{vmatrix} a_{12} & a_{14} \\ a_{32} & a_{34} \end{vmatrix} \begin{vmatrix} a_{21} & a_{23} \\ a_{41} & a_{43} \end{vmatrix} - \begin{vmatrix} a_{13} & a_{14} \\ a_{33} & a_{34} \end{vmatrix} \begin{vmatrix} a_{21} & a_{22} \\ a_{41} & a_{42} \end{vmatrix} \end{aligned}$$

EXERCISE 2.5 Deduce the expansion of a determinant in terms of its rows, equation (2.26), from Laplace's expansion (2.38).

EXERCISE 2.6 Prove Theorem 2.3 by using Laplace's expansion and the matrix equation

$$\begin{pmatrix} A & 0 \\ -I & B \end{pmatrix}\begin{pmatrix} I & B \\ 0 & I \end{pmatrix} = \begin{pmatrix} A & AB \\ -I & 0 \end{pmatrix}$$

[Use Theorem 2.2 (applied to columns) to show that multiplication by the second matrix on the left leaves the determinant of the first unchanged.]

Example 2.5

$$A = \begin{vmatrix} 1 & 1 & 0 & 0 \\ 2 & 1 & 3 & 1 \\ 0 & 1 & 1 & 3 \\ 2 & 1 & 1 & 2 \end{vmatrix}$$

$$= \begin{vmatrix} 1 & 1 \\ 2 & 1 \end{vmatrix}\begin{vmatrix} 1 & 3 \\ 1 & 2 \end{vmatrix} - \begin{vmatrix} 1 & 1 \\ 0 & 1 \end{vmatrix}\begin{vmatrix} 3 & 1 \\ 1 & 2 \end{vmatrix} + \begin{vmatrix} 1 & 1 \\ 2 & 1 \end{vmatrix}\begin{vmatrix} 3 & 1 \\ 1 & 3 \end{vmatrix}$$
$$+ \begin{vmatrix} 2 & 1 \\ 0 & 1 \end{vmatrix}\begin{vmatrix} 0 & 0 \\ 1 & 2 \end{vmatrix} - \begin{vmatrix} 2 & 1 \\ 2 & 1 \end{vmatrix}\begin{vmatrix} 0 & 0 \\ 1 & 3 \end{vmatrix} + \begin{vmatrix} 0 & 1 \\ 2 & 1 \end{vmatrix}\begin{vmatrix} 0 & 0 \\ 3 & 1 \end{vmatrix}$$

$$= -12$$

3 Matrix inversion

In Chapter 1, Section 7, matrix rings R^n over a ring R were discussed and it was noted that if R has a multiplicative identity then so has R^n. The identity in R^n is, in fact, the $n \times n$ matrix

$$I_n = \begin{pmatrix} 1 & 0 & \dots & 0 \\ 0 & 1 & \dots & 0 \\ \vdots & \vdots & & \vdots \\ 0 & 0 & \dots & 1 \end{pmatrix} \tag{3.1}$$

Now it is assumed that R is commutative and attention is turned to the units or regular elements of R^n, that is those with inverses, which (Chapter 1, Section 3) form a multiplicative group.

Let A be any matrix in R^n: then the *adjoint* or *adjunct* of A is the transpose of the matrix having as its elements the cofactors of the corresponding elements of A. The adjoint of A is denoted by adj A; thus adj $A = (A_{ji})$.

Example 3.1 Let

$$A = (a_{ij}) = \begin{pmatrix} 2 & 6 & 2 \\ 1 & 2 & 1 \\ 2 & 1 & 1 \end{pmatrix}$$

so that $|A| = 2$. Then the cofactor of a_{11} is

$$(-1)^{1+1}\begin{vmatrix} 2 & 1 \\ 1 & 1 \end{vmatrix} = 1$$

the cofactor of a_{12} is

$$(-1)^{1+2}\begin{vmatrix} 1 & 1 \\ 2 & 1 \end{vmatrix} = 1$$

and so on; giving

$$\text{adj } A = \begin{pmatrix} 1 & 1 & -3 \\ -4 & -2 & 10 \\ 2 & 0 & -2 \end{pmatrix}^{\mathrm{T}} = \begin{pmatrix} 1 & -4 & 2 \\ 1 & -2 & 0 \\ -3 & 10 & -2 \end{pmatrix}$$

It is now easily verified that

$$\begin{pmatrix} 2 & 6 & 2 \\ 1 & 2 & 1 \\ 2 & 1 & 1 \end{pmatrix}\begin{pmatrix} 1 & -4 & 2 \\ 1 & -2 & 0 \\ -3 & 10 & -2 \end{pmatrix} = \begin{pmatrix} 1 & -4 & 2 \\ 1 & -2 & 0 \\ -3 & 10 & -2 \end{pmatrix}\begin{pmatrix} 2 & 6 & 2 \\ 1 & 2 & 1 \\ 2 & 1 & 1 \end{pmatrix} = \begin{pmatrix} 2 & 0 & 0 \\ 0 & 2 & 0 \\ 0 & 0 & 2 \end{pmatrix}$$

and in the last matrix the diagonal elements are equal to $|A|$.

It follows from (2.28) and (2.30) that

$$(a_{ij})(A_{ji}) = (A_{ji})(a_{ij}) = \text{diag } (|A|) \tag{3.2}$$

where *diag* (d) means a matrix (d_{ij}) with $d_{ij} = d$, $i = j$ and $d_{ij} = 0$, $i \neq j$. Then provided that $|A|$ is a regular element of the ring R, with inverse $|A|^{-1}$,

$$(a_{ij})(|A|^{-1}A_{ji}) = (|A|^{-1}A_{ji})(a_{ij}) = I \tag{3.3}$$

Therefore, if $|A|$ is a regular element of R, the matrix A is a regular element of R^n and

$$A^{-1} = |A|^{-1} \text{ adj } A \tag{3.4}$$

Conversely, if $A, B \in R^n$ and $AB = I$, then by Theorem 2.3

$$|AB| = |A|\,|B| = |I| = 1$$

showing that $|A|$ and $|B|$ must be regular in the commutative ring R. It then follows as above that A has an inverse A^{-1}, and so

$$A^{-1}[A(B - A^{-1})] = A^{-1}0 = 0, \text{ and } B = A^{-1}.$$

EXERCISE 3.1 Prove that every $n \times n$ matrix over a field F has an inverse provided that its determinant is not zero.

Example 3.2 If R is the ring of integers then

$$A = \begin{pmatrix} 1 & 1 & 1 \\ 0 & -1 & 0 \\ 2 & 3 & 1 \end{pmatrix}$$

is invertible and

$$A^{-1} = \begin{pmatrix} -1 & 2 & 1 \\ 0 & -1 & 0 \\ 2 & -1 & -1 \end{pmatrix}$$

On the other hand

$$B = \begin{pmatrix} 1 & 2 & 1 \\ 3 & 1 & 1 \\ 0 & 0 & 1 \end{pmatrix}$$

is not invertible since $|B| = -5$ which is not regular in the ring of integers.

Example 3.3 If R is the ring of polynomials over the field of real numbers, then

$$A = \begin{pmatrix} 1 + x^2 & x \\ 2x & 2 \end{pmatrix}$$

is invertible and

$$A^{-1} = \begin{pmatrix} 1 & -\frac{1}{2}x \\ -x & \frac{1}{2}(1 + x^2) \end{pmatrix}$$

If on the other hand R is the ring of polynomials over the ring of integers (that is, R consists of polynomials with integer coefficients) then A has no inverse. [$|A| = 2$, which is regular in R in the first case, but not in the second.]

EXERCISE 3.2 If A and B are any two $n \times n$ matrices over a field F, prove the following results:

(a) $(AB)^{-1} = B^{-1}A^{-1}$

(b) $(A^{-1})^{\mathrm{T}} = (A^{\mathrm{T}})^{-1}$

4 Determinantal rank

Any square submatrix of a rectangular $n \times m$ matrix over a field F will have a determinant. The *determinantal rank* of A is defined as the order of the largest submatrix of A with non-zero determinant.

Example 4.1 Considered as a matrix over the ring of polynomials having real coefficients

$$A = \begin{pmatrix} 1 + x & 0 \\ 0 & 1 \end{pmatrix}$$

has determinantal rank 2 because $|A| = 1 + x$, which is not the zero polynomial. If x is given a numerical value, A can be regarded as a matrix over the field of real numbers: when $x = -1$ the determinantal rank of A is then 1.

If the rank of A is denoted by $\rho(A)$, it is obvious that $\rho(A) \leqslant n$ and $\rho(A) \leqslant m$. It is also evident from the expansions of a determinant that any matrix of rank r must contain at least one non-zero determinant of all orders less than r.

An important property of the determinantal rank of a matrix is that it is invariant under the following *elementary* (*row* or *column*) *operations*.

(a) Interchange row (column) i and row (column) j.
(b) Multiply row (column) i by a non-zero scalar a.
(c) Add row (column) i to row (column) j.

Invariance under (a) and (b) is obvious. Let the number of non-zero minors of any order k before the operation (c) be s. After the operation, each minor of order k will either be unchanged or will be the sum of two of the original minors: in the second case one of the two original minors involved in the sum persists unchanged. From this it follows that the number of non-zero minors of order k after the operation does not exceed $2s$, nor is it less than s if s is even, or $(s + 1)$ if s is odd. The invariance of determinantal rank under operation (c) follows at once.

5 Solution of systems of linear equations

Consider the system of m equations

$$\left.\begin{aligned} a_{11}x_1 + a_{12}x_2 + \cdots + a_{1n}x_n &= b_1 \\ a_{21}x_1 + a_{22}x_2 + \cdots + a_{2n}x_n &= b_2 \\ &\vdots \\ a_{m1}x_1 + a_{m2}x_2 + \cdots + a_{mn}x_n &= b_m \end{aligned}\right\} \tag{5.1}$$

in the n unknowns $x_1, x_2, \ldots, x_n$ with the a_{ij} and the b_i belonging to a field F. The system (5.1) can be written in matrix notation

$$Ax = b \tag{5.2}$$

where

$$A = (a_{ij}), \qquad x = \begin{pmatrix} x_1 \\ x_2 \\ \vdots \\ x_n \end{pmatrix} \qquad \text{and} \qquad b = \begin{pmatrix} b_1 \\ b_2 \\ \vdots \\ b_m \end{pmatrix}$$

With each system of equations (5.2) it is possible to associate the set

$$Ax = 0 \tag{5.3}$$

of *homogeneous* equations. It follows directly from (5.2) and (5.3) that if y and z are any two solutions of (5.2) then $y - z$ is a solution of (5.3). Again if y is a solution of (5.2) and z is a solution of (5.3) then $y + z$ is also a solution of (5.2). In general, therefore, all the solutions of (5.2) are obtained by adding the solutions of (5.3) to some particular solution of (5.2).

A system of linear equations is said to be *consistent* if it has at least one solution, and is *inconsistent* otherwise. If a consistent system has a unique solution then it is called *determinate*, and otherwise *indeterminate*.

EXERCISE 5.1 Prove that if a system of linear equations is indeterminate then it must have an infinite set of solutions. [There is at least one non-zero solution of (5.3), hence an infinite set of solutions.]

Suppose now that A in (5.2) is a regular square $n \times n$ matrix. Then

$$\begin{aligned} x &= A^{-1}b \\ &= |A|^{-1}(\text{adj } A)b \end{aligned}$$

so that

$$x_r = |A|^{-1} \sum_{i=1}^{n} A_{ir}b_i, \quad r = 1, 2, \ldots, n \tag{5.4}$$

The sum in (5.4), however, is just the cofactor expansion of the determinant of the matrix obtained by replacing the rth column of A by b. The result given in equation (5.4) is known as *Cramer's rule.*

Example 5.1 Solve the equations

$$\begin{aligned} 2x_1 + x_2 + x_3 &= 2 \\ x_1 + 2x_2 + x_3 &= 2 \\ 3x_1 + x_2 + x_3 &= 1 \end{aligned}$$

Here

$$A = \begin{pmatrix} 2 & 1 & 1 \\ 1 & 2 & 1 \\ 3 & 1 & 1 \end{pmatrix}, \qquad b = \begin{pmatrix} 2 \\ 2 \\ 1 \end{pmatrix}$$

therefore $|A| = -1 \neq 0$ and

$$x_1 = \frac{\begin{vmatrix} 2 & 1 & 1 \\ 2 & 2 & 1 \\ 1 & 1 & 1 \end{vmatrix}}{-1} = -1$$

$$x_2 = \frac{\begin{vmatrix} 2 & 2 & 1 \\ 1 & 2 & 1 \\ 3 & 1 & 1 \end{vmatrix}}{-1} = -1$$

$$x_3 = \frac{\begin{vmatrix} 2 & 1 & 2 \\ 1 & 2 & 2 \\ 3 & 1 & 1 \end{vmatrix}}{-1} = 5$$

The reader should note that although Cramer's rule is very useful theoretically, it is not an efficient numerical technique for solving equations owing to the difficulty of evaluating the determinants.

Attention is now given to the solution of (5.1) when A is rectangular.

Theorem 5.1 Equations (5.1) are consistent if and only if the determinantal rank of A and that of the matrix B given by

$$B = \begin{pmatrix} a_{11} & a_{12} & \dots & a_{1n} & b_1 \\ a_{21} & a_{22} & \dots & a_{2n} & b_2 \\ \vdots & \vdots & & \vdots & \vdots \\ a_{m1} & a_{m2} & \dots & a_{mn} & b_m \end{pmatrix} \tag{5.5}$$

are the same. The matrix A in the system of equations (5.2) is called the *coefficient matrix* of the system and the matrix B given by (5.5) is known as the *augmented matrix* of the system.

Proof The first step in the proof is to reduce the matrices A and B to a more manageable form by a process known as *Gauss reduction*. Consider the set of simultaneous equations (5.1), assuming that $a_{11} \neq 0$ (if this assumption does not hold, rearrange the equations so that the first coefficient of the first equation is not zero). Now divide the first equation by a_{11}, multiply the result by $a_{21}, a_{31}, \ldots, a_{m1}$ in turn and subtract the equations so formed from the 2nd, 3rd, ..., mth equations of (5.1) to give

$$\left.\begin{aligned} x_1 + a'_{12}x_2 + \ldots + a'_{1n}x_n &= b'_1 \\ a'_{22}x_2 + \ldots + a'_{2n}x_n &= b'_2 \\ \vdots \quad & \\ a'_{m2}x_2 + \ldots + a'_{mn}x_n &= b'_m \end{aligned}\right\} \tag{5.6}$$

Now suppose that $a'_{22} \neq 0$ (if not, rearrange) and divide the second of equations (5.6) by a'_{22}. Multiply the result by $a'_{12}, a'_{32}, \ldots, a'_{m2}$ successively and subtract the equations so formed from the 1st, 3rd, ..., mth equations of (5.6). This has the effect of eliminating x_2 and gives

$$\left.\begin{aligned} x_1 + a''_{13}x_3 + \ldots + a''_{1n}x_n &= b''_1 \\ x_2 + a''_{23}x_3 + \ldots + a''_{2n}x_n &= b''_2 \\ a''_{33}x_3 + \ldots + a''_{3n}x_n &= b''_3 \\ \vdots \quad & \\ a''_{m3}x_3 + \ldots + a''_{mn}x_n &= b''_m \end{aligned}\right\} \tag{5.7}$$

This process can obviously be repeated r times until either $r = n$, in which case the last equation contains only x_n on its left-hand side, or all the coefficients vanish on the left-hand sides of all the equations remaining after the rth. The final result after r steps of the process will be

$$A^*x = b^* \tag{5.8}$$

where

$$A^* = \begin{pmatrix} 1 & 0 & \ldots & 0 & a^*_{1,r+1} & \ldots & a^*_{1n} \\ 0 & 1 & \ldots & 0 & a^*_{2,r+1} & \ldots & a^*_{2n} \\ \vdots & \vdots & & \vdots & \vdots & & \vdots \\ 0 & 0 & \ldots & 1 & a^*_{r,r+1} & \ldots & a^*_{rn} \\ 0 & 0 & \ldots & 0 & 0 & \ldots & 0 \\ \vdots & \vdots & & \vdots & \vdots & & \vdots \\ 0 & 0 & \ldots & 0 & 0 & \ldots & 0 \end{pmatrix} \tag{5.9}$$

and

$$b^* = \begin{pmatrix} b_1^* \\ b_2^* \\ \vdots \\ b_m^* \end{pmatrix} \tag{5.10}$$

The final form of the augmented matrix B^* will be that of A^* with a column equal to b^* added at the extreme right.

The form of A^* shows directly that its largest non-vanishing determinant is of order r, so that $\rho(A^*) = r$. The rank of B^* will be $r + 1$ if one or more of b_{r+1}^*, $b_{r+2}^*, \ldots, b_m^*$ is non-zero, and will be r if these are all zero. Now notice that the non-vanishing of b_{r+i}^*, $1 \leqslant i \leqslant m - r$, involves the contradiction $0 = b_{r+i}^*$ so that the equations are consistent only if $b_{r+i}^* = 0$, $i = 1, 2, \ldots, m - r$: this condition is also sufficient as can be seen by putting $x_{r+1} = x_{r+2} = \cdots = x_n = 0$. Therefore the equations are consistent if and only if $\rho(A^*) = \rho(B^*) = r$.

To complete the proof it is only necessary to decide the relationship between $\rho(A)$ and $\rho(A^*)$, and between $\rho(B)$ and $\rho(B^*)$. The operations involved in obtaining A^* from A and B^* from B consist of interchanging two rows or columns, multiplication of a row by a non-zero constant, and adding one row to another. These are elementary operations, which by Section 4 cannot alter the determinantal rank of a matrix, so that $\rho(A) = \rho(A^*)$, $\rho(B) = \rho(B^*)$ and the proof is complete.

Suppose then that $\rho(A) = \rho(B) = r < n$. The equations (5.1) are certainly consistent and from the reduced form (5.8) it is seen that $n - r$ of the unknowns can be given arbitrary values and the remaining r unknowns found by solving the resulting r equations by Cramer's rule (or, of course, by Gauss reduction). Consequently, when $r < n$ an infinite number of solutions exist. If, however, $r = n$ there is a unique solution.

When the system (5.1) is homogeneous, that is, when $b = 0$, it follows that $\rho(A) = \rho(B)$ and the equations must always be consistent. It is obvious that $x = 0$ is a solution of $Ax = 0$ and from the above reasoning this trivial solution is the only solution if $\rho(A) = n$. On the other hand if $\rho(A) < n$ an infinite number of solutions exist and $n - r$ of the unknowns can be arbitrarily fixed.

EXERCISE 5.2 Interpret the above results for both the homogeneous and non-homogeneous cases in terms of the number of equations and number of unknowns.

EXERCISE 5.3 Prove that a square matrix is regular if and only if it can be reduced to the identity matrix by a sequence of elementary row and column operations.

The subject of systems of linear equations will be considered again in the next chapter.

Example 5.2 Solve the set of equations in Example 5.1 by Gauss reduction.

$$\begin{aligned} 2x_1 + x_2 + x_3 &= 2 \\ x_1 + 2x_2 + x_3 &= 2 \\ 3x_1 + x_2 + x_3 &= 1 \end{aligned}$$

The first step gives

$$\begin{aligned} x_1 + \tfrac{1}{2}x_2 + \tfrac{1}{2}x_3 &= 1 \\ 0 + \tfrac{3}{2}x_2 + \tfrac{1}{2}x_3 &= 1 \\ 0 - \tfrac{1}{2}x_2 - \tfrac{1}{2}x_3 &= -2 \end{aligned}$$

The second step gives

$$\begin{aligned} x_1 + 0 + \tfrac{1}{3}x_3 &= \tfrac{2}{3} \\ 0 + x_2 + \tfrac{1}{3}x_3 &= \tfrac{2}{3} \\ 0 + 0 - \tfrac{1}{3}x_3 &= -\tfrac{5}{3} \end{aligned}$$

The third step gives

$$\begin{aligned} x_1 + 0 + 0 &= -1 \\ 0 + x_2 + 0 &= -1 \\ 0 + 0 + x_3 &= 5 \end{aligned}$$

Example 5.3 Solve the equations

$$\begin{aligned} x + y + 2z &= -4 \\ x + 4y + 3z &= -11 \\ 2x + 5y + 5z &= 9 \\ 3x - 3y + 4z &= 2 \end{aligned}$$

by Gauss reduction.

The first step gives

$$\begin{aligned} x + y + 2z &= -4 \\ 0 + 3y + z &= -7 \\ 0 + 3y + z &= 17 \\ 0 - 6y - 2z &= 14 \end{aligned}$$

The second step gives

$$\begin{aligned} x + 0 + \tfrac{5}{3}z &= -\tfrac{5}{3} \\ 0 + y + \tfrac{1}{3}z &= -\tfrac{7}{3} \\ 0 + 0 + 0 &= 24 \\ 0 + 0 + 0 &= 0 \end{aligned}$$

Hence the equations are inconsistent and no solution exists.

Example 5.4 Solve the equations

$$\begin{aligned} x + y - 2z + w &= 5 \\ x + 2y + z - 3w &= 2 \\ 2x + y - z + 3w &= -1 \end{aligned}$$

The first step gives

$$\begin{aligned} x + y - 2z + w &= 5 \\ 0 + y + 3z - 4w &= -3 \\ 0 - y + 3z + w &= -11 \end{aligned}$$

The second step gives

$$\begin{aligned} x + 0 - 5z + 5w &= 8 \\ 0 + y + 3z - 4w &= -3 \\ 0 + 0 + 6z - 3w &= -14 \end{aligned}$$

Thus the system is consistent and one of the variables can be chosen arbitrarily provided the determinant of the coefficients of the remaining variables does not vanish. For example, if the value of w is taken to be 4 then the solution is $z = -\frac{1}{3}$, $y = 14$, $x = -13\frac{2}{3}$.

Further reading

Much of the work in Chapter 2 is dealt with in the references given at the end of Chapter 3. For a discussion of determinants, however, refer particularly to Mirsky (1955), Marcus and Minc (1965), and Ferrar (1941). Further work on the computational aspects of linear equations can be found in Faddeev and Faddeeva (1963) and Householder (1964).

References

FADDEEV, D. K. and V. N. FADDEEVA, 1963, *Computation methods of linear algebra*, W. H. Freeman, San Francisco.

FERRAR, W. L., 1941, *Algebra*, Oxford University Press, London.

HOUSEHOLDER, A. S., 1964, *The theory of matrices in numerical analysis*, Blaisdell, Waltham, Mass.
MARCUS, M. and H. MINC, 1965, *Introduction to linear algebra*, Macmillan, New York.
MIRSKY, L., 1955, *An introduction to linear algebra*, Clarendon Press, Oxford.
STEWARD, F. M., 1963, *Introduction to linear algebra*, Van Nostrand, Princeton, N.J.

Problems

1 If A and B are any two square, regular matrices show that

$$\begin{pmatrix} A & C \\ 0 & B \end{pmatrix}^{-1} = \begin{pmatrix} A^{-1} & -A^{-1}CB^{-1} \\ 0 & B^{-1} \end{pmatrix}$$

where C is an arbitrary matrix of suitable size. (The use of partitioning in this way can sometimes ease the computation of the inverse of a matrix.)

2 Show that the inverse of a quasi-diagonal matrix

$$C = \text{diag}\,(C_1, C_2, \ldots, C_n)$$

is given by

$$C^{-1} = \text{diag}\,(C_1^{-1}, C_2^{-1}, \ldots, C_n^{-1})$$

3 Calculate

$$\begin{pmatrix} 1 & 1 & 0 & 0 & 0 \\ 0 & 3 & 0 & 0 & 0 \\ 0 & 0 & 1 & 2 & 0 \\ 0 & 0 & 0 & 1 & 1 \\ 0 & 0 & 1 & 0 & 1 \end{pmatrix}^{-1}$$

4 Show that the determinant

$$\begin{vmatrix} 1 & x_1 & x_1^2 & \ldots & x_1^{n-1} \\ 1 & x_2 & x_2^2 & \ldots & x_2^{n-1} \\ \vdots & \vdots & \vdots & & \vdots \\ 1 & x_n & x_n^2 & \ldots & x_n^{n-1} \end{vmatrix}$$

is given by the product

$$\prod_{1 \leqslant i < j \leqslant n} (x_j - x_i)$$

(The above is called a *Vandermonde determinant* of order n.)

5 Prove that the determinant of a quasi-diagonal matrix is equal to the product of the determinants of the diagonal block matrices.

6 Show that for any $n \times n$ matrix, A,

$$\det\,(\text{adj}\,A) = (\det A)^{n-1}$$

7 If $a_2 \neq 0$ and $b_3 \neq 0$, give a necessary and sufficient determinantal condition for the two equations

$$a_0 + a_1x + a_2x^2 = 0$$
$$b_0 + b_1x + b_2x^2 + b_3x^3 = 0$$

to have a common root.

8 Prove that the determinant of a diagonal matrix or an upper (or lower) triangular matrix (that is, a matrix with all the elements below, or above, its principal diagonal equal to zero) is the product of the elements on the principal diagonal.

9 Use Gauss reduction to solve the set of linear equations

$$2x_1 + x_2 + x_3 = 1$$
$$x_1 + 2x_2 + x_3 = 3$$
$$x_1 + x_2 + 3x_3 = 2$$

10 Solve the matrix equation

$$A^{\mathrm{T}}B + BA = -C$$

for B when

$$A = \begin{pmatrix} -1 & 0 \\ 0 & -2 \end{pmatrix}$$

and

$$C = 3\begin{pmatrix} 1 & 2 \\ 2 & 4 \end{pmatrix}$$

3 Linear vector spaces

In this chapter the concept of a linear vector space is introduced and its relationship with matrices and linear transformations examined. A number of special types of matrix are discussed and this leads on to properties of eigenvalues, eigenvectors, and quadratic forms. As far as possible the properties of matrices are developed in conjunction with the properties of the associated linear transformations. Where this approach would have led too far from the objectives of the book, a direct approach has been taken.

1 Linear vector spaces

The reader will already be familar with geometric vectors in two and three dimensions. A three-dimensional vector is a set of three real numbers ξ_1, ξ_2, ξ_3, which represent the coordinates of the vector relative to some specified set of coordinate axes. A vector is represented by a single symbol as follows

$$x = (\xi_1, \xi_2, \xi_3)$$

Where no confusion can arise the commas are often omitted: $(\xi_1\ \xi_2\ \xi_3)$. Also the elements of x are often written x_1, x_2, x_3. The sum of two vectors is obtained by adding their coordinates, so that if

$$x = (\xi_1, \xi_2, \xi_3)$$
$$y = (\eta_1, \eta_2, \eta_3)$$

then

$$x + y = (\xi_1 + \eta_1, \xi_2 + \eta_2, \xi_3 + \eta_3)$$

The product of a vector by a real number is obtained by multiplying each coordinate by the real number,

$$\alpha x = (\alpha\xi_1, \alpha\xi_2, \alpha\xi_3)$$

A natural extension of these ideas is to consider the set of n-tuples $x=(\xi_1, \xi_2, \ldots, \xi_n)$

with the ξ_i elements of some field F and addition and scalar multiplication (by an element of F) defined as above. That is to say, if

$$x = (\xi_1, \xi_2, \ldots, \xi_n)$$
$$y = (\eta_1, \eta_2, \ldots, \eta_n)$$

then

$$x + y = (\xi_1 + \eta_1, \xi_2 + \eta_2, \ldots, \xi_n + \eta_n) \tag{1.1}$$

and

$$\alpha x = (\alpha\xi_1, \alpha\xi_2, \ldots, \alpha\xi_n) \tag{1.2}$$

where $\alpha \in F$.

Each of the n-tuples is called a *vector* and the set of all such vectors with addition and scalar multiplication defined as in (1.1) and (1.2) is called the *vector space $V_n(F)$ of n-tuples over the field F*. It is a simple matter to verify from the properties of a field given in Section 4 of Chapter 1 that $V_n(F)$ forms an additive group under addition and that scalar multiplication has the following properties:

$$\alpha(x + y) = \alpha x + \alpha y \tag{1.3}$$
$$(\alpha + \beta)x = \alpha x + \beta x \tag{1.4}$$
$$\alpha(\beta x) = (\alpha\beta)x \tag{1.5}$$
$$1x = x \tag{1.6}$$

where $\alpha, \beta \in F$ and 1 is the identity of F.

A further generalization is now made and an *abstract vector space over a field F* is defined as a set V of elements called vectors together with operations called addition and scalar multiplication such that V is an additive group under addition and the laws (1.3)–(1.6) above are satisfied for all $x, y \in V$ and $\alpha, \beta \in F$.

Example 1.1 It has already been demonstrated that the set of n-tuples $V_n(F)$ with elements in a field F forms a vector space.

Example 1.2 The set of all polynomials in an indeterminate x with coefficients in a field F forms a vector space. Addition and scalar multiplication are defined as follows: if

$$a = a_0 + a_1x + \ldots + a_nx^n$$
$$b = b_0 + b_1x + \cdots + b_mx^m, \quad m < n \text{ (say)}$$

then

$$a + b = (a_0 + b_0) + (a_1 + b_1)x + \cdots + (a_m + b_m)x^m + \cdots + a_nx^n$$

and

$$\alpha a = \alpha a_0 + \alpha a_1 x_1 + \cdots + \alpha a_n x^n$$

where $\alpha \in F$.

Example 1.3 Consider the set of all functions f mapping a field F into some arbitrary set S and define addition and scalar multiplication by the rules

$$(f_1 + f_2)(\alpha) = f_1(\alpha) + f_2(\alpha), \quad \alpha \in F$$

and

$$(\beta f_1)(\alpha) = \beta f_1(\alpha), \quad \alpha, \beta \in F$$

It is easy to verify that the functions, f, together with addition and scalar multiplication as defined, form a vector space V. The function which is identically zero on F is the zero element of V. Note that in the equation $f(x) = 0$, the zero is the zero of the field F, while in $f = 0$ the zero is the zero element of V. If for some reason the function (and not its value at x) is written $f(x)$, then to express $f = 0$ we write $f(x) \equiv 0$. (See Example 1.7.)

Example 1.4 The free motion of a dynamical system is governed by the differential equation

$$\ddot{x}(t) + 3\dot{x}(t) + 2x(t) = 0$$

having solutions x defined by

$$x(t) = c_1 e^{-t} + c_2 e^{-2t}$$

where c_1 and c_2 are arbitrary real constants. The solutions x form a vector space.

Exercise 1.1 Prove that any field is a vector space over any of its subfields. Give some examples.

Exercise 1.2 Show that the set of all possible sequences

$$x = (\xi_1, \xi_2, \ldots)$$

with elements in a field F and with only a finite number of non-zero elements forms a vector space over F when addition and scalar multiplication are defined as follows

$$x + y = (\xi_1 + \eta_1, \xi_2 + \eta_2, \ldots, \xi_n + \eta_n, \ldots)$$
$$\alpha x = (\alpha\xi_1, \alpha\xi_2, \ldots, \alpha\xi_n, \ldots)$$

What is the relationship of this vector space to that of Example 1.2?

EXERCISE 1.3 If x is any vector in a vector space V and 0 is the scalar zero, prove that

$$0.x = 0$$

(The zero on the right is, of course, the zero vector.)

1.1 Subspaces

A non-vacuous subset W of a vector space V over a field F is said to be a *subspace* of V if it is closed under addition and scalar multiplication. That is, for all $x, y \in W$ and $\alpha \in F$ we require $x + y \in W$ and $\alpha x \in W$. Obviously the set $\{0\}$ consisting of the zero vector, and V itself, satisfy these conditions and are therefore subspaces of V. They are called the trivial subspaces of V and any other subspace is said to be *proper* (compare Chapter 1, Section 2). Any subspace of a vector space is itself a vector space.

EXERCISE 1.4 Show that the set of all 5-tuples of the form $(\xi_1, \xi_2, 0, 0, 0)$ where $\xi_1, \xi_2 \in F$ is a subspace of $V_5(F)$. Generalize this result.

Now suppose that W is an arbitrary subset of a vector space V and consider the set sp (W) of all linear combinations of a finite number of elements of W. If $\sum_{i=1}^{n} \alpha_i x_i$ and $\sum_{i=1}^{m} \beta_i x_i$, with $m \leqslant n$ say, are two such combinations then their sum

$$\sum_{i=1}^{m} (\alpha_i + \beta_i)x_i + \sum_{i=m+1}^{n} \alpha_i x_i$$

and the product

$$\beta\left\{\sum_{i=1}^{n} \alpha_i x_i\right\} = \sum_{i=1}^{n} \beta\alpha_i x_i$$

are also in sp (W). It follows that sp (W) is a subspace of V: it is called the subspace of V *spanned* by W, or the *span* of W. If W has a finite number of elements $x, y, z, \ldots$ then sp (W) is written sp $(x, y, z, \ldots)$.

Let W_r be any collection of subspaces of a vector space V where the index r ranges over the elements of some subset R of the non-negative integers. Then if x and y are elements of the intersection $\bigcap_r W_r$ of all the members of the collection, it follows that $x, y \in W_r$ for every $r \in R$. But since each W_r is a subspace of V it is closed with respect to scalar multiplication and addition and hence the intersection is also closed with respect to these two operations, proving that $\bigcup_r W_r$ is also a subspace of V.

EXERCISE 1.5 Prove that the span sp (W) of any non-vacuous subset W of a vector space V is the smallest subspace containing all the vectors in W. [Prove that if any element of sp (W) is omitted the result is not a subspace containing all the vectors of W.]

1.2 Linear dependence

A set of vectors $x_1, x_2, \ldots, x_n$ is said to be *linearly dependent* if a set of scalars α_i can be found which are not all zero and for which

$$\sum_{i=1}^{n} \alpha_i x_i = 0$$

If no such set of scalars exists then the vectors are *linearly independent*. It follows that at least one of a set of linearly dependent vectors can be expressed as a linear combination of the others: it is then said to be linearly dependent on the others. Thus a vector which is linearly dependent on some set of vectors lies in their span. Notice that for a single vector to be linearly dependent it must be zero. If a subset of any set of vectors is linearly dependent it follows, by augmenting the set of scalars for the subset with sufficient zeros, that the set is itself linearly dependent.

A subset of a vector space V is called a *basis* for V provided that it is linearly independent and spans V. If a basis for V has a finite number of elements then V is said to be *finite-dimensional* and the number of elements in the basis is called the *dimension* of V and denoted by dim V. When a vector of V is expressed as a linear combination of some basis of V, the coefficients in the linear combination are known as the coordinates of the vector relative to the basis. Justification for the above terminology lies in the following theorem.

Theorem 1.1 If a finite basis with a certain number of elements is found for a vector space V, then any other basis has the same number of elements.

Proof Let $S_1 = (a_1, a_2, \ldots, a_n)$ be a set of vectors spanning V and let $b_1, b_2, \ldots, b_m$ be a linearly independent subset of V. Since S_1 spans V it is possible to write b_m (say) as a linear combination of the a_i, and certainly $b_m \cup S_1$ spans V. But $b_m \cup S_1$ is linearly dependent so that one of the a_i can be expressed as a linear combination of the rest of the elements in this set. If the result of deleting this element from S_1 is called S_2, then $b_m \cup S_2$ must span V. It now follows that $(b_{m-1}, b_m) \cup S_2$ spans V and is linearly dependent. One of the remaining a_i can again be expressed in terms of the other members of the set $(b_{m-1}, b_m) \cup S_2$ because b_{m-1} and b_m are linearly independent. If S_3 is the result of deleting this a_i from S_2 then $(b_{m-1}, b_m) \cup S_3$ spans V. This process can be repeated until a set $(b_1, b_2, \ldots, b_m) \cup S_{m+1}$ is obtained which spans V. The final set S_{m+1} will either be vacuous or will contain some of the a_i, proving that $n \geqslant m$. This shows that any set of vectors spanning V must have dimension greater than or equal to that of any linearly independent subset of V. As a direct consequence of the fact that a basis is a linearly independent set and spans V, it follows that the dimensions of any two bases must be equal. For if the dimensions of two bases are n and m, both $n \geqslant m$ and $m \geqslant n$.

For the rest of this chapter the discussion is confined to finite-dimensional vector spaces.

EXERCISE 1.6 Prove that any set of $n + 1$ vectors in a vector space of n dimensions must be linearly dependent.

EXERCISE 1.7 Prove that if V is an n-dimensional vector space then any set of $m \leqslant n$ linearly independent vectors can be augmented by a further $n - m$ vectors to form a basis for V. [Start with a basis and replace m of its vectors with the given vectors, as in the preceding proof.]

EXERCISE 1.8 Prove that the representation of any vector of V in terms of a basis of V is unique. [Assume the contrary and write down the difference between two representations of V.]

Example 1.5 Consider the vector spacc $V_n(F)$ of n-tuples over a field F. Then the set of vectors

$$e_i = (e_{ij}) = (\delta_{ij}); \quad i = 1, 2, \ldots, n; \quad j = 1, 2, \ldots, n$$

where the *Kronecker delta* δ_{ij} is defined by

$$\delta_{ij} = 1, \quad i = j$$
$$\delta_{ij} = 0, \quad i \neq j$$

forms a basis for $V_n(F)$. That the e_i are linearly independent is obvious; they also span $V_n(F)$ since any vector $x = (\xi_1, \xi_2, \ldots, \xi_n)$ can be written $x = \sum_{i=1}^{n} \xi_i e_i$, that is, the coefficients in the linear combination are just the elements of x.

This basis is known as the *standard basis* for $V_n(F)$. Notice that this has also shown that dim $V_n(F) = n$.

Example 1.6 The space defined by the function x in Example 1.4 has dimension 2. For it is easy to show that there are no non-zero α_1, α_2 such that $\alpha_1 e^{-t} + \alpha_2 e^{-2t}$ is the zero function†. The functions e^{-t} and e^{-2t} form a basis for the space.

Example 1.7 The functions† e^{-at}, where a is a real number, may be put into one-to-one correspondence (see Chapter 8) with the functions† $(s + a)^{-1}$ of the complex variable s, and this correspondence preserves sums and scalar

† These examples illustrate the loose but convenient usage mentioned in Section 8 of Chapter 1. More exactly, one would write, for example, 'there are no non-zero α_1, α_2 such that the function x, defined by $x(t) = \alpha_1 e^{-t} + \alpha_2 e^{-2t}$, is zero'.

multiples: that is, $\alpha_1 e^{-a_1 t} + \alpha_2 e^{-a_2 t}$ corresponds to $\alpha_1(s + a_1)^{-1} + \alpha_2(s + a)^{-1}$. Show from this, and also directly, that $(s + 1)^{-1}$ and $(s + 2)^{-1}$ span a two-dimensional space. [If there exist non-zero α_1, α_2 such that $\alpha_1(s + 1)^{-1} + \alpha_2(s + 2)^{-1} \equiv 0 \equiv 0(s + 1)^{-1} + 0(s + 2)^{-1}$, then $\alpha_1 e^{-t} + \alpha_2 e^{-2t} \equiv 0$.]

The preceding example is an illustration of the *isomorphism* of two spaces: spaces $V(F)$ and $W(F)$ are said to be isomorphic if there is a one-to-one correspondence between their vectors such that when v_1, v_2 correspond to w_1, w_2, $v_1 + v_2$ corresponds to $w_1 + w_2$ and αv_1 corresponds to αw_1 for any scalar α. It is easy to show from this definition that the zero vector in V corresponds to the zero vector in W; that a linearly independent set of vectors in V corresponds to a linearly independent set in W; and that a set of vectors spanning V corresponds to a set spanning W. Hence two finite-dimensional spaces which are isomorphic have the same dimension. Conversely, any vector space $V(F)$ of dimension n is easily shown to be isomorphic to $V_n(F)$, and hence to any other space $W(F)$ of dimension n.

EXERCISE 1.9 If S is a subspace of a vector space V, prove that $\dim S \leqslant \dim V$ with the equality sign holding only when $S = V$.

Example 1.8 The set $\{0\}$ consisting of the zero vector is a subspace of any space, but its dimension is undefined: the subspace is spanned by the vector 0, but this does not satisfy the definition of linear independence. It is usual to agree that $\{0\}$ is a subspace of dimension zero in order to achieve uniformity in results such as Theorem 1.2 below.

1.3 Direct sums

If all of $S_1, S_2, \ldots, S_n$ are subsets of a vector space V then the set of vectors formed by all possible sums of vectors, one taken from each subset, is called the *sum* or *join* of $S_1, S_2, \ldots, S_n$ and is written $S_1 + S_2 + \ldots + S_n$ or $\sum_{i=1}^{n} S_i$. The sum of a set of subspaces is itself a subspace of V. Notice that the union of a set of subspaces is not necessarily a subspace and hence is of little importance in the study of vector spaces.

Example 1.9 With the usual notation, the line $x = 0$ is a subspace S_1 of the plane, and the line $y = 0$ is another subspace S_2. The union $S_1 \cup S_2$ is the set of points lying on one or other of the coordinate axes: this set is not closed under addition and so is not a space. The sum $S_1 + S_2$ is the plane.

A vector space V is said to be the *direct sum* of subspaces $S_1, S_2, \ldots, S_n$ if every vector in V can be represented in one and only one way as a sum of vectors, one taken from each of the subspaces S_i. In this case V is written

$$V = S_1 \oplus S_2 \oplus \cdots \oplus S_n$$

Suppose that S_1 and S_2 are two subspaces of a vector space V and that $S_1 \cap S_2 = \{0\}$ (where 0 represents the zero vector). It follows that if any vector $x \in S_1 + S_2$ can be represented by two different sums $x_1 + x_2$ and $x_1' + x_2'$, with $x_1, x_1' \in S_1$ and $x_2, x_2' \in S_2$, then $x_1 - x_1' = x_2' - x_2$. Therefore both $x_1 - x_1'$ and $x_2' - x_2$ must lie in $S_1 \cap S_2$ and so must equal the zero vector. The representation of x as a sum of two vectors $x_1 \in S_1$ and $x_2 \in S_2$ is thus shown to be unique and the sum $S_1 + S_2$ is direct. When $V = S_1 \oplus S_2$ the subspaces S_1 and S_2 are said to be *complementary subspaces* of V. Extending these ideas to more than two subspaces, it is easily seen that

$$V = S_1 \oplus S_2 \oplus \ldots \oplus S_n$$

provided

$$V = S_1 + S_2 + \cdots + S_n$$

and

$$S_i \cap \left(\sum_{j \neq i} S_j \right) = \{0\}, \quad i = 1, 2, \ldots, n \tag{1.7}$$

Subspaces satisfying condition (1.7) are sometimes said to be *independent*.

Theorem 1.2 If S_1 and S_2 are subspaces of a vector space V then

$$\dim (S_1 + S_2) = \dim S_1 + \dim S_2 - \dim (S_1 \cap S_2) \tag{1.8}$$

Proof Assume that $S_1 \cap S_2 \neq \{0\}$ and select a basis $(s_1, s_2 \ldots, s_l)$ for $S_1 \cap S_2$. Since $S_1 \cap S_2 \subseteq S_1$ and $S_1 \cap S_2 \subseteq S_2$, this basis can be augmented (Exercise 1.7) to form bases $(s_1, s_2, \ldots, s_l, t_1, t_2, \ldots, t_m)$ and $(s_1, s_2, \ldots, s_l, u_1, u_2, \ldots, u_n)$ for S_1 and S_2 respectively. But every element of $S_1 + S_2$ is the sum of two elements, one from S_1 and one from S_2, so that the set $(s_1, s_2, \ldots, s_l, t_1, t_2, \ldots, t_m, u_1, u_2, \ldots, u_n)$ must span $S_1 + S_2$. Now suppose these vectors satisfy a linear relationship of the form

$$\sum_{i=1}^{l} \alpha_i s_i + \sum_{j=1}^{m} \beta_j t_j + \sum_{k=1}^{n} \gamma_k u_k = 0$$

giving

$$\sum_{i=1}^{l} \alpha_i s_i + \sum_{j=1}^{m} \beta_j t_j = - \sum_{k=1}^{n} \gamma_k u_k$$

the left-hand side of which is in S_1 and the right-hand side in S_2. This implies that both sides are in the intersection $S_1 \cap S_2$ and are therefore linear combinations of the basis $(s_1, s_2, \ldots, s_l)$. Since representation in terms of any basis is unique (Exercise 1.8), it follows directly that $\beta_j = 0, j = 1, 2, \ldots, m$. In exactly the same way it can be shown that $\gamma_k = 0$, $k = 1, 2, \ldots, n$ and then $\alpha_i = 0$, $i = 1, 2, \ldots, n$ results from the linear independence of the s_i. Consequently $(s_1, s_2, \ldots, s_l, t_1, t_2, \ldots, t_m, u_1, u_2, \ldots, u_n)$ is a basis for $S_1 + S_2$ with dimension $l + m + n$. But $\dim S_1 = l + m$, $\dim S_2 = l + n$, and $\dim (S_1 \cap S_2) = l$; and since $l + m + n = (l + m) + (l + n) - l$ the result is proved for $l > 0$. The case $l = 0$ is treated similarly.

EXERCISE 1.10 If V is a finite-dimensional vector space and $S_1, S_2, \ldots, S_n$ are subspaces of V, show that V is a direct sum $S_1 \oplus S_2 \oplus \ldots \oplus S_n$ if and only if

$$\dim (S_1 + S_2 + \ldots + S_n) = \sum_{i=1}^{n} \dim S_i = \dim V$$

EXERCISE 1.11 Prove that every subspace of a finite vector space V has a complement.

2 Inner products and euclidean spaces

So far no attempt has been made to generalize the concepts of distance and angle that are familiar in two- and three-dimensional vector spaces over a field of real numbers $\mathscr{R}$. To do this, the concept of inner product is introduced.

An *inner product* on a vector space $V(F)$, where F is the real or complex field†, is a mapping I of $V \times V$ into F (that is, a real number is associated with every ordered pair of vectors in V) with the following three properties:

(a) I is a *conjugate bilinear*: that is, for all vectors $x, x_1, x_2, y, y_1, y_2 \in V(F)$ and scalars $\alpha_1, \alpha_2 \in F$ the following relationships hold

$$I(\alpha_1 x_1 + \alpha_2 x_2, y) = \bar{\alpha}_1 I(x_1, y) + \bar{\alpha}_2 I(x_2, y) \tag{2.1}$$

$$I(x, \alpha_1 y_1 + \alpha_2 y_2) = \alpha_1 I(x, y_1) + \alpha_2 I(x, y_2) \tag{2.2}$$

where the bar denotes the complex conjugate.

(b) I is *conjugate symmetric*: that is, for all vectors $x, y \in V$

$$I(x, y) = I(\bar{y}, \bar{x}) \tag{2.3}$$

(c) I is *positive definite*: that is, for all vectors $x \in V$,

$$I(x, x) > 0,\ x \neq 0,\ I(0, 0) = 0 \tag{2.4}$$

A vector space $V(F)$ together with an inner product is called an *inner product space*. In the particular case when the field F is real an inner product space is

† See Jacobson (1953) for a more general definition of an inner product.

known as a *euclidean space*, and when F is the complex field the inner product space is known as a *unitary space*.

EXERCISE 2.1 Prove that equations (2.1) and (2.3) imply (2.2).

The *length* or *norm* of any vector $x \in V(F)$ is defined as the positive square root of the inner product $I(x, x)$ and denoted by $\|x\|$. Thus

$$\|x\| = \{I(x, x)\}^{1/2} \tag{2.5}$$

The norm of the zero vector is evidently itself zero. It is convenient to use the symbolism $I(x, y) = \langle x, y \rangle$ for the inner product. Other notations in use, particularly in physical texts, are $x.y$ and (x, y), and the inner product is often called the *scalar product*.

Example 2.1 Consider the vector space $V_n(\mathscr{R})$ of n-tuples of real numbers and define the *standard inner product* of two vectors $x = (\xi_1, \xi_2, \ldots, \xi_n)$ and $y = (\eta_1, \eta_2, \ldots, \eta_n)$ as

$$\langle x, y \rangle = \sum_{i=1}^{n} \xi_i \eta_i$$

The inner product space so formed is a euclidean space denoted by E_n. The length of a vector $x = (\xi_1, \xi_2, \ldots, \xi_n)$ in E_n is given by

$$\|x\| = \left(\sum_{i=1}^{n} \xi_i^2 \right)^{1/2}$$

Example 2.2 Consider the vector space $V_n(\mathscr{C})$ of n-tuples of complex numbers and define the *standard inner product* of two vectors $x = (\xi_1, \xi_2, \ldots, \xi_n)$ and $y = (\eta_1, \eta_2, \ldots, \eta_n)$ by

$$\langle x, y \rangle = \sum_{k=1}^{n} \bar{\xi}_k \eta_k$$

The resulting inner product space is denoted by C_n and the length of a vector $x = (\xi_1, \xi_2, \ldots, \xi_n)$ in C_n is

$$\|x\| = \left(\sum_{k=1}^{n} \bar{\xi}_k \xi_k \right)^{1/2}$$

EXERCISE 2.2 If P_n is the vector space of real polynomials (compare Example 1.2) of the form

$$a(x) = a_0 + a_1 x + \ldots + a_m x^m$$

with $m \leqslant n$, show that

$$\langle a, b \rangle = \int_0^1 a(x)b(x)\ dx$$

is an inner product for P_n.

2.1 Schwartz's inequality

Theorem 2.1 (Schwartz's inequality) Let x and y be any two vectors in an inner product space $V(F)$. Then

$$|\langle x, y \rangle| \leqslant \|x\|\ \|y\| \tag{2.6}$$

the equality sign holding if and only if x and y are linearly dependent (that is, one is a scalar multiple of the other). Since $z = \langle x, y \rangle$ is a scalar, the expression on the left-hand side of (2.6) is the absolute value of this scalar.

Proof Consider the inner product

$$\langle \alpha x - \beta y, \alpha x - \beta y \rangle \geqslant 0 \tag{2.7}$$

where α, β are scalars in F. Using the properties of the inner product, (2.7) can be expanded to give

$$|\alpha|^2\|x\|^2 - 2\ \text{Re}\ [\alpha\bar{\beta}\langle x, y \rangle] + |\beta|^2\|y\|^2 \geqslant 0 \tag{2.8}$$

For $\alpha = \|y\|^2$ and $\beta = \langle x, y \rangle$ the inequation (2.8) becomes

$$\|y\|^2(\|x\|^2\|y\|^2 - |\langle x, y \rangle|^2) \geqslant 0 \tag{2.9}$$

On dividing (2.9) by $\|y\|^2$ and taking the positive square root, the result $\|x\|\ \|y\| \geqslant |\langle x, y \rangle|$ follows.

Now suppose that x and y are linearly dependent, $x \neq 0$ (if $x = 0$, (2.6) is an equality) and $y = \gamma x$ with $\gamma \in F$. A direct calculation shows that

$$\|x\|\ \|y\| = \|x\|\ \|\gamma x\| = |\gamma|\ \|x\|^2$$

and

$$|\langle x, y \rangle| = |\langle x, \gamma x \rangle| = |\gamma|\ \|x\|^2$$

hence (2.6) is satisfied as an equality. On the other hand if (2.6) is an equality then equality must hold in (2.7) with the particular values that were selected for α and β: it follows that x and y are linearly dependent.

Example 2.3 When the Schwartz inequality is applied to E_n the following result is obtained. If $x = (\xi_1, \xi_2, \ldots, \xi_n)$ and $y = (\eta_1, \eta_2, \ldots \eta_n)$ are any two vectors in E_n, then

$$\left|\sum_{i=1}^{n} \xi_i\eta_i\right| \leqslant \left(\sum_{i=1}^{n} \xi_i^2\right)^{1/2}\left(\sum_{i=1}^{n} \eta_i^2\right)^{1/2}$$

EXERCISE 2.3 Prove that for any two vectors x, y in E_n or C_n the following inequality holds

$$\|x + y\| \leqslant \|x\| + \|y\|$$

This result is known as the *triangle inequality*. Give a geometric interpretation of the triangle inequality in E_2. [Expand $\langle x + y, x + y\rangle$ and use Theorem 2.1.]

2.2 Distance and angle

The *distance* $d(x, y)$ between two vectors x and y in E_n or C_n is defined to be the length $\|x - y\|$ of the vector $x - y$. For $x, y \in E_n$ the *angle* between x and y is the real number θ given by

$$\cos \theta = \frac{\langle x, y\rangle}{\|x\| \, \|y\|} \tag{2.10}$$

These definitions of distance and angle are directly motivated, of course, by the corresponding definitions for two- or three-dimensional geometric vectors.

For x, y in E_n or C_n, if $x, y \neq 0$ but $\langle x, y\rangle = 0$, x and y are said to be *perpendicular* or *orthogonal*. If S is a subset of V then $\{x \mid \langle x, y\rangle = 0, \text{ all } y \in S\}$ is known as the *orthogonal complement* of S and written $S^{\perp}$. Any vector of unit length is said to be a *unit vector* (sometimes it is said to be *normalized*). A set of unit vectors $x, y, z, \ldots$ which are mutually orthogonal is called an *orthonormal set.*

EXERCISE 2.4 Prove that the distance $d(x, y) = \|x - y\|$ between any two vectors x, y in E_n or C_n has the following properties.

D(i) $d(x, y) = d(y, x)$ (symmetry)

D(ii) $d(x, y) \geqslant 0$, with $d(x, y) = 0$, if and only if $x = y$ (positive)

D(iii) $d(x, y) \leqslant d(x, z) + d(z, y)$ (triangle inequality)

Let F be the field of real or of complex numbers, and let x and y belong to a space $V(F)$ not having an inner product. Then any $d(x, y) \in \mathscr{R}$ having the three properties listed in Exercise 2.4 is called a *distance function* on $V(F)$. The space $V(F)$ together with $d(x, y)$ is called a *metric space.*

EXERCISE 2.5 Prove that if x is a vector in E_n or C_n, the norm of x has the following properties.

N(i) $\|x\| \geqslant 0$, with $\|x\| = 0$ if and only if $x = 0$

N(ii) $\|\alpha x\| = |\alpha| \, \|x\|$

N(iii) $\|x + y\| \leqslant \|x\| + \|y\|$

If F is the field of real or complex numbers, and x belongs to a space $V(F)$ not having an inner product, any $n(x) \in \mathscr{R}$ having the three properties listed in Exercise 2.5 is called a *norm*. The space $V(F)$ together with $n(x)$ is called a *normed space*. In a normed space it is easy to show that $n(x - y)$ is a distance function. Inner product spaces are a special class of normed spaces, and normed spaces [with $n(x - y)$ as the distance function] are a special class of metric spaces. Some alternative norms are given in Section 2.3, but for the most part we shall be concerned only with E_n or C_n. For a more general account of metric spaces and normed spaces, see Dieudonné (1960).

2.3 Norms of vectors

The *euclidean* or *spherical* norm of any vector $x = (\xi_1, \xi_2, \ldots, \xi_n)$ in E_n has been defined as

$$\|x\|_e = \left(\sum_{k=1}^{n} \xi_k^2\right)^{1/2}$$

and shown to satisfy the conditions in Exercise 2.5. Two other norms are the *cubic norm*

$$\|x\|_c = \max_k |\xi_k|, \quad k = 1, 2, \ldots, n$$

and the *octahedral norm*

$$\|x\|_o = \sum_{k=1}^{n} |\xi_k|$$

It is obvious that both the cubic and octahedral norms satisfy the conditions of Exercise 2.5.

EXERCISE 2.6 Prove that $\|x\|_e$, $\|x\|_c$, and $\|x\|_o$ satisfy the following inequations

$$\|x\|_c \leqslant \|x\|_o \leqslant n\,\|x\|_c$$

$$\|x\|_c \leqslant \|x\|_e \leqslant \sqrt{n}\,\|x\|_c$$

$$\frac{1}{\sqrt{n}}\,\|x\|_o \leqslant \|x\|_e \leqslant \|x\|_o$$

2.4 Orthonormal bases

If $x_1, x_2, \ldots, x_n$ is a set of orthonormal vectors in an inner product space, and $\sum_{k=1}^{n} \alpha_k x_k = 0$, where the α_i are scalars, it follows that

$$\left\langle x_j, \sum_{k=1}^{n} \alpha_k x_k \right\rangle = \sum_{k=1}^{n} \alpha_k \langle x_k, x_j \rangle = \alpha_j = 0$$

and hence the set $x_1, x_2, \ldots, x_n$ is linearly independent.

Theorem 2.2 If $x_1, x_2, \ldots, x_n$ is a set of linearly independent vectors in an inner product space V, then it is possible to construct an orthonormal set $y_1, y_2, \ldots, y_n$ such that

$$\text{sp}\,(x_1, x_2, \ldots, x_n) = \text{sp}\,(y_1, y_2, \ldots, y_n)$$

Proof The construction is an inductive one. Suppose that $y_1, y_2, \ldots, y_r$, $1 \leqslant r \leqslant n$, is an orthonormal set with each y_k a linear combination of the vectors $x_1, x_2, \ldots, x_k$.

Consider the unit vector

$$y_{r+1} = \frac{z}{\|z\|}$$

where

$$z = x_{r+1} - \sum_{k=1}^{r} \langle x_{r+1}, y_k \rangle y_k$$

It is clearly orthogonal to y_k for $k = 1, 2, \ldots, r$, and it cannot vanish because this would imply a linear dependence of x_{r+1} on $y_1, y_2, \ldots, y_r$. The new vector y_{r+1} is a linear combination of $x_1, x_2, \ldots, x_{r+1}$ since by construction it is a linear combination of x_{r+1} and y_k, $k = 1, 2, \ldots, r$, with y_k a linear combination of $x_1, x_2, \ldots, x_k$ by assumption. If the inductive process is started by taking $y_1 = x_1/\|x_1\|$ then the resulting set of orthonormal vectors $y_1, y_2, \ldots, y_n$ has the required property.

The importance of this technique, known as the *Gram–Schmidt orthogonalization process*, is sufficient to warrant repeating its formulae in full.

$$y_1 = \frac{x_1}{\|x_1\|}$$

$$y_2 = \frac{x_2 - \langle x_2, y_1 \rangle y_1}{\|x_2 - \langle x_2, y_1 \rangle y_1\|}$$

$$y_3 = \frac{x_3 - \langle x_3, y_1 \rangle y_1 - \langle x_3, y_2 \rangle y_2}{\|x_3 - \langle x_3, y_1 \rangle y_1 - \langle x_3, y_2 \rangle y_2\|}$$

$$\vdots$$

$$y_k = \frac{x_k - \sum_{j=1}^{k-1} \langle x_k, y_j \rangle y_j}{\left\| x_k - \sum_{j=1}^{k-1} \langle x_k, y_j \rangle y_j \right\|}$$

EXERCISE 2.7 Prove that any finite-dimensional inner product space has an orthonormal basis.

Example 2.4 It is easily seen that the standard basis in E_n or C_n is orthonormal.

Example 2.5 Given three linearly independent vectors $x_1 = (1,1,1)$, $x_2 = (0, 1, 0)$, $x_3 = (1, 1, 0)$ in E_3 use the Gram–Schmidt process to construct an orthonormal basis.

$$y_1 = \frac{x_1}{\|x_1\|} = \left(\frac{1}{\sqrt{3}}, \frac{1}{\sqrt{3}}, \frac{1}{\sqrt{3}}\right)$$

$$y_2 = \frac{x_2 - \langle x_2, y_1\rangle y_1}{\|x_2 - \langle x_2, y_1\rangle y_1\|}$$

$$= \left(-\frac{1}{\sqrt{6}}, \frac{2}{\sqrt{6}}, -\frac{1}{\sqrt{6}}\right)$$

$$y_3 = \frac{x_3 - \langle x_3, y_1\rangle y_1 - \langle x_3, y_2\rangle y_2}{\|x_3 - \langle x_3, y_1\rangle y_1 - \langle x_3, y_2\rangle y_2\|}$$

$$= \left(\frac{1}{\sqrt{2}}, 0, -\frac{1}{\sqrt{2}}\right)$$

The reader should convince himself that y_1, y_2, y_3 are indeed orthonormal.

EXERCISE 2.8 If $V(F)$ is an inner product space and S is a subspace of V prove that $V = S \oplus S^{\perp}$.

3 Linear transformations

Let $V(F)$ and $W(F)$ be vector spaces over a field F. A mapping A of V into W is said to be linear if it has the following properties:

(a) $A(x + y) = A(x) + A(y)$
(b) $A(\alpha x) = \alpha A(x)$

for all $x, y \in V$, $\alpha \in F$. That is to say, the mapping preserves the operations of addition and multiplication. Notice that addition on the left of (a) is in V and on the right is in W, although the same symbol has been used for convenience. Any such mapping is known as a *linear transformation* of V into W and is often written $A\colon V \to W$. The mapping 0 which associates any vector $x \in V$ with the zero element of W is obviously a linear transformation. It is called the *zero transformation*. Another mapping I which is evidently linear is that which gives $I(x) = x$ for all $x \in V$: this is called the *identity transformation* of V.

EXERCISE 3.1 Prove that the conditions (a) and (b) above are equivalent to the single condition

$$A(\alpha x + \beta y) = \alpha A(x) + \beta A(y)$$

for all $x, y \in V$; $\alpha, \beta \in F$.

EXERCISE 3.2 Prove that any linear transformation $A: V \to W$ maps the zero of V into the zero of W and that linear transformations map linear combinations in V into the corresponding linear combinations in W and therefore preserve linear dependence.

Example 3.1 Let P_n be the vector space of polynomials in Exercise 2.2. Then the mapping defined by

$$D: a(x) \to \frac{da}{dx} = \sum_{k=0}^{m} ka_k x^{k-1}$$

is linear. To show this, let

$$a(x) = \sum_{k=0}^{m} a_k x^k, \qquad b(x) = \sum_{k=0}^{n} b_k x^k$$

and assume $n \geqslant m$. Then

$$a(x) + b(x) = \sum_{k=0}^{m} (a_k + b_k)x^k + \sum_{k=m+1}^{n} b_k x^k$$

and

$$D\big(a(x) + b\,(x)\big) = \sum_{k=0}^{m} k(a_k + b_k)x^{k-1} + \sum_{k=m+1}^{n} kb_k x^{k-1}$$

But

$$D\big(a(x)\big) = \sum_{k=0}^{m} ka_k x^{k-1}$$

$$D\big(b(x)\big) = \sum_{k=0}^{n} kb_k x^{k-1}$$

so that

$$D\big(a(x)\big) + D\big(b(x)\big) = D\big(a(x) + b(x)\big)$$

Again, for any $\alpha \in F$,

$$\alpha a(x) = \sum_{k=0}^{m} (\alpha a_k)x^k$$

and

$$D\big(\alpha a(x)\big) = \sum_{k=0}^{m} k(\alpha a_k)x^{k-1}$$

$$= \alpha \sum_{k=0}^{m} ka_k x^{k-1}$$

$$= \alpha D\big(a(x)\big)$$

As in Section 8 of Chapter 1, the image of an element $v \in V$ under the mapping $A: V \to W$ has been written $A(v)$. The space V is the domain of A, and $A(V)$ is the range of A. The set $\{v \in V \mid A(v) = 0\}$ is called the *null space* or *kernel* of A. Clearly, the range and kernel of A are subspaces of W and V respectively: when these subspaces are finite-dimensional their dimensions are called the *rank* and *nullity* of A. If A is both injective and surjective it is one-to-one (bijective) and has a one-to-one inverse mapping denoted by A^{-1}: A is then said to be *non-singular*; otherwise it is *singular*. A non-singular linear transformation $A: V \to W$ sets up an isomorphism between V and W as defined in Section 1.2.

Quite generally a subspace of V will map into a subspace of W and the inverse image set (Chapter 1, Section 8) of any subspace of W is a subspace of V. These two facts are almost self-evident; the proof of the second, for example, proceeds as follows. Let v_1 and v_2 be any two vectors in the inverse image set S of W_1, and α_1, α_2 any two scalars; then

$$A(\alpha_1 v_1 + \alpha_2 v_2) = \alpha_1 A(v_1) + \alpha_2 A(v_2) \in W_1$$

since W_1 is a subspace. Thus $\alpha_1 v_1 + \alpha_2 v_2 \in S$, showing that S is a subspace.

EXERCISE 3.3 Show that a necessary and sufficient condition for a linear transformation $A: V \to W$ to be one-to-one is that the null space of A is just the zero vector of V.

Example 3.2 Consider the mapping

$$A: V_n(\mathscr{R}) \to V_{n-1}(\mathscr{R})$$

where

$$A(\xi_1, \xi_2, \ldots, \xi_n) = (\xi_2, \xi_3, \ldots, \xi_n)$$

This mapping is evidently surjective, but it is not injective because $A(\eta, \xi_2, \ldots, \xi_n) = (\xi_2, \xi_3, \ldots \xi_n)$ for any $\eta \in \mathscr{R}$. It follows that range $A = V_{n-1}(\mathscr{R})$ and hence rank $A = n - 1$. A vector $x = (\xi_1, \xi_2, \ldots, \xi_n)$ in $V_n(\mathscr{R})$ can only map into the zero vector in $V_{n-1}(\mathscr{R})$ if $\xi_2 = \xi_3 = \cdots = \xi_n = 0$. Thus kernel A is just the space spanned by the unit vector $(1, 0, \ldots, 0)$ and therefore has unit dimension.

In the last example the sum of the range dimension and the null-space dimension is equal to that of $V_n(\mathscr{R})$ and, in general, the following theorem applies.

Theorem 3.1 For any linear transformation $A: V \to W$ the following result holds

$$\dim(\text{kernel } A) + \dim(\text{range } A) = \dim V \tag{3.1}$$

that is,

$$\text{nullity } A + \text{rank } A = \dim V \tag{3.2}$$

Proof Let $\dim V = n$, rank $A = r$ and nullity $A = k$. Since kernel $A \subseteq V$, a basis $(v_1, v_2, \ldots, v_k, \ldots, v_n)$ can be chosen for V in such a way that $(v_1, v_2, \ldots, v_k)$ is a basis for kernel A. Any vector $v \in V$ can be written as a linear combination $\sum_{j=1}^{n} \alpha_j v_j$ and the image of this element is, from the definition of null space,

$$\sum_{j=1}^{n} \alpha_j A(v_j) = \sum_{j=k+1}^{n} \alpha_j A(v_j)$$

Furthermore, the vectors $A(v_j), j = k+1, k+2, \ldots, n$ are linearly independent. For assuming, contrary to what is to be proved, that the vectors $A(v_j), j = k+1, k+2, \ldots, n$ are linearly dependent, there exist β_j, not all zero, such that

$$\sum_{j=k+1}^{n} \beta_j A(v_j) = A\left(\sum_{j=k+1}^{n} \beta_j v_j\right) = 0$$

and $\sum_{j=k+1}^{n} \beta_j v_j$ therefore lies in the null space of A. But since $(v_1, v_2. \ldots, v_k)$ is a basis for the null space of A there exist scalars γ_j such that

$$\sum_{j=1}^{k} \gamma_j v_j = \sum_{j=k+1}^{n} \beta_j v_j$$

contradicting the fact that $v_1, v_2, \ldots, v_n$ are linearly independent. Thus the $\beta_j = 0$ proving that $A(v_{k+1}), A(v_{k+2}), \ldots, A(v_n)$ is a basis for range A. It follows directly that $r = n - k$ and the proof is complete.

Example 3.3 The transformation A represented (see Section 4) by the matrix $\begin{pmatrix} 1 \\ 2 \\ 3 \end{pmatrix}$ has rank 1, nullity 0, and $\dim V = 1$. The transformation represented by (1, 2, 3) has rank 1, nullity 2, and $\dim V = 3$. Notice that range $A \subseteq W$ while kernel $A \subseteq V$. When A maps V into itself this distinction is obscured. Beware then of the fallacy: range A + kernel $A = V$, which is not true in general. For example, the transformation represented by $\begin{pmatrix} 0 & 1 \\ 0 & 0 \end{pmatrix}$ has rank 1, nullity 1; the range is spanned by $\begin{pmatrix} 1 \\ 0 \end{pmatrix}$ and so is the null space.

3.1 Operations with linear transformations

Let $A: V(F) \to W(F)$ and $B: V(F) \to W(F)$ be any two linear mappings. Their *sum* $A + B$ is defined by the equation

$$(A + B)(v) = A(v) + B(v)$$

for all vectors $v \in V$

The *scalar multiple* of the mapping A by an element $\alpha \in F$ is defined by the equation

$$(\alpha A)v = A(\alpha v)$$

for all $v \in V$.

EXERCISE 3.4 Show that the sum $A + B$ and scalar multiple αA defined above are also linear transformations and that the set of all linear transformations of $V(F)$ into $W(F)$, together with these operations of addition and scalar multiplication, is itself a vector space over F.

The product of two functions was defined quite generally in Section 8 of Chapter 1. Thus the product of two linear transformations $A: V(F) \rightarrow W(F)$, $B: W(F) \rightarrow Y(F)$ is the mapping defined by the equation

$$(BA)(v) = B(A(v))$$

It is again obvious that BA is a linear mapping of V into Y.

EXERCISE 3.5 By using the above definitions verify the following properties of the sum, scalar multiple, and product of linear transformations. If A and B are transformations of $V(F)$ into $W(F)$, C and D are transformations of $W(F)$ into $Y(F)$, and E is a transformation of $Y(F)$ into $Z(F)$, then

(i) $E(CA) = (EC)A$

(ii) $C(A + B) = CA + CB$

(iii) $(C + D)A = CA + DA$

(iv) $A0 = 0A = 0$

(v) $AI = IA = A$

(vi) $\alpha(A + B) = \alpha A + \alpha B$, all $\alpha \in F$

(vii) $\alpha(CA) = (\alpha C)A = C(\alpha A)$

(viii) If A is non-singular, $A^{-1}A = A\,A^{-1} = I$

(ix) If A and C are non-singular then so is CA and its inverse is $A^{-1}C^{-1}$.

From Exercise 3.2 and the fact that the range of A is a subspace of W it follows that the dimension of the range of A is less than or equal to both dim W and dim V. If now the product CA of the transformations $A: V \rightarrow W$ and $C: W \rightarrow Y$ is considered, it readily follows that the rank of CA is less than or equal to rank A or rank C. For range $CA = C(A(V)) \subseteq C(W)$ and so by the previous remark rank $CA \leqslant$ rank A, rank C.

4 Matrix representation of linear transformations

Let $V(F)$ and $W(F)$ be two vector spaces of dimensions n and m respectively and let $(v_1, v_2, \ldots, v_n)$ and $(w_1, w_2, \ldots, w_m)$ be bases for V and W. Let A be a linear transformation of V into W: then $A(v_j) \subseteq W$ and can therefore be written as a linear combination of the w_i. Thus

$$A(v_j) = \sum_{i=1}^{m} \alpha_{ij} w_i, \quad j = 1, 2, \ldots, n$$

where the α_{ij} are scalars in F. Any vector $v \in V$ is, however, a unique linear combination of the v_j, and since linear transformations preserve linear combinations (Exercise 3.2) it follows that $w = A(v)$ is known provided that the bases $(v_1, v_2, \ldots, v_n)$ and $(w_1, w_2, \ldots, w_m)$ and the matrix (α_{ij}) are known. The matrix (α_{ij}) is known as the *matrix representation* of the linear transformation A with respect to the two given bases.

EXERCISE 4.1 Prove that the correspondence between the set of linear transformations of $V(F)$ into $W(F)$ and the set of their matrix representations with respect to two fixed bases is one-to-one.

Example 4.1 Let $V(F)$ be a vector space over a field F, and A the linear transformation defined by $A(v) = \alpha v$, for $v \in V$, where α is a scalar in F. This transformation is known as a scalar transformation. If $(v_1, v_2, \ldots, v_n)$ is any basis whatever for V it follows that $A(v_j) = \alpha v_j$ and therefore the matrix representation of A with respect to any basis is just the $n \times n$ diagonal matrix with the scalar α on its diagonal. (This, of course, is just αI_n.)

It follows directly from this example that the matrix representation of the identity transformation with respect to any basis is the identity matrix I_n.

Example 4.2 Let A be the linear transformation $A:(\alpha_1, \alpha_2, \alpha_3) \rightarrow (\alpha_2, \alpha_3)$ of $V_3(\mathscr{R})$ into $V_2(\mathscr{R})$. Take the standard bases in $V_3(\mathscr{R})$ and $V_2(\mathscr{R})$ as e_i and e_i' respectively; then

$$(1, 0, 0) \rightarrow (0, 0) = 0e_1' + 0e_2'$$
$$(0, 1, 0) \rightarrow (1, 0) = 1e_1' + 0e_2'$$
$$(0, 0, 1) \rightarrow (0, 1) = 0e_1' + 1e_2'$$

and the matrix representation of the mapping with respect to (e_1, e_2, e_3) and (e_1', e_2') is

$$A_1 = \begin{pmatrix} 0 & 1 & 0 \\ 0 & 0 & 1 \end{pmatrix}$$

The image of any vector in $V_3(\mathscr{R})$ is found by multiplying the vector by A_1. For example, $(2,3,5) \rightarrow (3, 5)$ agreeing with

$$\begin{pmatrix} 0 & 1 & 0 \\ 0 & 0 & 1 \end{pmatrix} \begin{pmatrix} 2 \\ 3 \\ 5 \end{pmatrix} = \begin{pmatrix} 3 \\ 5 \end{pmatrix}$$

Now take $(1, 2, 0)$, $(-1, 1, 3)$, $(0, 2,4)$ as a basis in $V_3(\mathscr{R})$ and $(1, 1)$, $(0, 1)$ as a basis in $V_2(\mathscr{R})$. Then

$$\begin{aligned} (1, 2, 0) \rightarrow (2, 0) &= 2(1, 1) - 2(0, 1) \\ (-1, 1, 3) \rightarrow (1, 3) &= 1(1, 1) + 2(0, 1) \\ (0, 2, 4) \rightarrow (2, 4) &= 2(1, 1) + 2(0, 1) \end{aligned}$$

and the matrix representation with respect to these bases is

$$A_2 = \begin{pmatrix} 2 & 1 & 2 \\ -2 & 2 & 2 \end{pmatrix}$$

To find the image of $(2, 3, 5)$, it is first written in terms of the second basis for $V_3(\mathscr{R})$, which gives $(-\frac{1}{3}, -\frac{7}{3}, 3)$, and then multiplied by A_2:

$$\begin{pmatrix} 2 & 1 & 2 \\ -2 & 2 & 2 \end{pmatrix} \begin{pmatrix} -\frac{1}{3} \\ -\frac{7}{3} \\ 3 \end{pmatrix} = \begin{pmatrix} 3 \\ 2 \end{pmatrix}$$

Notice that $(3, 2)$ is $3(1, 1) + 2(0, 1) = (3, 5)$ in the standard basis, as would be expected.

Now consider the two linear transformations

$$A: V(F) \rightarrow W(F) \qquad \text{and} \qquad B: W(F) \rightarrow X(F)$$

Let $(v_1, v_2, \ldots, v_n)$, $(w_1, w_2, \ldots, w_m)$ and $(x_1, x_2, \ldots, x_r)$ be bases for V, W, and X respectively. If (a_{ij}) is the matrix representation of A with respect to the given bases, (b_{ij}) that of B, and (c_{ij}) that of $C = BA$, then

$$A(v_j) = \sum_{k=1}^{m} a_{kj} w_k, \quad j = 1, 2, \ldots, n \tag{4.1}$$

$$B(w_k) = \sum_{i=1}^{r} b_{ik} x_i, \quad k = 1, 2, \ldots, m \tag{4.2}$$

It follows that

$$\begin{aligned} C(v_j) &= BA(v_j) \\ &= \sum_{k=1}^{m} a_{kj} B(w_k), \quad j = 1, 2, \ldots, n \end{aligned}$$

$$= \sum_{i=1}^{r} \left\{ \sum_{k=1}^{m} a_{kj} b_{ik} \right\} x_i, \quad j = 1, 2, \ldots, n$$

whence

$$c_{ij} = \sum_{k=1}^{m} b_{ik} a_{kj} \tag{4.3}$$

This shows that the matrix representation of the product transformation $C = BA$ is the product of the matrices B and A as defined in Chapters 1 and 2. It can be shown in a similar way that the sum of two linear transformations and the scalar multiple of a linear transformation are given by the corresponding sum and scalar multiple, again as previously defined, of their matrix representations. It is these facts, of course, that motivated the previous definitions for operations on matrices.

Example 4.3 In Example 4.2, call the standard basis in $V_3(\mathscr{R})$, b_1, and the second basis in $V_3(\mathscr{R})$, b_2. The identity transformation in $V_3(\mathscr{R})$ may, if we wish, be regarded as a transformation from a space $V_3(\mathscr{R})$ with basis b_2 to another space $V_3(\mathscr{R})$ with basis b_1. This transformation will be written $I_{b_2 b_1}$, and its matrix representation is easily seen to be

$$\begin{pmatrix} 1 & -1 & 0 \\ 2 & 1 & 2 \\ 0 & 3 & 4 \end{pmatrix}$$

Similarly, the transformation in $V_2(\mathscr{R})$ from bases b_1' to b_2', written $I_{b'_1 b'_2}$, is clearly $I_{b'_2 b'_1}^{-1}$ and has matrix

$$\begin{pmatrix} 1 & 0 \\ 1 & 1 \end{pmatrix}^{-1}$$

Then the transformation A from $V_3(\mathscr{R})$ with basis b_2 to $V_2(\mathscr{R})$ with basis b_2' can be written as a product of three transformations,

$$A = I_{b'_1 b'_2} \hat{A} I_{b_2 b_1}$$

where $\hat{A}$ on the right-hand side is regarded as a transformation from $V_3(\mathscr{R})$ with basis b_1 to $V_2(\mathscr{R})$ with basis b_1': this, of course, is the same linear transformation as A. The matrix representation of the last equation is

$$A_2 = \begin{pmatrix} 1 & 0 \\ 1 & 1 \end{pmatrix}^{-1} A_1 \begin{pmatrix} 1 & -1 & 0 \\ 2 & 1 & 2 \\ 0 & 3 & 4 \end{pmatrix}$$

which will be found to agree with the result in Example 4.2.

4.1 Row and column rank of a matrix

Let $A: V(F) \to W(F)$ be a linear transformation and (a_{ij}) its matrix representation with respect to the bases $(v_1, v_2, \ldots, v_n)$ and $(w_1, w_2, \ldots, w_m)$. Denote the rows of (a_{ij}) by

$$A_i = (a_{i1}, a_{i2}, \ldots, a_{in}), \quad i = 1, 2, \ldots, m$$

and the columns by

$$A^j = \begin{pmatrix} a_{1j} \\ a_{2j} \\ \vdots \\ a_{mj} \end{pmatrix}, \quad j = 1, 2, \ldots, n$$

A_i is generally called a *row vector* and A^j a *column vector* in the spaces $V_n(F)$ and $V_m(F)$ respectively. Column vectors are often written horizontally for typographical convenience.

The image of the base vector v_j is

$$A(v_j) = \sum_{i=1}^{m} a_{ij} w_i, \quad j = 1, 2, \ldots, n$$

so that the coordinates of $A(v_j)$ with respect to the w_i are given by the column vector A^j. Consider now the range $A(V)$ of A; any vector in V is a linear combination of the v_j and therefore any vector in $A(V)$ must bea linear combination of $A(v_j), j = 1, 2, \ldots, n$, so that $A(V)$ is spanned by $A(v_j), j = 1, 2, \ldots, n$. The rank of A is dim $A(V)$, but this is just the largest set of linearly independent vectors taken from $A(v_1), A(v_2), \ldots, A(v_n)$ and consequently must also equal the greatest number of linearly independent columns of (a_{ij}). This number is called the *column rank* of (a_{ij}) and the space spanned by the linearly independent columns is called the *column space* of (a_{ij}).

Similarly, it is possible to define the *row space* of a matrix, and its *row rank*.

Theorem 4.1 The number of linearly independent rows of any matrix is equal to the number of linearly independent columns.

Proof Suppose the column rank of (a_{ij}) is r: then r of the A^j must be linearly independent; call them $B^1, B^2, \ldots, B^r$. It is now possible to write any column of (a_{ij}) as a linear combination of $B^1, B^2, \ldots, B^r$ as follows

$$A^j = \sum_{i=1}^{r} \alpha_{ij} B^i, \quad j = 1, 2, \ldots, n$$

so that

$$a_{kj} = \sum_{i=1}^{r} \alpha_{ij} b_{ki}, \quad j = 1, 2, \ldots, n; \quad k = 1, 2, \ldots, m \tag{4.4}$$

where b_{ki} is the kth element of the column B^i. Equation (4.4) is easily reorganized in terms of the rows A_i to give

$$A_k = \sum_{i=1}^{r} b_{ki}\phi_i$$

where ϕ_i is the ith row of the $r \times n$ matrix (α_{ij}). The row space of (a_{ij}) is therefore spanned by the set of r n-vectors, ϕ_i, and hence its dimension is less than or equal to r. If the above reasoning is applied in exactly the same way to the transpose of (a_{ij}), the result will be that the dimension of the column space is less than or equal to that of the row space. It follows that the two dimensions must be equal and consequently the row rank and column rank are equal.

In Chapter 2 the determinantal rank of a matrix (a_{ij}) was defined as the order of the largest submatrix of (a_{ij}) with non-vanishing determinant.

Theorem 4.2 The determinantal rank of a matrix is equal to its row (and column) rank.

Proof Let r be the determinantal rank of (a_{ij}): then it will have a square submatrix Q of order r with non-zero determinant. The columns of Q are linearly independent, for by Section 5 of Chapter 2 there exists no non-zero r-vector, q, such that $Qq = 0$. The columns of (a_{ij}) which correspond with the r linearly independent columns of Q must also be linearly independent and therefore column rank (a_{ij}) = row rank $(a_{ij}) \geqslant r$.

If the row rank of (a_{ij}) is p, then p of the rows of (a_{ij}) are linearly independent and form a $p \times n$ matrix from which p linearly independent columns can be selected (row rank = column rank) to give a $p \times p$ submatrix of (a_{ij}) with non-zero determinant. Thus $r \geqslant p$, proving that row rank (or column rank) and determinantal rank are the same; the common symbol $\rho(a_{ij})$ will be used for them in future.

4.2 Inverse transformations and matrix inversion

It has already been seen that a one-to-one linear transformation $A: V(F) \to W(F)$ has a unique inverse A^{-1} such that $A^{-1}A = AA^{-1} = I$, where I is the identity transformation (Exercise 3.5). The next step in relating the properties of a linear transformation to its matrix representation lies in the following theorem.

Theorem 4.3 A is non-singular if and only if its matrix representation is regular; furthermore the matrix representation of A^{-1} is the inverse of that of A.

Proof First suppose that A is non-singular; then if (a_{ij}) is the matrix representation of A with respect to fixed bases in V and W it follows that

$$I = A^{-1}A$$

and

$$I_n = (b_{kl})(a_{ij})$$

where (b_{kl}) is the matrix representation of A^{-1}. Then by Exercise 3.2 of Chapter 2, $(b_{kl}) = (a_{ij})^{-1}$. Again, suppose that (a_{ij}) is regular and let B be the linear transformation represented by its inverse, then

$$I_n = (a_{ij})^{-1}(a_{ij})$$

and

$$I = BA$$

showing that $B = A^{-1}$ and A is non-singular.

By the preceding argument we are justified in calling a regular matrix non-singular. It also follows immediately that the transformation A is non-singular if and only if its rank is equal to the dimension of both V and W. In particular if $V = W$, A is non-singular if and only if $\rho(A) = \dim V$.

Theorem 4.4 The rank of a matrix is unchanged if it is multiplied by any non-singular matrix.

Proof From the corresponding result for linear transformations the rank of the product of two matrices is less than or equal to the rank of either factor. Let the matrix A have rank r_1; then $\rho(BA) = r_2 \leqslant r_1$ and provided B is non-singular $\rho\big(B^{-1}(BA)\big) = \rho(A) = r_1 \leqslant r_2$, proving that $\rho(BA) = \rho(A)$. Similarly $\rho(AB) = \rho(A)$.

5 Similar matrices

From now on, single capitals will be used for matrices (as in previous chapters) where no confusion can arise between a linear transformation and its matrix representation with respect to different sets of bases. Any two $n \times n$ matrices A and B over a field F are said to be *similar* if there exists a non-singular (regular) matrix H such that

$$A = HBH^{-1} \tag{5.1}$$

EXERCISE 5.1 Prove that similarity of matrices as defined above is an equivalence relationship.

The concept of similarity can be interpreted in terms of linear transformations. Let $T: V(F) \to V(F)$ be a linear transformation of the vector space $V(F)$, over the field F, into itself. Let $b_1 = (v_1, v_2, \ldots, v_n)$ and $b_2 = (u_1, u_2, \ldots, u_n)$ be two distinct bases for $V(F)$, and let A and B be the matrix representations of T with respect to b_1 and b_2 respectively. By Example 4.3, it follows that

$$A = \text{rep}\, I_{b_2 b_1} B \,\text{rep}\, I_{b_1 b_2} \tag{5.2}$$

where rep $I_{b_2b_1}$ denotes the matrix representation of the identity transformation on V with respect to the bases b_2 and b_1. On putting $H = \text{rep } I_{b_2b_1}$, equation (5.2) gives (5.1).

Now assume that a non-singular matrix $H = (h_{ij})$ exists and satisfies (5.1). If b_1 is a basis for V and A is the matrix representation of T with respect to b_1, is it possible to find another basis b_2 such that B is the matrix representation of T with respect to b_2? Consider as a candidate for b_2 the set of n vectors

$$u_i = \sum_{j=1}^{n} h_{ij}v_j, \quad i = 1, 2, \ldots, n \tag{5.3}$$

If they are linearly dependent there exists a set of scalars α_i such that

$$\sum_{i=1}^{n} \alpha_i u_i = 0$$

Therefore from (5.3)

$$\sum_{j=1}^{n} \sum_{i=1}^{n} \alpha_i h_{ij} v_j = 0$$

and since the v_j are linearly independent

$$\sum_{i=1}^{n} \alpha_i h_{ij} = 0, \quad j = 1, 2, \ldots, n$$

But this last expression is just $\alpha H = 0$, where α is the row vector $(\alpha_1, \alpha_2, \ldots, \alpha_n)$ and since H is non-singular $\alpha HH^{-1} = \alpha = 0$ showing that b_2 is indeed a basis for V (Exercise 1.7).

Finally, notice from (5.3) that

$$\text{rep } I_{b_2b_1} = H$$

so that as before

$$\text{rep } I_{b_1b_2} = H^{-1}$$

and

$$\begin{aligned} \text{rep } T_{b_2b_2} &= \text{rep } I_{b_1b_2} A \text{ rep } I_{b_2b_1} \\ &= H^{-1}AH \\ &= B \end{aligned}$$

The above result can be summarized by saying that a necessary and sufficient condition for two matrices to be similar is that they represent the same linear transformation with respect to different bases. This, of course, means that similar matrices have equal rank: this also follows directly from the invariance of rank under multiplication by a non-singular matrix.

EXERCISE 5.2 Prove that, if A and B are similar matrices, the powers A^n and B^n are similar and if A and B are also non-singular then A^{-1} and B^{-1} are similar.

6 Equivalent matrices

Two matrices A and B over a field F are said to be *equivalent* (sometimes *associative*) if one can be obtained from the other by a sequence of elementary row and column operations (Chapter 2, Section 4).

Theorem 6.1 A necessary and sufficient condition for A and A_1 to be equivalent is that two non-singular matrices P and Q exist for which $A_1 = PAQ$. (In fact this latter relationship is often used to define equivalence.)

Proof The proof relies on the fact that any elementary row operation can be carried out by pre-multiplication by a suitable regular matrix. In fact it is easily verified by direct computation that the latter is equal to the matrix obtained by carrying out the same elementary operation on the identity matrix of appropriate order: such transformed identity matrices are known as *elementary matrices.* Also, any column transformation can be achieved by post-multiplying by an appropriate elementary matrix.

Now suppose that the $m \times n$ matrix A is equivalent to A_1 and let $E_1, E_2, \ldots, E_r$ and $F_1, F_2, \ldots, F_s$ be the elementary matrices representing the row and column operations (in their order of application) needed to transform A into A_1. Then

$$A_1 = PAQ$$

where

$$P = E_r E_{r-1} \ldots E_1$$

and

$$Q = F_1 F_2 \ldots F_s$$

On the other hand if $A_1 = PAQ$, where P and Q are non-singular, then A and A_1 are equivalent, since any non-singular matrix can be written as the product of a set of elementary matrices (Exercise 5.3, Chapter 2).

EXERCISE 6.1 Prove that equivalence of matrices is an equivalence relation as defined in Chapter 1, Section 2.1.

It is an immediate consequence of the Gauss reduction technique (Chapter 2, Section 5) that any $m \times n$ matrix A of rank r is equivalent to a matrix of the form

$$B = \begin{pmatrix} I_r & 0_{r,n-r} \\ 0_{m-r,r} & 0_{m-r,n-r} \end{pmatrix} \tag{6.1}$$

where I_r is the $r \times r$ identity matrix and 0_{ij} is an $i \times j$ matrix of zeros. It is also obvious that two $m \times n$ matrices are equivalent if and only if they have the same rank. The 'if' part follows from (6.1) and the 'only if' from the invariance of rank under multiplication by a non-singular matrix.

6.1 Linear equations

It is convenient at this point to give a brief discussion of the geometrical interpretation of the results of Chapter 2, Section 5. Any $m \times n$ matrix A can be thought of as a representation of a mapping of $V_n(F)$ into $V_m(F)$, so that a solution of the homogeneous linear system $Av = 0$ must lie in the kernel of A. It follows that the set of all solutions of $Ax = 0$ is just the kernel of A and hence is a subspace of $V_n(F)$, known as the *solution space* of A, with dimension equal to the nullity of A. From Section 3, the dimension of the solution space is seen to be $n - r$, where r is the rank of A. Notice that when $V_n(F)$ is euclidean the solution space is just the orthogonal complement of the rows of A.

Now consider the non-homogeneous system

$$Av = w \tag{6.2}$$

As was seen in Chapter 2, Section 5, any two solutions x and y of (6.2) must satisfy

$$A(x - y) = 0$$

and so must lie in the kernel of A. The most general solution of (6.2) is obtained by adding any solution of $Av = 0$ to a particular solution of (6.2). Provided that a solution of (6.2) exists it is therefore unique if and only if the kernel of A is the zero vector or, in other words, A is of rank n.

Suppose S is the solution space of $Av = 0$ and x is a particular solution of (6.2). Then the set S^1 of solutions of (6.2) is

$$S^1 = x + S \tag{6.3}$$

and S^1 is called a *plane parallel to S*. The set S^1 is a subspace of $V_n(F)$ only if $x \in S$, but in any case its dimension is defined to be that of S.

Equation (6.2) can be written in vector form as follows:

$$v_1 a_1 + v_2 a_2 + \cdots + v_n a_n = w \tag{6.4}$$

in which $(v_1, v_2, \ldots, v_n)^{\mathrm{T}} = v$ and $a_1, a_2, \ldots, a_n$ are the columns of A. For (6.2) to have a solution it follows that w must lie in sp $(a_1, a_2, \ldots, a_n)$. Let the matrix A have rank r: then r of the columns a_i are linearly independent and form a basis for sp $(a_1, a_2, \ldots, a_n)$. If the r linearly independent columns are taken as $a_1, a_2, \ldots, a_r$ (for convenience of notation) then (6.4) can be written

$$v_1 a_1 + v_2 a_2 + \cdots + v_r a_r = w - v_{r+1} a_{r+1} - v_{r+2} a_{r+2} - \cdots - v_n a_n \tag{6.5}$$

and it is immediately seen that provided $w \in \text{sp}(a_1, a_2, \ldots, a_n)$ then, if arbitrary values are given to $v_{r+1}, v_{r+2}, \ldots, v_n$, the values $v_1, v_2, \ldots, v_r$ are uniquely determined, because $a_1, a_2, \ldots, a_r$ is a basis for $\text{sp}(a_1, a_2, \ldots, a_n)$.

Finally, consider the vector spaces $V_n(F)$ and $V_m(F)$ and let $Av = 0$ be a set of homogeneous linear equations with A $m \times n$ and $v \in V_n(\mathscr{R})$. Suppose $m < n$. Then the columns of A form a set of n vectors in the m-dimensional space $V_m(\mathscr{R})$ and are therefore dependent, showing that $Av = 0$ has a non-trivial solution.

EXERCISE 6.2 Any $(n - 1)$-dimensional subspace of an n-dimensional vector space V is called a *hyperplane of* V. Show that the intersection of $m < n$ hyperplanes in $V_n(\mathscr{R})$ is a non-vacuous subspace of $V_n(\mathscr{R})$. [The null space of a $1 \times n$ matrix q (a row vector) is a hyperplane. Assemble m such row vectors into an $m \times n$ matrix and interpret the intersection of hyperplanes as a solution space.]

7 Eigenvalues and eigenvectors

An important aspect of similarity transformations is the possibility of reducing a matrix to a diagonal or nearly diagonal form which will facilitate computation. It is convenient to approach this problem using linear transformations.

Suppose $V(F)$ is a vector space over the field F. Then any non-zero vector $v \in V$ is said to be an *eigenvector*† of the linear transformation $A: V \to V$ if

$$A(v) = \lambda v \tag{7.1}$$

for some scalar $\lambda \in F$. The scalar λ is called an *eigenvalue* of A. Notice that if (7.1) is satisfied for any given vector $v \in V$, it is also satisfied for any scalar multiple of v, that is, the subspace of V spanned by v is mapped into itself by A. More generally, if W is any subspace of V and $A(w) \in W$ for all $w \in W$, then W is said to be an *invariant subspace* of V under A.

Now let $b_1 = (v_1, v_2, \ldots, v_n)$ be a basis in V and A_1 the matrix representation of A with respect to this basis. If each vector of b_1 is an eigenvector of A then

$$A(v_i) = \lambda_i v_i, \quad i = 1, 2, \ldots, n$$

and

$$A_1 = \text{diag}(\lambda_i)$$

Furthermore if $A_1 = \text{diag}(\lambda_i)$ then

$$A(v_j) = \sum_{i=1}^{n} a_{ij} v_i = \lambda_j v_j, \quad j = 1, 2, \ldots, n$$

Therefore the matrix representation A_1 is diagonal if and only if each vector in the basis b_1 is an eigenvector of A.

† Synonyms for eigenvector are *proper vector*, *latent vector*, and *characteristic vector*.

Theorem 7.1 A set of eigenvectors which all correspond to different eigenvalues are linearly independent.

Proof Let $\lambda_1, \lambda_2, \ldots, \lambda_r$ be a set of distinct eigenvalues of the transformation $A: V(F) \to V(F)$ and let $h_1, h_2, \ldots, h_r$ be the corresponding eigenvectors. It will now be shown that the assumption that the set $\{h_1, h_2, \ldots, h_r\}$ is dependent involves a contradiction.

Assume, then, that $\{h_1, h_2, \ldots, h_r\}$ is a dependent set and that it has been rearranged (if necessary) so that $h_1, h_2, \ldots, h_k$ are independent and $h_{k+1}, h_{k+2}, \ldots, h_r$ depend on them. It is therefore possible to write

$$h_r = \sum_{i=1}^{k} \alpha_i h_i \tag{7.2}$$

where not all of the α_i are zero ($h_r \neq 0$). Applying the transformation A to both sides of (7.2) gives

$$\lambda_r h_r = \sum_{i=1}^{k} \alpha_i \lambda_i h_i \tag{7.3}$$

If $\lambda_r = 0$ then (7.3) implies that $h_1, h_2, \ldots, h_k$ are dependent since no other λ_i, $i = 1, 2, \ldots, k$, can also be zero: this is a contradiction. If $\lambda_r \neq 0$ then (7.3) can be written

$$h_r = \sum_{i=1}^{k} \alpha_i \frac{\lambda_i}{\lambda_r} h_i \tag{7.4}$$

and, since $\lambda_i \neq \lambda_r$, equation (7.4) gives a different representation for h_r from equation (7.2): again, a contradiction. Thus $h_1, h_2, \ldots, h_r$ must be independent.

EXERCISE 7.1 Prove that if the linear transformation $A: V(F) \to V(F)$ has finite dimension n then it has, at most, n distinct eigenvalues.

The problem now arises of how to determine the eigenvalues and eigenvectors of a given linear transformation $A: V(F) \to V(F)$. Let $b_1 = (v_1, v_2, \ldots, v_n)$ be a basis for V and suppose that the coordinates of an eigenvector $h \in V$ are $\alpha_i \in F$, $i = 1, 2, \ldots, n$, so that

$$h = \sum_{i=1}^{n} \alpha_i v_i$$

Then, since h is an eigenvector

$$A(h) = \lambda h$$

and

$$A_1 a = \lambda a \tag{7.5}$$

where a is the column vector $(\alpha_1, \alpha_2, \ldots, \alpha_n)$ and $A_1 = \text{rep } A_{b_1 b_1}$.

The matrix equation (7.5) can be rewritten

$$(\lambda I_n - A_1)a = 0 \tag{7.6}$$

showing that a lies in the null space of the matrix $(\lambda I_n - A_1)$, which is called the *characteristic matrix* of A_1. Furthermore, (7.6) is a system of homogeneous linear equations which has a non-zero solution if and only if

$$|\lambda I_n - A_1| = 0 \tag{7.7}$$

Equation (7.7) is called the *characteristic equation* of A_1 and the determinant can be expanded into a polynomial of the form

$$p(\lambda) = \lambda^n + \beta_{n-1}\lambda^{n-1} + \cdots + \beta_0 \tag{7.8}$$

with coefficients $\beta_i \in F$. This polynomial is called the *characteristic polynomial* of A_1 and its roots are the *eigenvalues* of A_1. Any non-zero vector which satisfies (7.6) is an *eigenvector* of A_1. Of course, the eigenvalues of A_1 are identical with those of the transformation A.

EXERCISE 7.2 Prove that similar matrices have the same characteristic equation and prove by counter example that the converse is not true.

EXERCISE 7.3 Show that every $n \times n$ matrix over $\mathscr{C}$ has n eigenvalues, and corresponding to each distinct eigenvalue is at least one eigenvector.

It is important to appreciate that although similar matrices have the same set of eigenvalues, the corresponding eigenvectors are not necessarily equal.

Example 7.1 Find the eigenvalues and eigenvectors of the matrix

$$A = \begin{pmatrix} 3 & 2 & 2 \\ 1 & 4 & 1 \\ -2 & -4 & -1 \end{pmatrix}$$

The characteristic equation for this matrix is

$$\begin{vmatrix} \lambda - 3 & -2 & -2 \\ -1 & \lambda - 4 & -1 \\ 2 & 4 & \lambda + 1 \end{vmatrix} = 0$$

or

$$(\lambda - 3)(\lambda - 4)(\lambda + 1) + 6(\lambda - 3) = 0$$

The eigenvalues are therefore 1, 2, 3.

The eigenvector corresponding to the eigenvalue 1 must satisfy the equation

$$Ah = h$$

or if $h = (h_1\ h_2\ h_3)$,

$$\begin{aligned} h_1 + \ \ h_2 + h_3 &= 0 \\ h_1 + 3h_2 + h_3 &= 0 \\ h_1 + 2h_2 + h_3 &= 0 \end{aligned}$$

Thus $h_2 = 0$, $h_1 = -h_3$ and the eigenvector representing the first eigenvalue is $(-1, 0, 1)$, up to a scalar multiple.

Similarly, the eigenvector corresponding to the second eigenvalue 2 must satisfy the set of equations

$$\begin{aligned} h_1 + 2h_2 + 2h_3 &= 0 \\ h_1 + 2h_2 + \ \ h_3 &= 0 \\ 2h_1 + 4h_2 + 3h_3 &= 0 \end{aligned}$$

from which $h_3 = 0$, $h_1 = -2h_2$, and the corresponding eigenvector is $(-2, 1, 0)$. In exactly the same way, the third eigenvector is seen to be $(0, 1, -1)$.

The next result shows how an $n \times n$ matrix A with n linearly independent eigenvectors over a field F can be reduced to diagonal form by means of a similarity transformation. First, suppose that $H^{-1}AH = \operatorname{diag}(\lambda_i)$, $i = 1, 2, \ldots, n$, so that

$$AH = H \operatorname{diag}(\lambda_i) \tag{7.9}$$

Now let $h_1, h_2, \ldots, h_n$ be the linearly independent columns of H, which is non-singular. Then from (7.9)

$$Ah_i = \lambda_i h_i$$

showing that the columns of H are just the eigenvectors of A. Moreover, if $(v_1, v_2, \ldots, v_n)$ are assumed to be n linearly independent eigenvectors of A, then $Av_i = \lambda_i v_i$, $i = 1, 2, \ldots, n$ and if H is taken to be the matrix with columns $h_i = v_i$, then $H^{-1}AH = \operatorname{diag}(\lambda_i)$. This proves the following theorem.

Theorem 7.2 A necessary and sufficient condition for A to be similar to a diagonal matrix is that it should have n linearly independent eigenvectors.

It follows directly from Theorem 7.1 that a sufficient condition for A to be similar to a diagonal matrix is that it should have n distinct eigenvalues.

Example 7.2 Find a similarity transformation that will diagonalize the matrix of Example 7.1.

If the transformation is $H^{-1}AH$, then the columns of H are the eigenvectors $(-1, 0, 1)$, $(-2, 1, 0)$, $(0, 1, -1)$, that is

$$H = \begin{pmatrix} -1 & -2 & 0 \\ 0 & 1 & 1 \\ 1 & 0 & -1 \end{pmatrix}$$

$$H^{-1} = \frac{1}{-1}\begin{pmatrix} -1 & 1 & -1 \\ -2 & 1 & -2 \\ -2 & 1 & -1 \end{pmatrix}^T = \begin{pmatrix} 1 & 2 & 2 \\ -1 & -1 & -1 \\ 1 & 2 & 1 \end{pmatrix}$$

This gives

$$H^{-1}A = \begin{pmatrix} 1 & 2 & 2 \\ -1 & -1 & -1 \\ 1 & 2 & 1 \end{pmatrix}\begin{pmatrix} 3 & 2 & 2 \\ 1 & 4 & 1 \\ -2 & -4 & -1 \end{pmatrix} = \begin{pmatrix} 1 & 2 & 2 \\ -2 & -2 & -2 \\ 3 & 6 & 3 \end{pmatrix}$$

and finally

$$H^{-1}AH = \begin{pmatrix} 1 & 2 & 2 \\ -2 & -2 & -2 \\ 3 & 6 & 3 \end{pmatrix}\begin{pmatrix} -1 & -2 & 0 \\ 0 & 1 & 1 \\ 1 & 0 & -1 \end{pmatrix} = \begin{pmatrix} 1 & 0 & 0 \\ 0 & 2 & 0 \\ 0 & 0 & 3 \end{pmatrix}$$

Example 7.3 If h is an eigenvector of a non-singular matrix A with eigenvalue λ, show that h is also an eigenvector of A^{-1} with eigenvalue $1/\lambda$.

Since $Ah = \lambda h$,

$$A^{-1}Ah = h = A^{-1}(\lambda h) = \lambda A^{-1}h$$

$$A^{-1}h = (1/\lambda)h$$

proving the result.

EXERCISE 7.4 A matrix (a_{ij}) is called *upper triangular* if all the entries below the principal diagonal are zero, and *lower triangular* if all entries above the principal diagonal are zero. Prove that the eigenvalues of any triangular matrix are just the principal diagonal entries.

EXERCISE 7.5 Prove that a necessary and sufficient condition for a matrix A to have a zero eigenvalue is that A be singular.

7.1 Cayley–Hamilton theorem

Theorem 7.3 (Caley–Hamilton) Every square matrix satisfies its own characteristic equation.

Proof Let the characteristic equation of the $n \times n$ matrix A be

$$|\lambda I_n - A| = \sum_{k=0}^{n} \beta_k \lambda^k = 0 \qquad (7.10)$$

where $\beta_n = 1$.

Then adj $(\lambda I_n - A)$ is also an $n \times n$ matrix with entries that are polynomials in λ of degree not exceeding $n - 1$, and so it can be split up to give

$$\text{adj } (\lambda I_n - A) = \sum_{k=0}^{n-1} B_k \lambda^k \tag{7.11}$$

where the B_k are $n \times n$ matrices. But

$$(\lambda I_n - A) \text{ adj } (\lambda I_n - A) = |\lambda I_n - A| \, I_n \tag{7.12}$$

and using (7.10) and (7.11) in (7.12) gives

$$(\lambda I_n - A) \sum_{k=0}^{n-1} B_k \lambda^k = \sum_{k=0}^{n} \beta_k \lambda^k \, I_n \tag{7.13}$$

Now equate coefficients of like powers of λ on either side of (7.13) to obtain

$$\begin{aligned} -AB_0 &= \beta_0 I_n \\ -AB_1 + B_0 &= \beta_1 I_n \\ -AB_2 + B_1 &= \beta_2 I_n \\ &\vdots \\ -AB_{n-1} + B_{n-2} &= \beta_{n-1} I_n \\ B_{n-1} &= I_n \end{aligned} \tag{7.14}$$

Finally, if the equations (7.14) are pre-multiplied by $I_n, A, \ldots, A^n$ respectively (from the top) and are added, the result

$$\beta_0 I_n + \beta_1 A + \beta_2 A^2 + \cdots + A^n = 0 \tag{7.15}$$

follows. The expression on the left-hand side of (7.15) is called a *polynomial in the matrix A*, or *matrix polynomial.*

Example 7.4 Over the field of integers modulo 3, the polynomial $p(x) = x^3 + 2x + 1$ is irreducible by Exercise 6.3 of Chapter 1. The matrix

$$A = \begin{pmatrix} 0 & 2 & 0 \\ 0 & 2 & 2 \\ 2 & 0 & 1 \end{pmatrix}$$

has $p(\lambda)$ for its characteristic polynomial, and $A^3 + 2A + I = 0$. Notice that the characteristic roots of A do not lie in the base field.

EXERCISE 7.6 Prove that adj $A = (-1)^{n-1}(\beta_1 I_n + \beta_2 A + \cdots + A^{n-1})$. [From (7.11), adj $A = (-1)^{n-1}\beta_0$.]

By putting $\lambda = 0$ in equation (7.10) it is easily seen that $\beta_0 = (-1)^n \, |A|$.

EXERCISE 7.7 Show that $-\beta_{n-1}$ in equation (7.10) is equal to $a_{11} + a_{22} + \cdots + a_{nn}$. [Use the definition of a determinant directly.]

The negative of the coefficient β_{n-1} in the characteristic equation, that is $a_{11} + a_{22} + \cdots + a_{nn}$, is called the *trace* or *spur* of the matrix A.

EXERCISE 7.8 If λ_i, $i = 1, 2, \ldots, n$ are the characteristic roots of a matrix A, prove that $p(\lambda_i)$ are the characteristic roots of any matrix polynomial $p(A)$.

7.2 Minimal polynomials

Any polynomial p such that $p(A) = 0$ for the $n \times n$ matrix A is said to be an *annihilating polynomial* for A. In particular, the monic polynomial of least degree which annihilates A is called the *minimal polynomial* of A. Since any square matrix satisfies its own characteristic equation, it follows that the characteristic polynomial annihilates A and that the degree of the minimal polynomial must be less than or equal to n. It is a direct consequence of Exercise 5.2 that similar matrices have the same minimal polynomial.

Suppose that $m(\lambda)$ is the minimal polynomial of any square matrix A and that $f(\lambda)$ is any other polynomial that annihilates A. Then by the remainder theorem (Chapter 1, Section 6) there exist two polynomials $a(\lambda)$ and $b(\lambda)$ such that

$$f(\lambda) = a(\lambda)m(\lambda) + b(\lambda) \tag{7.16}$$

where $\deg b(\lambda) < \deg m(\lambda)$, or $b(\lambda) \equiv 0$.

But since $f(A) = 0$, so must $b(A) = 0$, and $b(\lambda) \equiv 0$ otherwise there is a contradiction with regard to the definition of a minimal polynomial. It has been shown, therefore, that the minimal polynomial of A divides every polynomial which annihilates A and, in particular, the minimal polynomial divides the characteristic polynomial.

EXERCISE 7.9 Prove that the minimal polynomial of a matrix is unique.

The relationship between the characteristic polynomial and the minimal polynomial of a given $n \times n$ matrix A will now be discussed. The elements of $\operatorname{adj}(\lambda I_n - A)$ are polynomials in λ of degree $n - 1$ at most; let their monic greatest common divisor be $g(\lambda)$ and let $f(\lambda)$ be the characteristic polynomial of A and $m(\lambda)$ its minimal polynomial. Write $\operatorname{adj}(\lambda I_n - A)$ in the form $g(\lambda)\Gamma(\lambda)$ where Γ is a matrix with elements having no polynomial common factor (Γ is sometimes called a *reduced adjoint matrix*). Then

$$g(\lambda)\Gamma(\lambda)(\lambda I_n - A) = f(\lambda)I_n \tag{7.17}$$

so that $g(\lambda)$ divides $f(\lambda)$ without remainder. If $s(\lambda) = f(\lambda)/g(\lambda)$ then (7.17) becomes

$$\Gamma(\lambda)(\lambda I_n - A) = s(\lambda)I_n \tag{7.18}$$

On comparing (7.18) and (7.12) it is obvious that the same argument can be used as in the proof of the Cayley–Hamilton theorem to show that $s(\lambda)$ annihilates A. This means that $m(\lambda)$ divides $s(\lambda)$. It will now be shown that $s(\lambda)$ divides $m(\lambda)$, so that in fact $m(\lambda) = s(\lambda)$ because both $s(\lambda)$ and $m(\lambda)$ are monic.

First notice that because

$$\lambda^r I_n - A^r = (\lambda I_n - A)(\lambda^{r-1} I_n + \lambda^{r-2} A + \cdots + A^{r-1})$$

$$m(\lambda I_n) - m(A) = (\lambda I_n - A)B \tag{7.19}$$

where B is a polynomial in λI_n and A. Since $m(A) = 0$, multiplying (7.19) by adj $(\lambda I_n - A)$ gives

$$\text{adj}\,(\lambda I_n - A)m(\lambda I_n) = \text{adj}\,(\lambda I_n - A)(\lambda I_n - A)B$$

or

$$m(\lambda)\,\text{adj}\,(\lambda I_n - A) = f(\lambda)B \tag{7.20}$$

But adj $(\lambda I_n - A) = g(\lambda)\Gamma(\lambda)$ and $f(\lambda) = s(\lambda)g(\lambda)$ so that (7.20) can be written

$$m(\lambda)\Gamma(\lambda) = s(\lambda)B \tag{7.21}$$

From (7.21), and the fact that the elements of Γ have no common polynomial factors, $s(\lambda)$ must divide $m(\lambda)$, and the result follows. Thus the minimal polynomial of any square matrix A is the ratio of its characteristic polynomial and the greatest common divisor of the elements of the adjoint of its characteristic matrix $(\lambda I_n - A)$.

EXERCISE 7.10 In Chapter 8 the transfer function $G(s)$ of a certain linear system will be defined by

$$G(s) = C(sI_n - A)^{-1}B + D$$

where A, B, C, D are real matrices, respectively $n \times n, n \times l, m \times n$, and $m \times l$, and s is a complex number. Show that $G(s)$ can be written

$$G(s) = C\Gamma(s)B/m(s) + D$$

where $\Gamma(s)$ is the reduced adjoint matrix and $m(s)$ is the minimal polynomial of A.

EXERCISE 7.11 Show that the characteristic polynomial of

$$A = \begin{pmatrix} -1 & 1 & 0 \\ 0 & -1 & 0 \\ 0 & 0 & -1 \end{pmatrix}$$

is $(\lambda + 1)^3$, the minimal polynomial is $(\lambda + 1)^2$, and

$$\Gamma(s) = \begin{pmatrix} s+1 & -1 & 0 \\ 0 & s+1 & 0 \\ 0 & 0 & s+1 \end{pmatrix}$$

If $B = \begin{pmatrix} 0 \\ 0 \\ 1 \end{pmatrix}$, $C = (0, 0, 1)$, $D = 0$ show that $G(s) = \dfrac{1}{(s+1)}$.

[Notice that the denominator of $G(s)$ can have lower degree than $m(s)$.]

Let (7.19) be written in the form

$$m(\lambda)I_n = (\lambda I_n - A)B \tag{7.22}$$

and take determinants on both sides of (7.22) to give

$$(m(\lambda))^n = f(\lambda)|B| \tag{7.23}$$

Together with the fact that $m(\lambda)$ divides $f(\lambda)$, this last equation implies that the sets of distinct roots of $f(\lambda)$ and $m(\lambda)$ are the same. In other words, all the distinct characteristic roots of a matrix are roots of its minimal polynomial. In the case where the characteristic equation has no repeated roots it follows that the characteristic polynomial and the minimal polynomial are the same.

7.3 Invariant polynomials and elementary divisors

In Section 6 the equivalence of matrices over a field F was considered. The ideas involved will now be extended to matrices over a polynomial ring $F[x]$. Matrices having their elements in a polynomial ring are known simply as *polynomial matrices*. In the elementary transformation involving the multiplication of a row or column by a non-zero constant, the multipliers will now be restricted to be non-zero elements of the field F. An analogue of the Gauss reduction process will first be established.

Theorem 7.4 Any $n \times n$ polynomial matrix P of rank r is equivalent to a matrix of the form

$$D = \text{diag}(p_1, p_2, \ldots, p_r, 0, \ldots, 0) \tag{7.24}$$

where each p_i, $i = 1, 2, \ldots, r$ is a monic polynomial and p_i is a factor of p_{i+1}, $i = 1, 2, \ldots, r - 1$.

Proof The proof is very similar to the corresponding proof in Chapter 2, but complications arise because the elements of P are polynomials. It is obviously possible to ignore the case in which P is the null matrix and so to take $r > 0$. Start by bringing a non-zero element of P with least degree to the top left-hand position (1, 1) by appropriate elementary operations. Write $p_{12} = p_{11}g_{12} + s_{12}$, where s_{12} is zero or has degree less than p_{11}, and subtract g_{12} times column 1 from column 2. Proceed similarly with $p_{13}, p_{14}, \ldots, p_{1n}$ and $p_{21}, p_{31}, \ldots, p_{n1}$. Unless all the remainders s_{ij} produced in this way are zero, an element of lower degree than before can be brought to position (1, 1). In this way the degree of the leading element can be continually reduced until the process terminates with all elements in the first row and first column zero except the leading element.

If any one of columns 2, 3, ..., n now contains an element not divisible (without remainder) by the leading element, add this column to the first column. Then again the degree of the leading element can be reduced. When the process terminates, a matrix Q equivalent to P has been obtained with

$$Q = \begin{pmatrix} p_1 & 0 & 0 & \cdots & 0 \\ 0 & q_{22} & q_{23} & \cdots & q_{2n} \\ 0 & q_{32} & q_{33} & \cdots & q_{3n} \\ \vdots & \vdots & \vdots & & \vdots \\ 0 & q_{n2} & q_{n3} & \cdots & q_{nn} \end{pmatrix} \tag{7.25}$$

In (7.25) it can be assumed that p_1 is monic; and it is a factor of every other element of Q.

Now suppose that not every $q_{ij} = 0$ $(i, j = 2, 3, \ldots, n)$. Exactly the same procedure can be used on the submatrix obtained from Q by deleting its first row and first column to obtain a second matrix R equivalent to P and given by

$$R = \begin{pmatrix} p_1 & 0 & 0 & 0 & \cdots & 0 \\ 0 & p_2 & 0 & 0 & \cdots & 0 \\ 0 & 0 & r_{33} & r_{34} & \cdots & r_{3n} \\ 0 & 0 & r_{43} & r_{44} & \cdots & r_{4n} \\ \vdots & \vdots & \vdots & \vdots & & \vdots \\ 0 & 0 & r_{n3} & r_{n4} & \cdots & r_{nn} \end{pmatrix} \tag{7.26}$$

Again in (7.26) p_2 can be made monic; and it is a factor of every other element of R except possibly p_1. Since R is obtained from Q by a sequence of elementary operations it is also apparent that p_1 will still divide every other element of R.

Repeated application of this process will result in a matrix of the form given in (7.24), and so the theorem is proved.

EXERCISE 7.12 Extend the theorem to rectangular polynomial matrices. [Given $P(x)$, $m \times n$, with $n < m$, consider $(P(x), 0)$, $m \times m$.]

Let $D_k(x)$ be the monic polynomial which is the greatest common divisor of the set of all k-rowed minors of P for $1 \leqslant k \leqslant r$, where r is the rank of P, and define $D_0(x) = 1$. Then the $D_k(x)$ are called the *determinantal divisors* of P. The next theorem concerns the determinantal divisors of equivalent polynomial matrices.

Theorem 7.5 Equivalent polynomial matrices have the same determinantal divisors.

Proof Clearly if the result can be shown to hold for a single row transformation, say, then it also holds for a single column transformation of the same type and for a series of row and column transformations. If two rows are interchanged or a row is multiplied by an element of the field F, the result is obvious since each k-rowed minor of the transformed matrix is a constant multiple of the corresponding original minor. This leaves the operation of adding a multiple, by a polynomial in $F[x]$, of one row to another.

Suppose then that P_1 is obtained from P by adding $g(x)$ times the jth row of P to the ith row of P. Any k-rowed minor of P which does not involve the ith row of P, or does involve both the ith and jth rows of P, is unchanged under the transformation. If a k-rowed minor of P involves the ith row but does not involve the jth row of P, then the corresponding minor of P_1 is just a linear sum of k-rowed minors of P. This shows that the determinantal divisors of P divide those of P_1. Interchanging the roles of P_1 and P shows that the determinantal divisors of P_1 divide those of P, thus completing the proof of the theorem.

Now notice that the determinantal divisors of D in equation (7.24) are just

$$\begin{aligned} &D_0(x) = 1, \qquad D_1(x) = p_1(x), \qquad D_2(x) = p_1(x)p_2(x), \;\ldots, \\ &D_r(x) = p_1(x)p_2(x)\ldots p_r(x) \end{aligned} \tag{7.27}$$

so that

$$p_k(x) = \frac{D_k(x)}{D_{k-1}(x)}, \quad k = 1, 2, \ldots, r \tag{7.28}$$

The polynomials $p_k(x)$ in equation (7.28) are called† the *invariant polynomials* of the polynomial matrix P. It has therefore been shown that any polynomial matrix P is equivalent to a unique diagonal matrix D, given by (7.24), where p_i are the invariant polynomials of P. The matrix D is called the *Smith canonical form* of P. Notice that it has also been demonstrated that equivalent polynomial matrices have equal rank.

Exercise 7.13 Prove that a necessary and sufficient condition for two polynomial matrices P_1 and P over $F[x]$ to be equivalent is that

$$P_1 = MPN \tag{7.29}$$

† The polynomials are sometimes numbered in the reverse order.

where M and N are polynomial matrices with their determinants non-zero and independent of x (such matrices are said to be *unimodular*).

EXERCISE 7.14 Let $G(s)$ be an $m \times m$ transfer function matrix (Chapter 8, Section 5.2) with rational polynomial entries. Write $G(s) = N(s)/d(s)$, where d(s) is a polynomial and $N(s)$ is a polynomial matrix, and hence show that $G(s) = LQR$, where L, R are unimodular polynomial matrices,

$$Q = \text{diag}\,(\epsilon_1/\psi_1, \epsilon_2/\psi_2, \ldots, \epsilon_r/\psi_r, 0, \ldots, 0)$$

the ϵ_i and ψ_i are relatively prime polynomials, $i = 1, 2, \ldots, r$, and ϵ_i divides ϵ_{i+1}, ψ_{i+1} divides ψ_i, $i = 1, 2, \ldots, r - 1$. Q is called the *McMillan form*, or *Smith–McMillan* form of $G(s)$.

EXERCISE 7.15 Prove that a necessary and sufficient condition for two polynomial matrices to be equivalent is that they have the same invariant polynomials.

Now suppose that the invariant polynomials $p_k(x)$ of a matrix P are decomposed into their monic irreducible factors over the field F to give

$$p_k(x) = [\phi_1(x)]^{j_{k1}}[\phi_2(x)]^{j_{k2}} \ldots [\phi_s(x)]^{j_{ks}}, \quad k = 1, 2, \ldots, r \tag{7.30}$$

Then those factors ϕ^j of (7.30) with non-zero exponents are called the *elementary divisors* of P. Because of the divisibility properties of the $p_k(x)$, it is evident that

$$j_{rl} \geqslant j_{r-1,l} \geqslant \cdots \geqslant j_{1l} \geqslant 0 \quad \text{for } l = 1, 2, \ldots, s$$

(Notice that all of the distinct irreducible factors of all the $p_k(x)$, $k = 1, 2, \ldots, r$, have been included in the factorization of each $p_k(x)$ so that some of the exponents j_{kl} may be zero.)

EXERCISE 7.16 Prove that a necessary and sufficient condition for two polynomial matrices to be equivalent is that they have the same rank and elementary divisors. [The invariant polynomials of a matrix obviously determine its elementary divisors: show, by using the divisibility properties of the invariant polynomials, that a knowledge of the rank and elementary divisors will determine the invariant polynomials.]

Two theorems concerning quasi-diagonal matrices that will be needed later in this section will now be proved.

Theorem 7.6 The invariant polynomials of the quasi-diagonal matrix

$$C(x) = \begin{pmatrix} C_1(x) & 0 \\ 0 & C_2(x) \end{pmatrix} \tag{7.31}$$

are the union of the invariant polynomials of $C_1(x)$ and $C_2(x)$, provided that every invariant polynomial of $C_1(x)$ divides every invariant polynomial of $C_2(x)$.

Proof The result follows directly from the fact that $C(x)$ is equivalent to $\text{diag}\,(p_1, p_2, \ldots, p_r, q_1, q_2, \ldots, q_s, 0, \ldots, 0)$, where $p_1, p_2, \ldots, p_r$ are the invariant polynomials of $C_1(x)$ and $q_1, q_2, \ldots, q_s$ are the invariant polynomials of $C_2(x)$.

Theorem 7.7 The elementary divisors of the quasi-diagonal matrix $C(x)$ given by equation (7.31) are the union of those of $C_1(x)$ and $C_2(x)$.

Proof Let the invariant polynomials of $C_1(x)$ and $C_2(x)$ be given by

$$p_k(x) = [\phi_1(x)]^{j_{k1}}[\phi_2(x)]^{j_{k2}} \ldots [\phi_t(x)]^{j_{kt}}, \quad k = 1, 2, \ldots, r \tag{7.32}$$

$$q_k(x) = [\phi_1(x)]^{l_{k1}}[\phi_2(x)]^{l_{k2}} \ldots [\phi_t(x)]^{l_{kt}}, \quad k = 1, 2, \ldots, s \tag{7.33}$$

Now suppose that these are arranged in ascending order of powers of $\phi_1(x)$ so that, by a suitable sequence of elementary operations, it can be seen that $C(x)$ is equivalent to the matrix

$$D = \text{diag}\,\big(d_1(x), d_2(x), \ldots, d_\alpha(x), g_1(x){\phi_1}^{k'_1}(x), g_2(x){\phi_1}^{k'_2}(x), \\ \ldots, g_\beta(x){\phi_1}^{k'_\beta}(x), 0, \ldots, 0\big) \tag{7.34}$$

where $d_1(x), d_2(x), \ldots, d_\alpha(x), g_1(x), g_2(x), \ldots, g_\beta(x)$ are all prime relative to $\phi_1(x)$, $k'_1 \leqslant k'_2 \leqslant \cdots \leqslant k'_\beta$, and $\alpha + \beta = r + s = m$. Let the determinantal divisors and invariant polynomials of D, and hence of $C(x)$, be represented by D_k and p'_k, $k = 1, 2, \ldots, m$. It follows from (7.34) that D_m is the product of a polynomial prime to $\phi_1(x)$ with ${\phi_1}^{k'_1+k'_2+\cdots+k'_\beta}(x)$, D_{m-1} is the product of a polynomial prime to $\phi_1(x)$ with ${\phi_1}^{k'_1+k'_2+\cdots+k'_{\beta-1}}(x)$ and so on. Consequently p'_m is the product of a polynomial prime to $\phi_1(x)$ with ${\phi_1}^{k'_\beta}(x)$, p'_{m-1} is a similar product with ${\phi_1}^{k'_{\beta-1}}$, and so on.

From this it is clear that all the elementary divisors of $C_1(x)$ and $C_2(x)$ which involve $\phi_1(x)$ are also elementary divisors of $C(x)$. By treating $\phi_2(x), \phi_3(x), \ldots, \phi_t(x)$ in exactly the same way the theorem is proved.

Given any $n \times n$ matrix A over a field F, then its characteristic matrix $(\lambda I_n - A)$ is a polynomial matrix of rank n over $F[\lambda]$ and the preceding results apply. Notice that $D_n(\lambda)$ is the characteristic polynomial of A, $D_{n-1}(\lambda)$ is the polynomial $g(\lambda)$ in equation (7.17), and $p_n(\lambda)$ is $s(\lambda)$ in equation (7.18), which was shown to be the minimal polynomial of A.

It is evident from (7.28) that the product $\prod_{k=1}^{n} p_k(\lambda)$ is equal to $D_n(\lambda)$ so that the $p_k(\lambda)$ are factors of the characteristic polynomial. The invariant polynomials and elementary divisors of $(\lambda I_n - A)$ are said to be the *invariant polynomials* and *elementary divisors* of A.

Finally we give an important theorem for matrices over a field F.

Theorem 7.8 A necessary and sufficient condition for two square matrices to be similar is that they have the same invariant polynomials (or elementary divisors).

Proof If A and B are similar there exists a non-singular matrix H such that

$$B = HAH^{-1} \tag{7.35}$$

It follows that

$$\begin{aligned}(\lambda I_n - B) &= \lambda I_n - HAH^{-1}\\ &= H(\lambda I_n - A)H^{-1}\end{aligned} \tag{7.36}$$

so $\lambda I_n - A$ and $\lambda I_n - B$ are equivalent (Exercise 7.13), and therefore have the same invariant polynomials (Exercise 7.15).

On the other hand, if $\lambda I_n - A$ and $\lambda I_n - B$ have the same invariant polynomials they are equivalent and there exist unimodular polynomial matrices M and N such that

$$(\lambda I_n - A) = M(\lambda I_n - B)N \tag{7.37}$$

Let $R = N^{-1}$ (this exists since N is unimodular) so that

$$(\lambda I_n - A)R = M(\lambda I_n - B) \tag{7.38}$$

Now R can be written $R(\lambda) = R_0 + R_1\lambda + \cdots + R_q\lambda^q$, and because I_n has an inverse it is easy to see that the division algorithm (Chapter 1, Section 6) can be used with multiplication on the appropriate side to give

$$R = Q_1(\lambda I_n - B) + Q_0 \tag{7.39}$$

and similarly

$$M = (\lambda I_n - A)S_1 + S_0 \tag{7.40}$$

where Q_0, S_0 are constant matrices (that is, are independent of λ), and Q_1, S_1 are polynomial matrices.

Using (7.39) and (7.40) in (7.38) gives

$$\begin{aligned}&(\lambda I_n - A)Q_1(\lambda I_n - B) + (\lambda I_n - A)Q_0\\ &\qquad = (\lambda I_n - A)S_1(\lambda I_n - B) + S_0(\lambda I_n - B)\end{aligned} \tag{7.41}$$

which can be rearranged as follows

$$(\lambda I_n - A)(Q_1 - S_1)(\lambda I_n - B) = S_0(\lambda I_n - B) - (\lambda I_n - A)Q_0 \tag{7.42}$$

The degree of the right-hand side of (7.42) is one, but unless $Q_1 - S_1 = 0$ the degree of the left-hand side is at least two: this implies that $Q_1 = S_1$ and hence

$$(\lambda I_n - A)Q_0 = S_0(\lambda I_n - B) \tag{7.43}$$

Again, using the division algorithm it is possible to write

$$N = T_1(\lambda I_n - A) + T_0 \tag{7.44}$$

where T_0 is constant. Hence

$$I_n = NR = T_1(\lambda I_n - A)R + T_0R \tag{7.45}$$

and, from (7.38), (7.39), and (7.45),

$$\begin{aligned} I_n &= T_1M(\lambda I_n - B) + T_0Q_1(\lambda I_n - B) + T_0Q_0 \\ &= (T_1M + T_0Q_1)(\lambda I_n - B) + T_0Q_0 \end{aligned} \tag{7.46}$$

Since the degree of I_n is zero it follows from (7.46) that $T_1M + T_0Q_1 = 0$ and therefore $T_0Q_0 = I_n$, so that Q_0 is non-singular and $Q_0^{-1} = T_0$. Thus equation (7.43) becomes

$$(\lambda I_n - A) = S_0(\lambda I_n - B)T_0 \tag{7.47}$$

and it has been shown that the polynomial matrices M and N in (7.37) can be replaced by constant matrices S_0 and T_0.

The rest of the proof now follows directly from (7.47), because this can be written

$$\lambda I_n - A = \lambda S_0T_0 - S_0BT_0 \tag{7.48}$$

implying that $S_0T_0 = I_n$, $A = S_0BT_0 = S_0BS_0^{-1}$, so that A and B are similar. This ends the proof.

It should be noted that this theorem means that a necessary and sufficient condition for similarity of the matrices A and B is that their characteristic matrices should be equivalent.

EXERCISE 7.17 If M and N are written as matrix polynomials in the form $\sum_{k=0}^{m} \lambda^k M_k$ and $\sum_{k=0}^{n} N_k\lambda^k$ respectively, prove that

$$S_0 = \sum_{k=0}^{m} A^k M_k \tag{7.49}$$

$$T_0 = \sum_{k=0}^{n} N_k A^k \tag{7.50}$$

[Use the division algorithm.]

EXERCISE 7.18 If $A = A_1\lambda + A_0$ and $B = B_1\lambda + B_0$, with B_1 non-singular, prove that a necessary and sufficient condition for the equivalence of A and B is that there exist two constant non-singular matrices P and Q such that $PAQ = B$. [By considering the degree of $|A|$ show that A_1 is non-singular: then use the remark following the proof of Theorem 7.8.]

7.4 Companion matrix and normal forms

Consider the polynomial

$$f(\lambda) = \lambda^m + \alpha_{m-1}\lambda^{m-1} + \cdots + \alpha_0$$

Is it possible to construct a matrix with $f(\lambda)$ as its minimal polynomial? A direct answer to this question is obtained by considering the characteristic matrix of the *companion matrix* C of $f(\lambda)$, given† by

$$C = \begin{pmatrix} 0 & 0 & \dots & 0 & -\alpha_0 \\ 1 & 0 & \dots & 0 & -\alpha_1 \\ 0 & 1 & \dots & 0 & -\alpha_2 \\ \vdots & \vdots & & \vdots & \vdots \\ 0 & 0 & \dots & 0 & -\alpha_{m-1} \end{pmatrix} \tag{7.51}$$

The characteristic determinant of C is

$$|\lambda I_n - C| = \begin{vmatrix} \lambda & 0 & 0 & \dots & 0 & \alpha_0 \\ -1 & \lambda & 0 & \dots & 0 & \alpha_1 \\ 0 & -1 & \lambda & \dots & 0 & \alpha_2 \\ \vdots & \vdots & \vdots & & \vdots & \vdots \\ 0 & 0 & 0 & \dots & -1 & \alpha_{m-1} + \lambda \end{vmatrix} \tag{7.52}$$

and expansion by the last column easily verifies that $|\lambda I_n - C| = f(\lambda)$. Since the minor of α_0 in (7.52) is ± 1 it follows that $D_{m-1} = 1$ and hence $f(\lambda)$ is the minimal polynomial of C.

Now let A be an $n \times n$ matrix with elements in some field F and let its invariant polynomials be

$$i_1(\lambda) \equiv i_2(\lambda) \equiv \cdots \equiv i_r(\lambda) \equiv 1, \quad i_{r+1}(\lambda), \quad i_{r+2}(\lambda), \dots, i_n(\lambda) \tag{7.53}$$

where $i_{r+1}, i_{r+2}, \dots, i_n$ have degree greater than zero, the sum of the degrees of these polynomials is n, and i_{r+k} divides i_{r+k+1}, $k = 1, 2, \dots, n - r - 1$. Each of the polynomials $i_k(\lambda)$ will have a companion matrix C_k and it is possible to form the $n \times n$ matrix

$$C = \begin{pmatrix} C_{r+1} & & & & \\ & C_{r+2} & & 0 & \\ & & \cdot & & \\ & & & \cdot & \\ & 0 & & & \cdot \\ & & & & C_n \end{pmatrix} \tag{7.54}$$

By Theorem 7.6, the invariant polynomials of C are just the set of all the invariant polynomials of the C_k. Thus the invariant polynomials of C are just those of A, and A and C must be similar by Theorem 7.8.

It has been shown that any $n \times n$ matrix A is similar to a matrix C having the

† The three matrices, similar to C, obtained by reflections in the two diagonals are also called companion matrices.

companion matrices of the invariant polynomials of A on its principal diagonal and zero elsewhere. This matrix C is called the *first natural form*† of A.

Similarly, it is possible to show that A is similar to the matrix D given by

$$D = \begin{pmatrix} D_1 & & & & \\ & D_2 & & 0 & \\ & & \cdot & & \\ & & & \cdot & \\ & 0 & & & \cdot \\ & & & & D_s \end{pmatrix} \tag{7.55}$$

where $D_1, D_2, \ldots, D_s$ are the companion matrices of the elementary divisors of A. The matrix D is called the *second natural normal form* of A.

Up to the present the nature of the field F has been left quite general; attention is now turned to the special case where F is the field of complex numbers. In this case the characteristic polynomial has linear factors only (Chapter 1, Section 6) and the elementary divisors of A are of the form

$$(\lambda - \lambda_1)^{k_1}, \quad (\lambda - \lambda_2)^{k_2}, \quad \ldots, \quad (\lambda - \lambda_r)^{k_r}$$

with $\sum_{i=1}^{r} k_i = n$; where not all of the λ_i need be distinct. If the matrix J is formed with

$$J = \begin{pmatrix} J_1 & & & & \\ & J_2 & 0 & & \\ & & \cdot & & \\ & & & \cdot & \\ & 0 & & & \cdot \\ & & & & J_r \end{pmatrix} \tag{7.56}$$

where the $k_i \times k_i$ matrix J_i is

$$J_i = \begin{pmatrix} \lambda_i & 1 & 0 & \ldots & 0 \\ 0 & \lambda_i & 1 & \ldots & 0 \\ \vdots & \vdots & \vdots & & \vdots \\ 0 & 0 & 0 & \ldots & 1 \\ 0 & 0 & 0 & \ldots & \lambda_i \end{pmatrix}, \quad i = 1, 2, \ldots, r$$

then it is easily seen that the elementary divisors of J_i are just $(\lambda - \lambda_i)^{k_i}$ and hence the elementary divisors of J are those of A. It follows (Theorem 7.8) that any matrix over the field of complex numbers is similar to a matrix of the same form as J, which is known as the *Jordan canonical form*.

† Though it is more usual to arrange the companion matrices in the reverse order to that shown in (7.54).

EXERCISE 7.19 Prove that the nth power of the *Jordan block*

$$J = \begin{pmatrix} \lambda & 1 & 0 & \dots \\ 0 & \lambda & 1 & \dots \\ 0 & 0 & \lambda & \dots \\ . & . & . & . \end{pmatrix}$$

is given by

$$J^n = \begin{pmatrix} \lambda^n & \binom{n}{1}\lambda^{n-1} & \binom{n}{2}\lambda^{n-2} & \dots \\ 0 & \lambda^n & \binom{n}{1}\lambda^{n-1} & \dots \\ 0 & 0 & \lambda^n & \dots \\ . & . & . & . \end{pmatrix}$$

where $\binom{n}{k} = \dfrac{n!}{(n-k)!k!}$. [Use induction.]

EXERCISE 7.20 In Chapter 6 the $n \times n$ transition matrix $\Phi(\tau)$ will be defined for certain linear systems by the equation, valid for all t,

$$x(t + \tau) = \Phi(\tau)x(t)$$

By writing $\Phi(\tau) = HJH^{-1}$ where J is as in equation (7.56) show how the result of Exercise 7.19 can be used to obtain $x(t + n\tau)$.

8 Some special matrices

In this section some special forms of matrices over the field of complex numbers are considered and their properties are developed.

Symmetric and skew-symmetric matrices, the transpose of a matrix, and its adjoint have been defined in Chapter 2. The *conjugate* of a matrix A is denoted by $\bar{A}$ and is the matrix having as its ijth element the complex conjugate of the ijth element of A, that is, $\bar{A} = (\bar{a}_{ij})$. The *conjugate transpose* of A is $(\bar{A})^{\mathrm{T}}$ and is denoted by A^*. Notice that $(\bar{A})^{\mathrm{T}} = \overline{(A^{\mathrm{T}})}$, $(A^*)^* = A$, and $(AB)^* = B^*A^*$.

A matrix is called *normal* if $AA^* = A^*A$, *hermitian* if $A^* = A$ (symmetric when F is real), *skew-hermitian* if $A^* = -A$ (skew-symmetric when F is real), and *unitary* if $AA^* = A^*A = I_n$ (orthogonal when F is real). It is easily verified that hermitian, skew-hermitian, and unitary matrices are all normal.

EXERCISE 8.1 Prove that any $n \times n$ matrix with elements in the field of complex numbers can be expressed uniquely as the sum of a hermitian and a skew-hermitian matrix.

EXERCISE 8.2 Prove that the inverse of a unitary matrix is unitary and the product of two unitary matrices is unitary.

8.1 Normal matrices

Normal matrices are the most general of the types of matrix defined above so a number of important results are first shown for normal matrices.

Any two matrices A and B are said to be *unitarily similar* (or *orthogonally similar* when F is real) when there exists a unitary matrix U for which $B = U^*AU$.

Theorem 8.1 A necessary and sufficient condition for a matrix A to be normal is that it be unitarily similar to a diagonal matrix.

Proof First, let D be a diagonal matrix: then it is obvious that $DD^* = D\bar{D} = \bar{D}D = D^*D$ so that any diagonal matrix is normal. Now let U be any unitary matrix, $B = U^*CU$, and assume that C is normal. Then

$$\begin{aligned} B^*B &= U^*C^*UU^*CU \\ &= U^*C^*I_nCU \\ &= U^*CC^*U \\ &= U^*CUU^*C^*U \\ &= BB^* \end{aligned}$$

so that any matrix unitarily similar to a normal matrix is itself normal: the sufficiency of the conditions in the theorem follows directly.

To prove necessity is rather more troublesome and it is first necessary to show that a unitary matrix may be constructed with a given column vector as its first column. Let x_1 be the given column vector and augment it with $v_2, v_3, \ldots, v_n$ to form a basis of the vector space of n-tuples over F, that is, of $V_n(F)$. Now use the Gram–Schmidt orthogonalization process (with standard inner product) to form an orthonormal set $(x_1, x_2, \ldots, x_n)$ from $(x_1, v_2, \ldots, v_n)$ and form the matrix U with the x_i as its columns. Then the ijth element of U^*U is easily seen to be $\langle x_j, x_i\rangle$ and this is equal to the Kronecker delta, δ_{ij}, because the x_i are an orthonormal set. Thus $U^*U = I_n$ and U is unitary.

Suppose that x_1 is an eigenvector of A with λ_1 as the corresponding eigenvalue and let U be a unitary matrix with x_1 as its first column. Then the element of U^*AU in position $i1$ is given by

$$\begin{aligned} (U^*AU)_{i1} &= \sum_{k=1}^{n} (U^*)_{ik}\,(AU)_{k1} \\ &= \sum_{k=1}^{n} (U^*)_{ik}(Ax_1)_k \end{aligned}$$

$$= \lambda_1 \sum_{k=1}^{n} (U^*)_{ik}(x_1)_k$$

$$= \lambda_1 \delta_{i1}$$

If $B = U^*AU$, then B can be written in the form

$$B = \left(\begin{array}{c|c} \lambda_1 & z^{\mathrm{T}} \\ \hline 0 & B_1 \end{array}\right) \tag{8.1}$$

where B_1 is an $(n - 1) \times (n - 1)$ matrix.

From (8.1), on taking the conjugate transpose

$$B^* = \left(\begin{array}{c|c} \bar{\lambda}_1 & 0 \\ \hline \bar{z} & B_1^* \end{array}\right) \tag{8.2}$$

therefore

$$B^*B = \left(\begin{array}{c|c} \bar{\lambda}_1\lambda_1 & \bar{\lambda}_1 z^{\mathrm{T}} \\ \hline \lambda_1 \bar{z} & \bar{z}z^{\mathrm{T}} + B_1^* B_1 \end{array}\right) \tag{8.3}$$

and

$$BB^* = \left(\begin{array}{c|c} \lambda_1\bar{\lambda}_1 + z^{\mathrm{T}}\bar{z} & z^{\mathrm{T}}B_1^* \\ \hline B_1\bar{z} & B_1 B_1^* \end{array}\right) \tag{8.4}$$

But the matrix B is unitarily similar to A so if A is assumed normal, B must also be normal. Comparing (8.3) and (8.4):

$$\bar{\lambda}_1\lambda_1 = \lambda_1\bar{\lambda}_1 + z^{\mathrm{T}}\bar{z}$$

so that $z = 0$, and $B_1 B_1^* = B_1^* B_1$. It is now possible to write (8.1) in the form

$$B = \left(\begin{array}{c|c} \lambda_1 & 0 \\ \hline 0 & B_1 \end{array}\right) \tag{8.5}$$

where B_1 is a normal $(n - 1) \times (n - 1)$ matrix.

The necessity of the conditions in the theorem now follows by induction on n; for if the theorem holds for $n - 1$ there will be a diagonal matrix D_1 and a unitary matrix U_1 for which $B_1 U_1 = U_1 D_1$. This means that

$$\begin{pmatrix} \lambda_1 & 0 \\ 0 & B_1 \end{pmatrix}\begin{pmatrix} 1 & 0 \\ 0 & U_1 \end{pmatrix} = \begin{pmatrix} 1 & 0 \\ 0 & U_1 \end{pmatrix}\begin{pmatrix} \lambda_1 & 0 \\ 0 & D_1 \end{pmatrix} \tag{8.6}$$

Let

$$V = \begin{pmatrix} 1 & 0 \\ 0 & U_1 \end{pmatrix}$$

then V is obviously unitary if U_1 is unitary; also let

$$D = \begin{pmatrix} \lambda_1 & 0 \\ 0 & D_1 \end{pmatrix}$$

then

$$BV = VD \tag{8.7}$$

and

$$V^*BV = D$$

or

$$(V^*U^*)A(UV) = D \tag{8.8}$$

and the proof is complete.

Since all unitary, hermitian and skew-hermitian matrices are certainly normal it follows that they also are unitarily similar to diagonal matrices.

It was shown in Theorem 7.2 that a necessary and sufficient condition for an $n \times n$ matrix A to be similar to a diagonal matrix is that it should have n linearly independent eigenvectors; it follows from this and Theorem 8.1 that A is normal if and only if it has n orthonormal eigenvectors. Consequently hermitian, skew-hermitian, and unitary matrices all have n orthonormal eigenvectors.

If H is any hermitian (resp. skew-hermitian) matrix, then since it is normal it is unitarily similar to a diagonal matrix having as its principal diagonal elements the eigenvalues of H. In addition, since this diagonal matrix is also hermitian (resp. skew-hermitian), the eigenvalues of H are obviously real (resp. pure imaginary). Notice, in particular, that any real symmetric matrix must have real eigenvalues.

EXERCISE 8.3 Prove that any real symmetric $n \times n$ matrix is orthogonally similar to a diagonal matrix. [Show that the eigenvectors are real.]

EXERCISE 8.4 Prove that any normal matrix with real eigenvalues must be hermitian.

If U is a unitary matrix then, since it is normal, it is unitarily similar to a diagonal matrix having as its principal diagonal elements the eigenvalues of U. This diagonal matrix must also be unitary and therefore the eigenvalues of U have unit modulus.

EXERCISE 8.5 Prove that any normal matrix with eigenvalues of unit modulus must be unitary.

EXERCISE 8.6 Prove directly from the definition that if λ is a non-zero eigenvalue of a unitary (resp. orthogonal) matrix, so is $1/\bar{\lambda}$ (resp. $1/\lambda$).

EXERCISE 8.7 Prove that a real matrix A is orthogonal if and only if

$$\langle Ax, Ay \rangle = \langle x, y \rangle$$

for all real vectors x and y. [Note the interpretation that if x is a vector of currents and y is a vector of voltages, $\langle x, y \rangle$ is a power. Orthogonal transformations (and only orthogonal transformations) keep this power invariant.]

8.2 Jacobi matrices

A real tridiagonal matrix

$$J = \begin{pmatrix} b_1 & c_1 & 0 & \dots & 0 & 0 \\ a_1 & b_2 & c_2 & \dots & 0 & 0 \\ 0 & a_2 & b_3 & \dots & 0 & 0 \\ \vdots & \vdots & \vdots & & \vdots & \vdots \\ 0 & 0 & 0 & \dots & a_{n-1} & b_n \end{pmatrix} \tag{8.9}$$

is called a *Jacobi matrix* if

$$a_i c_i > 0, \quad i = 1, 2, \dots, n - 1 \tag{8.10}$$

Such matrices are of importance in the numerical solution of certain partial differential equations.

Let D be a real diagonal matrix with elements d_i, $i = 1, 2, \dots, n$, and consider

$$R = DJD^{-1} \tag{8.11}$$

It is easily seen, by direct calculation, that

$$r_{ij} = b_j, \quad i = j$$
$$r_{ij} = 0, \quad |i - j| \geqslant 2$$

$$\left.\begin{aligned} r_{ij} &= \frac{d_i}{d_{i+1}} c_i \\ r_{ji} &= \frac{d_{i+1}}{d_i} a_i \end{aligned}\right\} j = i + 1$$

where $i, j = 1, 2, \dots, n$. It follows from (8.10) that the d_i can be taken so that $d_i/d_{i+1} = (a_i/c_i)^{1/2}$, making $r_{ij} = r_{ji}$ and R symmetric. Thus J is similar to a symmetric matrix and so must have real eigenvalues.

Example 8.1

$$\begin{pmatrix}1 & 0 & 0\\ 0 & 2 & 0\\ 0 & 0 & 6\end{pmatrix}\begin{pmatrix}1 & 4 & 0\\ 1 & 1 & 9\\ 0 & 1 & 1\end{pmatrix}\begin{pmatrix}1 & 0 & 0\\ 0 & \frac{1}{2} & 0\\ 0 & 0 & \frac{1}{6}\end{pmatrix} = \begin{pmatrix}1 & 4 & 0\\ 2 & 2 & 18\\ 0 & 6 & 6\end{pmatrix}\begin{pmatrix}1 & 0 & 0\\ 0 & \frac{1}{2} & 0\\ 0 & 0 & \frac{1}{6}\end{pmatrix}$$

$$= \begin{pmatrix}1 & 2 & 0\\ 2 & 1 & 3\\ 0 & 3 & 1\end{pmatrix}$$

EXERCISE 8.8 Show that the characteristic polynomial $\phi_n(\lambda)$ of an $n \times n$ Jacobi matrix can be obtained from the recurrence formula

$$\phi_r(\lambda) = (\lambda - b_r)\phi_{r-1} - c_{r-1}a_{r-1}\phi_{r-2}, \quad 1 < r \leqslant n$$
$$\phi_1(\lambda) = \lambda - b_1$$
$$\phi_0(\lambda) = 1$$

[Expand the characteristic determinant in terms of the last row.]

8.3 Some properties of eigenvalues

It has already been shown that the eigenvalues of any real symmetric matrix or any hermitian matrix must be real. Any diagonal matrix has its eigenvalues along its main diagonal and so has any triangular matrix. The first part of this statement is obvious; the second is proved as follows. Let T be any upper triangular matrix. Then $(\lambda I - T)$ is also upper triangular and $|\lambda I - T| = \prod_{i=1}^{n} (\lambda - t_{ii})$, so that the eigenvalues of T are just the main diagonal elements of T. The result follows in a similar manner for a lower triangular matrix.

EXERCISE 8.9 Prove that the eigenvalues of any unitary matrix lie on the unit circle and that those of any skew-hermitian matrix are pure imaginary. [Consider the unitarily similar diagonal matrices.]

Results such as those in Exercise 8.9 restrict the eigenvalues of certain special types of matrix to particular regions in the complex plane. For general matrices an interesting and useful result of similar type is the following theorem due to Gershgorin. (See Marcus and Minc, 1964, for other results on the location of eigenvalues.)

Theorem 8.2 All the eigenvalues of an arbitrary matrix $A = (a_{ij})$ over the complex field are located in the union of the set of circular disks defined by

$$|z - a_{ii}| \leqslant r_i, \quad i = 1, 2, \ldots, n \tag{8.12}$$

where

$$\sum_{\substack{j=1\\j\neq i}}^{n} |a_{ij}| = r_i$$

and also in the union of the set of disks defined by

$$|z - a_{jj}| \leqslant \rho_j, \quad j = 1, 2, \ldots, n$$

where

$$\sum_{\substack{i=1\\i\neq j}}^{n} |a_{ij}| = \rho_j$$

Proof Suppose that λ is an eigenvalue of (a_{ij}) with $v = (v_1, v_2, \ldots, v_n)^{\mathrm{T}}$ as the corresponding eigenvector. Then

$$Av = \lambda v$$

or

$$\sum_{j=1}^{n} a_{ij}v_j = \lambda v_i, \quad i = 1, 2, \ldots, n$$

On subtracting $a_{ii}v_i$ from both sides

$$\sum_{\substack{j=1\\j\neq i}}^{n} a_{ij}v_j = (\lambda - a_{ii})v_i$$

and if $v_i \neq 0$

$$(\lambda - a_{ii}) = \sum_{\substack{j=1\\j\neq i}}^{n} a_{ij}\,\frac{v_j}{v_i} \tag{8.13}$$

Now assume that i has been selected so that

$$|v_i| \geqslant |v_j|, \quad j = 1, 2, \ldots, n$$

Taking moduli on both sides of (8.13) gives

$$|\lambda - a_{ii}| \leqslant \sum_{\substack{j=1\\j\neq i}}^{n} |a_{ij}| \left|\frac{v_j}{v_i}\right| \leqslant r_i$$

showing that each eigenvalue lies in a disk with centre a_{ii} and radius r_i. Thus all the eigenvalues lie in the union of these disks. The second part is obtained by noticing that the eigenvalues of A^{T} are those of A.

Example 8.2 Given

$$A = \begin{pmatrix} 4 & 0 & 2 \\ -1 & 3 & 0 \\ 1 & 2 & 5 \end{pmatrix}$$

the first set of Gershgorin disks is

$$|z - 4| \leqslant 2$$
$$|z - 3| \leqslant 1$$
$$|z - 5| \leqslant 3$$

so that all the eigenvalues of A lie in the union of these three disks, that is to say, in the last of them. Notice that all the circles lie in the positive half-plane and so the eigenvalues of A all have positive real parts. The second set of disks is

$$|z - 4| \leqslant 2$$
$$|z - 3| \leqslant 2$$
$$|z - 5| \leqslant 2$$

and all the eigenvalues of A lie also in the union of these disks. The eigenvalues consequently lie in the intersection of the two unions, which gives a slightly sharper result, as shown in Fig. 1.

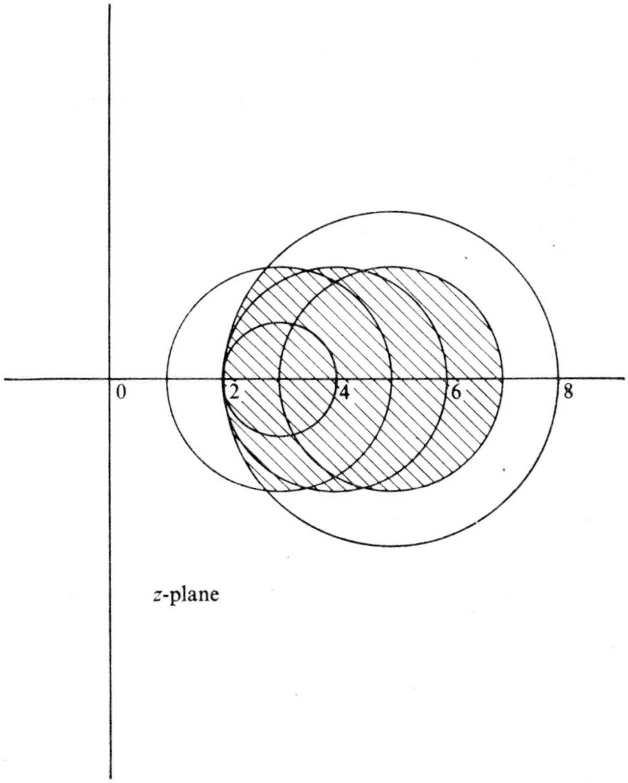

FIG. 1

EXERCISE 8.10 It will be shown in Chapter 6 that the solution $x = 0$ of the differential equation $\dot{x} = Ax$ is asymptotically stable if every eigenvalue of A has a negative real part: the matrix A is then called a *stability matrix*. Show that the following are stability matrices

$$\begin{pmatrix} -3 & 0 & 2 \\ 1 & -2 & 0 \\ 1 & 0 & -2 \end{pmatrix}, \quad \begin{pmatrix} -1 & 1 & 0 & \dots & 0 & 0 \\ 1 & -2 & 1 & \dots & 0 & 0 \\ 0 & 1 & -2 & \dots & 0 & 0 \\ \vdots & \vdots & \vdots & & \vdots & \vdots \\ 0 & 0 & 0 & \dots & -2 & 1 \\ 0 & 0 & 0 & \dots & 1 & -\alpha \end{pmatrix}$$

where the second matrix is $n \times n$ and $\alpha > 1$. [Choose β so that $1 < \beta^{2(n-2)} < \max\{\alpha, \frac{3}{2}\}$ and multiply the columns of the second matrix by $1, \beta, \beta^3, \beta^7, \dots, \beta^{2(n-1)-1}$ and the rows by $1, \beta^{-1}, \beta^{-3}, \beta^{-7}, \dots, \beta^{1-2(n-1)}$. Notice that $1 - 2\gamma^2 + \gamma^3 < 0$ if $1 < \gamma < \frac{3}{2}$, and apply Gershgorin's theorem.]

9 Bilinear functions and forms

Let $V(F)$ be a vector space over a field F. Then a function $f(v, w)$ of $V \times V$ into F is said to be *bilinear* if it is linear in both v and w. Thus if $v, w, x \in V$ and $\alpha \in F$, a bilinear function $f(v, w)$ has the properties

$$f(v, w + x) = f(v, w) + f(v, x)$$
$$f(v + w, x) = f(v, x) + f(w, x)$$
$$f(\alpha v, w) = f(v, \alpha w) = \alpha f(v, w)$$

In addition, a bilinear function $f(v, w)$ is said to be *symmetric* if $f(v, w) = f(w, v)$ and *skew-symmetric* if $f(v, w) = -f(w, v)$ for all $v, w \in V$. A symmetric bilinear function is usually called a *bilinear form*.

EXERCISE 9.1 Prove that the standard inner product in $V_n(\mathscr{R})$ is a bilinear form.

Suppose that $(v_1, v_2, \dots, v_n)$ is a basis for $V(F)$ and f is a bilinear function on V. Then for all pairs $1 \leqslant i, j \leqslant n$ the set of scalars $a_{ij} = f(v_i, v_j)$ form a matrix $A = (a_{ij})$ which is known as the *matrix of f relative to the basis* $(v_1, v_2, \dots, v_n)$. If the coordinates of two vectors w and x in $V(F)$ are α_i and β_i, $i = 1, 2, \dots, n$ respectively, then

$$\begin{aligned} f(w, x) &= f\left(\sum_{i=1}^{n} \alpha_i v_i, \ \sum_{j=1}^{n} \beta_j v_j\right) \\ &= \sum_{i=1}^{n} \sum_{j=1}^{n} \alpha_i \beta_j f(v_i, v_j) \end{aligned}$$

$$= \sum_{i=1}^{n} \sum_{j=1}^{n} \alpha_i \beta_j a_{ij} \tag{9.1}$$

In terms of matrices, (9.1) can be written

$$f(w, x) = a^{\mathrm{T}} A b \tag{9.2}$$

where a and b are the column vectors $(\alpha_1, \alpha_2, \ldots, \alpha_n)^{\mathrm{T}}$ and $(\beta_1, \beta_2, \ldots, \beta_n)^{\mathrm{T}}$.

EXERCISE 9.2 Find the matrix of the standard inner product of $V_n(\mathscr{R})$ relative to the standard basis for $V_n(\mathscr{R})$.

Denote the basis $(v_1, v_2, \ldots, v_n)$ of $V(F)$ by b_1 and let $b_2 = (w_1, w_2, \ldots, w_n)$ be a second basis for $V(F)$ with $w_j = \sum_{i=1}^{n} c_{ij} v_i$. Then if f is a bilinear function with matrix $A = (a_{ij})$ relative to b_1,

$$f(w_i, w_j) = f\left(\sum_{k=1}^{n} c_{ki} v_k, \quad \sum_{h=1}^{n} c_{hj} v_h\right)$$

$$= \sum_{k=1}^{n} \sum_{h=1}^{n} c_{ki} c_{hj} f(v_k, v_h)$$

$$= \sum_{k=1}^{n} \sum_{h=1}^{n} c_{ki} a_{kh} c_{hj} \tag{9.3}$$

It is easily seen that in terms of matrices (9.3) can be written

$$(b_{ij}) = (c_{ij})^{\mathrm{T}} (a_{ij}) (c_{ij})$$

or

$$B = C^{\mathrm{T}} A C \tag{9.4}$$

where $C = (c_{ij})$, and $B = (b_{ij})$ is the matrix of f relative to b_2.

The transformation represented by (9.4) is known as a *congruence transformation* and the matrices A and B are said to be *congruent*. It is evident, because C is non-singular, that A and B must have equal rank, and therefore the rank of a matrix representing f relative to any basis for $V(F)$ is known as the *rank of the bilinear form f*. Notice that a congruence transformation is necessarily an equivalence transformation. When F is real, a congruence transformation is also a similarity transformation if C is orthogonal. Directly from the definition of A, and (9.2), it follows that a bilinear function is symmetric if and only if its matrix is symmetric, and is skew-symmetric if and only if its matrix is skew-symmetric.

EXERCISE 9.3 Prove that any bilinear function f on a vector space $V(F)$ can be expressed uniquely as the sum of a symmetric and a skew-symmetric bilinear function on $V(F)$, provided that $1 + 1 \neq 0$ in F.

Example 9.1 The integers modulo 2 form a field (Chapter 1, Section 4). Over this field the bilinear function with matrix $\begin{pmatrix} 1 & 1 \\ 0 & 1 \end{pmatrix}$ cannot be expressed as the sum of a symmetric and skew-symmetric bilinear function.

Example 9.2 Over the field of integers modulo 3 the matrix in Example 9.1 can be written

$$\begin{pmatrix} 1 & 1 \\ 0 & 1 \end{pmatrix} = \begin{pmatrix} 1 & 2 \\ 2 & 1 \end{pmatrix} + \begin{pmatrix} 0 & 2 \\ 1 & 0 \end{pmatrix}$$

where the last matrix is skew-symmetric. Note that the inverse of 2 in this field is 2, and

$$\begin{pmatrix} 1 & 2 \\ 2 & 1 \end{pmatrix} = 2\left\{\begin{pmatrix} 1 & 1 \\ 0 & 1 \end{pmatrix} + \begin{pmatrix} 1 & 0 \\ 1 & 1 \end{pmatrix}\right\} \text{mod } 3$$

$$\begin{pmatrix} 0 & 2 \\ 1 & 0 \end{pmatrix} = 2\left\{\begin{pmatrix} 1 & 1 \\ 0 & 1 \end{pmatrix} - \begin{pmatrix} 1 & 0 \\ 1 & 1 \end{pmatrix}\right\} \text{mod } 3$$

9.1 Quadratic forms

Given any bilinear function f on an n-dimensional vector space $V(F)$, it is possible to associate with f a function q of $V(F)$ into F by setting $q(v) = f(v, v)$ for all $v \in V$. Such a function $q(v)$ is known as the *quadratic form associated with f.*

This definition does not, as it stands, give a unique bilinear form corresponding to each quadratic form since, if f is written as the sum of its symmetric and skew-symmetric parts, it is immediately apparent that the quadratic form associated with f depends only on the symmetric part. Because of this fact the bilinear form f is taken to be symmetric in the definition of the quadratic form. There is then a one-to-one correspondence between quadratic forms and their associated symmetric bilinear forms.

Note that in applications a quadratic form may arise from an unsymmetric matrix A in the guise of $x^{\mathrm{T}}Ax$. This is to be replaced by $\frac{1}{2}x^{\mathrm{T}}(A + A^{\mathrm{T}})x$ in order to bring it within the scope of the definition.

Consider a quadratic form $q(v)$ on $V(F)$. Then if $v, w \in V$

$$\begin{aligned} q(w + v) &= f(w + v, w + v) \\ &= f(w, w) + 2f(w, v) + f(v, v) \\ &= q(w) + 2f(w, v) + q(v) \end{aligned}$$

where the symmetry of f has been used. Therefore (if $1 + 1 \neq 0$ in F)

$$f(w, v) = 2^{-1}\{q(w + v) - q(w) - q(v)\} \qquad (9.5)$$

Equation (9.5) shows how the bilinear form can be found if the quadratic form is given: $f(w, v)$ is often called the *polar form* of $q(v)$.

Now let $V(F)$ be the vector space of n-tuples with elements in F, that is $V_n(F)$, and let $A = (a_{ij})$ be an $n \times n$ symmetric matrix with elements in F. Then, with respect to the standard basis, (a_{ij}) represents a symmetric bilinear form f with

$$f(x, y) = x^{\mathrm{T}}Ay \tag{9.6}$$

and since A is symmetric, the quadratic form associated with f is

$$q(x) = x^{\mathrm{T}}Ax \tag{9.7}$$

Example 9.3 Find the symmetric matrix representing the quadratic form

$$q(x) = \xi_1^2 + 4\xi_1\xi_2 + \xi_2^2 + 2\xi_1\xi_3 + \xi_3^2$$

and write down the polar form of q.

If A is the matrix representing $q(x)$ then $q(x) = x^{\mathrm{T}}Ax$. But

$$(\xi_1, \xi_2, \xi_3)\begin{pmatrix} a_{11} & a_{12} & a_{13} \\ a_{12} & a_{22} & a_{23} \\ a_{13} & a_{23} & a_{33} \end{pmatrix}\begin{pmatrix} \xi_1 \\ \xi_2 \\ \xi_3 \end{pmatrix}$$

gives

$$a_{11}\xi_1^2 + a_{22}\xi_2^2 + a_{33}\xi_3^2 + 2a_{23}\xi_2\xi_3 + 2a_{13}\xi_1\xi_3 + 2a_{12}\xi_1\xi_2$$

and by direct comparison with $q(x)$

$$a_{11} = a_{22} = a_{33} = 1, \qquad 2a_{12} = 4, \qquad 2a_{23} = 0, \qquad 2a_{13} = 2$$

giving

$$A = \begin{pmatrix} 1 & 2 & 1 \\ 2 & 1 & 0 \\ 1 & 0 & 1 \end{pmatrix}$$

The polar form of q is therefore

$$\begin{aligned}(\xi_1, \xi_2, \xi_3)&\begin{pmatrix} 1 & 2 & 1 \\ 2 & 1 & 0 \\ 1 & 0 & 1 \end{pmatrix}\begin{pmatrix} \eta_1 \\ \eta_2 \\ \eta_3 \end{pmatrix} \\ &= \xi_1\eta_1 + 2(\xi_2\eta_1 + \xi_1\eta_2) + \xi_2\eta_2 + \xi_1\eta_3 + \xi_3\eta_1 + \xi_3\eta_3\end{aligned}$$

Example 9.4 Write down the polar form of the quadratic form

$$\xi_1^2 + 2\xi_1\xi_2 + 4\xi_1\xi_3 + 3\xi_2^2 + \xi_2\xi_3 + 7\xi_3^2$$

From (9.1), $2f(w, v) = q(w + v) - q(w) - q(v)$

$$
\begin{aligned}
2f(w, v) = {} & (\xi_1 + \eta_1)^2 + 2(\xi_1 + \eta_1)(\xi_2 + \eta_2) \\
& + 4(\xi_1 + \eta_1)(\xi_3 + \eta_3) + 3(\xi_2 + \eta_2)^2 \\
& + (\xi_2 + \eta_2)(\xi_3 + \eta_3) + 7(\xi_3 + \eta_3)^2 \\
& - \xi_1^2 - 2\xi_1\xi_2 - 4\xi_1\xi_3 - 3\xi_2^2 - \xi_2\xi_3 - 7\xi_3^2 \\
& - \eta_1^2 - 2\eta_1\eta_2 - 4\eta_1\eta_3 - 3\eta_2^2 - \eta_2\eta_3 - 7\eta_3^2
\end{aligned}
$$

which on simplification gives

$$
\begin{aligned}
2f(w, v) = {} & 2\xi_1\eta_1 + 2(\xi_1\eta_2 + \eta_1\xi_2) + 4(\xi_1\eta_3 + \eta_1\xi_3) \\
& + 6\xi_2\eta_2 + (\xi_2\eta_3 + \eta_2\xi_3) + 14\xi_3\eta_3
\end{aligned}
$$

9.2 Diagonalization of quadratic forms

Let $V(F)$ be a vector space over any field for which $1 + 1 \neq 0$, and $q(v)$ a quadratic form on V. If A is the matrix of $q(v)$ relative to some given basis $b_1 = (v_1, v_2, \ldots, v_n)$ for $V(F)$ then the matrix of $q(v)$ relative to a second basis for $V(F)$ is given by a congruence transformation C^TAC. If the second basis is such that C^TAC is diagonal, the quadratic form takes on a particularly simple form and it is important to find such bases.

Theorem 9.1 A congruence transformation can be found which will reduce a symmetric matrix to diagonal form.

Proof This theorem can be proved directly using elementary operations (Mirksy, 1955) in a manner similar to that used for equivalence transformations; it is also possible to proceed as follows.

Let $f(v, w)$ be the bilinear form corresponding to the quadratic form $q(v)$. Then from (9.5) it can be seen that unless $f(v, w) \equiv 0$, in which case $A = 0$ is already diagonal, $q(w_1) \neq 0$ for some $w_1 \in V$; let $q(w_1) = \alpha_1$. Now assume that the result concerning diagonalization holds for $V(F)$ of dimension $n - 1$ (it certainly holds for $n = 1$) and use induction on n. If w is held fixed then w^TAv can be thought of as a linear transformation T of $V(F)$ into F (considered as a vector space). The transformation T is not identically zero since $w_1^TAw_1 = q(w_1) = \alpha_1 \neq 0$: the dimension of its range is 1 and therefore the nullity of T is $n - 1$ (see Section 3).

Now consider the subspace of V defined by

$$S = \{v \in V(F) \mid w_1^TAv = 0\}$$

It has just been shown that its dimension is $n - 1$, hence by the induction hypothesis there exists a basis $(w_2, w_3, \ldots, w_n)$ relative to which the matrix A_1 of f on S is diagonal. But $f(w_1, w_1) = \alpha_1$ and $f(w_1, w_j) = f(w_j, w_1) = 0$ for $2 \leqslant j \leqslant n$ and therefore the matrix of $f(v, w)$ relative to the basis $b_2 = (w_1, w_2, \ldots, w_n)$ is also diagonal. It follows that if P is the matrix representing the change of basis from b_1 to b_2 and if r is the rank of A

$$P^{\mathrm{T}}AP = \left(\begin{array}{c|c} \operatorname{diag}(\alpha_1, \alpha_2, \ldots, \alpha_r) & 0_{r,n-r} \\ \hline 0_{n-r,r} & 0_{n-r,n-r} \end{array}\right) \tag{9.8}$$

EXERCISE 9.4 Show that it is possible to choose another basis for $V(F)$ in such a way that the diagonal representation of a quadratic form has its main diagonal elements multiplied by the squares of arbitrary non-zero scalars from F, and hence when F is the field of complex numbers that these elements can be made to take the value 1.

Example 9.5 Find a congruence transformation that will reduce the matrix

$$A = \begin{pmatrix} 2 & 1 & 1 \\ 1 & 2 & 1 \\ 1 & 1 & 2 \end{pmatrix}$$

to diagonal form.

Let $w_1 = e_1 = (1, 0, 0)^{\mathrm{T}}$; then $w_1^{\mathrm{T}}Aw_1 = 2 \neq 0$ and $w_1^{\mathrm{T}}A = (2, 1, 1)$. A vector that $w_1^{\mathrm{T}}A = (2, 1, 1)$ will annihilate is $(0, 1, -1)^{\mathrm{T}} = w_2$, and $w_2^{\mathrm{T}}Aw_2 = 2$ with $w_2^{\mathrm{T}}A = (0, 1, -1)$. Finally, a vector that $w_2^{\mathrm{T}}A = (0, 1, -1)$ and $w_1^{\mathrm{T}}A = (2, 1, 1)$ will both annihilate is $(1, -1, -1)^{\mathrm{T}} = w_3$ (either by inspection or solving some linear equations) and $w_3^{\mathrm{T}}A = (0, -2, -2)$ with $w_3^{\mathrm{T}}Aw_3 = 4$. Hence

$$\begin{pmatrix} 1 & 0 & 0 \\ 0 & 1 & -1 \\ 1 & -1 & -1 \end{pmatrix}\begin{pmatrix} 2 & 1 & 1 \\ 1 & 2 & 1 \\ 1 & 1 & 2 \end{pmatrix}\begin{pmatrix} 1 & 0 & 1 \\ 0 & 1 & -1 \\ 0 & -1 & -1 \end{pmatrix}$$

$$= \begin{pmatrix} 2 & 1 & 1 \\ 0 & 1 & -1 \\ 0 & -2 & -2 \end{pmatrix}\begin{pmatrix} 1 & 0 & 1 \\ 0 & 1 & -1 \\ 0 & -1 & -1 \end{pmatrix} = \begin{pmatrix} 2 & 0 & 0 \\ 0 & 2 & 0 \\ 0 & 0 & 4 \end{pmatrix}$$

9.3 Real quadratic forms

When F is the field of real numbers, -1 is not a square and so it is only possible to reduce the main diagonal elements of the diagonal form of the matrix representation of a quadratic form q to ± 1. If this is done, however, then the numbers of positive and negative terms do not vary with different diagonal reductions. The number of positive elements less the number of negative elements is therefore an invariant; it is called the *signature* of the quadratic form.

Theorem 9.2 If a real quadratic form is diagonalized in any way its signature remains unchanged.

Proof Let $b_1 = (v_1, v_2, \ldots, v_n)$ and $b_2 = (w_1, w_2, \ldots, w_n)$ be two bases for $V(F)$ each reducing the matrix representation of q to diagonal form. It is obviously possible to arrange for all the positive terms to appear before the negative terms on the diagonal for each representation: suppose that there are l positive terms in the representation with respect to b_1, and m in that with respect to b_2, with $l > m$.

Let the coordinates of any non-zero vector v in sp $(v_1, v_2, \ldots, v_l)$ be $(\alpha_1, \alpha_2, \ldots, \alpha_l)$; then

$$q(v) = \sum_{i=1}^{l} \alpha_i^2 > 0$$

Let the coordinates of any non-zero vector w in sp $(w_{m+1}, w_{m+2}, \ldots, w_n)$ be $(\beta_{m+1}, \beta_{m+2}, \ldots, \beta_n)$; then

$$q(w) = -\sum_{i=m+1}^{n} (\beta_i)^2 \leqslant 0$$

But

$$\dim \operatorname{sp}(v_1, v_2, \ldots, v_l) = l$$

and

$$\dim \operatorname{sp}(w_{m+1}, w_{m+2}, \ldots, w_n) = n - m$$

showing that these two subspaces must have a common non-zero vector since $l > m$. For this vector $q > 0$ and also $q < 0$, which is impossible, so that $l \leqslant m$. Similarly $m \leqslant l$, so that $m = l$. If r is the rank of the form and p the number of positive elements, then the signature $s = p - (r - p) = 2p - r$.

If the signature s of a quadratic form q is equal to n, the dimension of $V(F)$, the form is said to be *positive definite* and $q(v) > 0$ for all non-zero $v \in V(F)$. If $s = r$ then q is said to be *positive semi-definite* and $q(v) \geqslant 0$ for all $v \in V(F)$. *Negative definiteness* and *negative semi-definiteness* are similarly defined. If q can take both positive and negative values it is said to be *indefinite*.

9.4 Hermitian forms

Let $V(\mathscr{C})$ be a vector space over the field of complex numbers. Then a function $f(v, w)$ of $V \times V$ into $\mathscr{C}$ is said to be *hermitian symmetric* if $f(v, w) = \overline{f(w, v)}$. A function $f(v, w)$ which is hermitian symmetric and linear in the second variable is called a *hermitian form*. If $f(v, w)$ is a hermitian form, $v_1, v_2, w \in V(\mathscr{C})$, and $\alpha_1, \alpha_2 \in \mathscr{C}$ then

$$\begin{aligned} f(\alpha_1 v_1 + \alpha_2 v_2, w) &= \overline{f(w, \alpha_1 v_1 + \alpha_2 v_2)} \\ &= \overline{\alpha_1 f(w, v_1) + \alpha_2 f(w, v_2)} \end{aligned}$$

$$= \overline{\alpha_1}\overline{f(w, v_1)} + \overline{\alpha_2}\overline{f(w, v_2)}$$
$$= \overline{\alpha_1} f(v_1, w) + \overline{\alpha_2} f(v_2, w)$$

Because of this relationship $f(v, w)$ is said to be *complex linear* with respect to its first variable.

From the definition of a hermitian form it follows that the matrix of any hermitian form relative to a given basis is hermitian. Also, the transformation of the matrix A of a hermitian form relative to one basis to its matrix representation B relative to another basis is given by $B = C^*AC$, where C is the matrix of transition between the bases [equation (9.4)]. The transformation $B = C^*AC$ is said to be *hermitian congruent.* As with a real quadratic form a *hermitian quadratic form* $q(v)$ is obtained from the associated hermitian form $f(v, w)$ by setting $q(v) = f(v, v)$. In terms of matrices, any hermitian quadratic form can be represented by v^*Av where A is a hermitian matrix, and conversely any hermitian matrix determines a unique hermitian quadratic form. It is easily seen that the value of any hermitian quadratic form is real, since

$$\begin{aligned} \overline{v^*Av} &= v^{\mathrm{T}}\bar{A}\bar{v} \\ &= (v^{\mathrm{T}}\bar{A}\bar{v})^{\mathrm{T}} \quad \text{(since it is scalar)} \\ &= v^*A^*v \\ &= v^*Av \end{aligned}$$

The rank of a hermitian form is the same as that of its matrix representation with respect to any basis. It is also possible to define signature and positive and negative definiteness of a hermitian quadratic form in exactly the same way as for a real quadratic form. The reader should have no difficulty in proving the two following theorems on hermitian quadratic forms which are left as exercises.

EXERCISE 9.5 Show that any hermitian matrix can be reduced to diagonal form by a non-singular hermitian congruence transformation, and that any positive definite hermitian matrix can be reduced to the unit matrix by such a transformation.

EXERCISE 9.6 Show that any hermitian quadratic form x^*Ax can be reduced to the form $\sum_{i=1}^{r} \lambda_i \bar{x}_i x_i$, where λ_i are the non-zero eigenvalues of A, by a suitable unitary transformation. Give the corresponding theorem when x^*Ax is a real quadratic form. [Use Theorem 8.1.]

9.5 Determination of positive and negative definiteness

From Exercise 9.6 it follows that a hermitian or a real quadratic form is positive (negative) definite if and only if all the eigenvalues of its matrix are positive (negative). Similarly, positive (negative) semi-definiteness occurs if and only if all the eigenvalues are non-negative (non-positive); indefiniteness arises if and only if there is at least one positive and one negative eigenvalue.

Positive and negative definiteness of forms have been defined in terms of their rank and signature, and the above assertions give the corresponding results in terms of the eigenvalues of the matrices representing the forms. Conditions are now developed for positive and negative definiteness in terms of the elements of these matrices. A matrix is defined to have the same sign definiteness as the form it represents and therefore the conditions given below apply equally to the matrices and to the forms. The proof is given for hermitian forms but applies with very little alteration to real quadratic forms.

The fact that a matrix A is positive (negative) definite is commonly expressed by writing $A > 0$ $(A < 0)$. Similarly, positive (negative) semi-definiteness is expressed by $A \geqslant 0$ $(A \leqslant 0)$.

Theorem 9.3 The hermitian form v^*Av (where $A^* = A$) is positive definite if and only if its *leading minors* satisfy

$$a_{11} > 0, \quad \begin{vmatrix} a_{11} & a_{12} \\ a_{21} & a_{22} \end{vmatrix} > 0, \quad \ldots, \quad \begin{vmatrix} a_{11} & \cdots & a_{1n} \\ \vdots & & \vdots \\ a_{n1} & \cdots & a_{nn} \end{vmatrix} > 0 \tag{9.9}$$

Proof The necessity of conditions (9.9) follows from the fact that the determinant of the matrix A of any positive definite hermitian form is positive. To see this, recall that (from Exercise 9.5) there exists a non-singular matrix C such that

$$C^*AC = I \tag{9.10}$$

and taking the determinant of both sides of (9.10)

$$|\overline{C}|\,|C|\,|A| = \overline{|C|}\,|C|\,|A| = 1 \tag{9.11}$$

and hence $|A| > 0$. Assume now that v^*Av is positive definite; then $\sum_{i,j=1}^{m} a_{ij}\bar{x}_i x_j$ must also be positive definite for $1 \leqslant m \leqslant n$ proving the necessity of (9.9).

To prove sufficiency, induction is used on n. The theorem holds trivially for $n = 1$; assume that it holds for $n - 1$ where $n \geqslant 2$, and that conditions (9.9) are satisfied. Now $a_{11} > 0$, so

$$\begin{aligned}\sum_{i,j=1}^{n} a_{ij}\bar{v}_i v_j &= a_{11}\bar{v}_1 v_1 + \sum_{i=2}^{n} (a_{1i}\bar{v}_1 v_i + a_{i1}\bar{v}_i v_1) + \sum_{i,j=2}^{n} a_{ij}\bar{v}_i v_j \\ &= a_{11}\Bigg[\left(\bar{v}_1 + \frac{a_{21}}{a_{11}}\bar{v}_2 + \cdots + \frac{a_{n1}}{a_{11}}\bar{v}_n\right)\left(v_1 + \frac{a_{12}}{a_{11}}v_2 + \cdots + \frac{a_{1n}}{a_{11}}v_n\right) \\ &\qquad - \sum_{i,j=2}^{n} \frac{a_{i1}}{a_{11}}\bar{v}_i \cdot \frac{a_{1j}}{a_{11}} v_j\Bigg] \\ &\qquad + \sum_{i,j=2}^{n} a_{ij}\bar{v}_i v_j \\ &= a_{11}\left|v_1 + \frac{a_{12}}{a_{11}}v_2 + \cdots + \frac{a_{1n}}{a_{11}}v_n\right|^2 + \sum_{i,j=2}^{n}\left(a_{ij} - \frac{a_{i1}a_{1j}}{a_{11}}\right)\bar{v}_i v_j\end{aligned} \tag{9.12}$$

Since

$$\overline{\left(a_{ij} - \frac{a_{i1}a_{1j}}{a_{11}}\right)} = a_{ji} - \frac{a_{j1}a_{1i}}{a_{11}}$$

the sum over i and j in (9.12) is a hermitian form, ϕ say, on a space of dimension $n - 1$. Also

$$a_{11}\left|\left(a_{ij} - \frac{a_{i1}a_{1j}}{a_{11}}\right)\right| = |(a_{kl})| > 0, \qquad \begin{aligned} i, j &= 2, 3, \ldots, m \\ k, l &= 1, 2, \ldots, m \\ m &= 2, 3, \ldots, n \end{aligned}$$

by subtraction of a_{1p}/a_{11} times the first column from the pth column of (a_{kl}), $p = 2, 3, \ldots, m$, so that by the induction hypothesis $\phi > 0$ for $v \neq 0$.

Suppose $(v_1, v_2, \ldots, v_n)$ makes $v^*Av \leqslant 0$; then since ϕ is positive definite and $a_{11} > 0$ it follows from (9.12) that $v_2 = v_3 = \cdots = v_n = 0$ and $\sum_{i=1}^{n} a_{1i}v_i = 0$, so that $v_1 = 0$. Hence v^*Av is positive definite.

By applying this theorem to $-v^*Av$ it is clear that the hermitian form v^*Av is negative definite if and only if

$$a_{11} < 0, \quad \begin{vmatrix} a_{11} & a_{12} \\ a_{21} & a_{22} \end{vmatrix} > 0, \quad \begin{vmatrix} a_{11} & a_{12} & a_{13} \\ a_{21} & a_{22} & a_{23} \\ a_{31} & a_{32} & a_{33} \end{vmatrix} < 0, \quad \ldots \tag{9.13}$$

that is, the successive minors are alternately negative and positive.

Exercise 9.7 Show that the matrix of any definite quadratic or hermitian form must be non-singular. What can be said in this respect regarding indefinite forms?

Exercise 9.8 Show that Theorem 9.3 holds for any suitable set of principal minors of the matrix A; that is, the theorem does not depend on the labelling of the a_{ij}.

Exercise 9.9 Show that the hermitian form v^*Av is positive (negative) semi-definite if and only if all its principal minors of all orders are non-negative (its even-order principal minors are non-negative and those of odd order are non-positive). [Note that it is not enough for the leading minors to be non-negative: compare $\begin{pmatrix} 0 & 0 \\ 0 & -1 \end{pmatrix}$.]

10 Norms of matrices

A matrix norm is a real-valued function $\|A\|$ of the elements of a matrix A which has the four properties

M (i) $\|A\| > 0 \quad \text{if } A \neq 0$

M (ii) $\|\alpha A\| = |\alpha|\,\|A\|, \quad \alpha$ any scalar

M (iii) $\|A + B\| \leqslant \|A\| + \|B\|$

M (iv) $\|AB\| \leqslant \|A\|\,\|B\|$

Two commonly used matrix norms are

$$\|A\|_e = \left(\sum_{i,j} |a_{ij}|^2\right)^{1/2}$$

and

$$\|A\|_c = n \max_{i,j} |a_{ij}|$$

The first of these is obtained by considering A as an n^2-dimensional vector and using the euclidean vector norm. The second is a direct extension of the cubic norm for vectors. Notice that

$$\|A\|_e = (\text{tr } A^*A)^{1/2}$$

That the two matrix norms defined above satisfy conditions M (i), M (ii), and M (iii) is obvious from their connection with vector norms. Since the ijth element of AB is $\sum_{k=1}^{n} a_{ik}b_{kj}$ it follows that

$$\begin{aligned}
\|AB\|_c &= n \max_{i,j} \left|\sum_{k=1}^{n} a_{ik}b_{kj}\right| \\
&\leqslant n \max_{i,j} \sum_{k=1}^{n} |a_{ik}|\,|b_{kj}| \\
&\leqslant n \sum_{k=1}^{n} \frac{1}{n}\|A\|_c \frac{1}{n}\|B\|_c \\
&= \|A\|_c\,\|B\|_c
\end{aligned}$$

and M (iv) holds for the cubic matrix norm.

Again,

$$\begin{aligned}
\|AB\|_e^2 &= \sum_{i,j} \left|\sum_{k=1}^{n} a_{ik}b_{kj}\right|^2 \\
&\leqslant \sum_{i,j} \left\{\sum_{k=1}^{n} |a_{ik}|^2 \sum_{l=1}^{n} |b_{lj}|^2\right\} \\
&= \sum_{i,k} |a_{ik}|^2 \sum_{l,j} |b_{lj}|^2 \\
&= \|A\|_e^2\,\|B\|_e^2
\end{aligned}$$

showing that $\|AB\|_e \leqslant \|A\|_e \|B\|_e$, and M (iv) holds for the euclidean matrix norm.

If the condition $\|Ax\| \leqslant \|A\| \|x\|$ holds for any vector x and matrix A, the matrix norm $\|A\|$ and the vector norm $\|x\|$ are said to be *consistent* (*compatible*, by some authors).

EXERCISE 10.1 Show that $\|A\|_c$ is consistent with $\|x\|_e$, $\|x\|_c$, and $\|x\|_o$.

It is possible to set up a matrix norm which is consistent with a given vector norm in many different ways. If $\|Ax\|$ is any vector norm, one such way is to define a matrix norm $\|A\|_s$ by

$$\|A\|_s = \max_{\|x\|=1} \|Ax\|$$

The matrix norm $\|A\|_s$ is said to be *subordinate* to the given vector norm.

EXERCISE 10.2 Prove that $\|A\|_s$ satisfies conditions M (i), M (ii), M (iii), and M (iv) and that it is consistent with $\|x\|$.

An interesting property of a subordinate norm is that it is less than or equal to any other matrix norm which is consistent with the same vector norm. The proof is as follows: let $\|x\|$ be any vector norm, $\|A\|$ any matrix norm consistent with $\|x\|$, and $\|A\|_s$ a matrix norm subordinate to $\|x\|$. By definition there exists a vector y with $\|y\| = 1$ and

$$\|A\|_s = \|Ay\|$$

but by M (iv),

$$\|Ay\| \leqslant \|A\| \|y\| = \|A\|$$

hence

$$\|A\|_s \leqslant \|A\|$$

Now suppose that

$$\|A\|_s = \max_{\|x\|_e=1} \|Ax\|_e$$

that is, $\|A\|_s$ is subordinate to the euclidean vector norm. Since

$$\begin{aligned}\|Ax\|_e^2 &= \langle Ax, Ax\rangle \\ &= \langle x, A^*Ax\rangle\end{aligned}$$

it follows (Exercise 8.17 in Chapter 4: notice that A^*A is hermitian) that

$$\max_{\|x\|_e=1} \langle x, A^*Ax\rangle$$

is just the largest eigenvalue λ_m of A^*A and therefore

$$\|A\|_s = \sqrt{\lambda_m}$$

Again let

$$\|A\|_s = \max_{\|x\|_o = 1} \|Ax\|_o$$

that is, in this case $\|A\|_s$ is subordinate to the octahedral vector norm. If x, with elements ξ_i, is any vector with unit octahedral norm,

$$\begin{aligned} \|Ax\|_o &= \sum_{i=1}^{n} \left| \sum_{j=1}^{n} a_{ij}\xi_j \right| \\ &\leqslant \sum_{i=1}^{n} \sum_{j=1}^{n} |a_{ij}|\ |\xi_j| \\ &= \sum_{j=1}^{n} |\xi_j| \left\{ \sum_{i=1}^{n} |a_{ij}| \right\} \\ &\leqslant \left\{ \max_j \sum_{i=1}^{n} |a_{ij}| \right\} \left\{ \sum_{k=1}^{n} |\xi_k| \right\} \\ &= \max_j \sum_{i=1}^{n} |a_{ij}| \end{aligned}$$

since

$$\sum_{k=1}^{n} |\xi_k| = 1$$

It is easy to see that $\|Ax\|_o$ can achieve the upper bound established above, and so the matrix norm subordinate to the octahedral vector norm is

$$\|A\|_s = \max_j \sum_{i=1}^{n} |a_{ij}|$$

EXERCISE 10.3 Prove that the matrix norm subordinate to the cubic vector norm is

$$\max_i \sum_{j=1}^{n} |a_{ij}|$$

11 Pseudo or generalized inverse

The matrix equation

$$AX = I_n \tag{11.1}$$

has the unique solution $X = A^{-1}$ provided that A is non-singular. When A is square and singular, or rectangular, A^{-1} does not exist and (11.1) has no solu tion in the above sense. It is still of interest, however, to investigate the 'best approximate solution' of (11.1) in these circumstances.

The matrix X_0 is defined as the *best approximate solution* of (11.1) if, for all other matrices X, either

$$\|AX - I_n\|_e > \|AX_0 - I_n\|_e \tag{11.2}$$

or

$$\|AX - I_n\|_e = \|AX_0 - I_n\|_e \quad \text{and} \quad \|X\|_e > \|X_0\|_e \tag{11.3}$$

The matrix X_0 so defined is called the *generalized* or *pseudo inverse* of A and is denoted by $A^\dagger$; obviously $A^\dagger = A^{-1}$ when A is non-singular.

The pseudo inverse can also be defined as the matrix $A^\dagger$ satisfying the four equations

$$AA^\dagger A = A \tag{11.4}$$

$$A^\dagger AA^\dagger = A^\dagger \tag{11.5}$$

$$(AA^\dagger)^* = AA^\dagger \tag{11.6}$$

$$(A^\dagger A)^* = A^\dagger A \tag{11.7}$$

and it is possible to show that the two definitions are consistent (Penrose, 1956).

The following properties of the pseudo inverse (which will not be used again in this book) are stated without proof:

(i) $A^*AA^\dagger = A^* = A^\dagger AA^*$ (11.8)

(ii) $(A^\dagger)^\dagger = A$ (11.9)

(iii) $A^*(A^\dagger)^*A^\dagger = A^\dagger = A^\dagger(A^\dagger)^*A^*$ (11.10)

(iv) $A^\dagger = A^{-1}$ if A is non-singular (11.11)

(v) $A^\dagger = (A^*A)^{-1}A^*$ if (A^*A) is non-singular (11.12)

Further reading

The whole of the work of this chapter is covered in Gantmacher (1959). This is probably the best modern treatment of the subject. Halmos (1958) has a highly individual approach and is recommended for the mathematically inclined reader. Another individual and interesting treatment of the basic theory is that of Bellman (1960). Other important general texts are Mirsky (1955), Jacobson (1951, 1953), Birkhoff and MacLane (1965), and Hodge and Pedoe (1947). The treatment by Pease (1965) is specifically oriented to applications of matrix theory.

There are a large number of recent books dealing with linear algebra and matrix theory at the undergraduate level. Of these we particularly mention Nering (1963), Marcus and Minc (1965), Nomizu (1966), Steward (1963), and Thrall and Tornheim (1957). The geometric aspects of linear algebra are emphasized in Martin and Mizel (1966).

For the computational aspects of matrix theory the reader is referred to Householder (1964), Wilkinson (1965), Faddeev and Faddeeva (1936), and Fox (1964).

Useful general works on algebra are Archbold (1961), Bocher (1936), and Ferrar (1941).

Finally we recommend the very comprehensive text by Marcus and Minc (1964) for general reference purposes.

References

ARCHBOLD, J. W., 1961, *Algebra*, Pitman, London.

BELLMAN, R., 1960, *Introduction to matrix analysis*, McGraw-Hill, New York.

BIRKHOFF, G. and S. MACLANE, 1965, *A survey of modern algebra*, Allyn and Bacon, Boston.

BOCHER, M., 1936, *Introduction to higher algebra*, Macmillan, New York.

DIEUDONNÉ, J., 1960, *Foundations of modern analysis*, Academic Press, New York.

FADDEEV, D. K. and V. N. FADDEEVA, 1963, *Computational methods of linear algebra*, W. H. Freeman, San Francisco.

FERRAR, W. L., 1941, *Algebra*, Oxford University Press, London.

FOX, L., 1964, *An introduction to numerical linear algebra*, Clarendon Press, Oxford.

GANTMACHER, F. R., 1959, *The theory of matrices*, vols. 1 and 2, Chelsea Publishing Co., New York.

HALMOS, P. R., 1958, *Finite dimensional vector spaces*, Van Nostrand, Princeton, N.J.

HODGE, W. V. D. and D. PEDOE, 1947, *Methods of algebraic geometry*, vol. 1, Cambridge University Press, London.

HOUSEHOLDER, A. S., 1964, *The theory of matrices in numerical analysis*, Blaisdell, Waltham, Mass.

JACOBSON, N., 1951, *Lectures in abstract algebra*, vol. 1, *Basic concepts*, Van Nostrand, Princeton, N.J.

JACOBSON, N., 1953, *Lectures in abstract algebra*, vol. 2, *Linear algebra*, Van Nostrand, Princeton, N.J.

MARCUS, M. and H. MINC, 1964, *A survey of matrix theory and matrix inequalities*, Allyn and Bacon, Boston.

MARCUS, M. and H. MINC, 1965, *Introduction to linear algebra*, Macmillan, New York.

MARTIN, A. D. and V. J. MIZEL, 1966, *Introduction to linear algebra*, McGraw-Hill, New York.

MIRSKY, L., 1955, *An introduction to linear algebra*, Oxford University Press, London.

NERING, E. D., 1963, *Linear algebra and matrix theory*, John Wiley, New York.

NOMIZU, K., 1966, *Fundamentals of linear algebra*, McGraw-Hill, New York.

PEASE, M. C., 1965, *Methods of matrix algebra*, Academic Press, New York.

PENROSE, R., 1956, On best approximate solutions of linear matrix equations. *Proc. Camb. Phil Soc.*, **52**, 17–19.

STEWARD, F. M., 1963, *Introduction to linear algebra*, Van Nostrand, Princeton, N.J.

THRALL, R. M. and L. TORNHEIM, 1957, *Vector spaces and matrices*, John Wiley, New York.

WILKINSON, J. H., 1965, *The algebraic eigenvalue problem*, Oxford University Press, London.

Problems

1 Prove that the set of even complex polynomials and the set of odd complex polynomials are complementary subspaces of the space of all complex polynomials $C[\lambda]$. [$a(\lambda)$ is even if $a(\lambda) = a(-\lambda)$ and $a(\lambda)$ is odd if $a(\lambda) = -a(-\lambda)$.]

2 The *Kronecker product* $A \otimes B$ of an $m \times n$ matrix A and a $p \times q$ matrix B is defined by

$$A \otimes B = \begin{vmatrix} a_{11}B & a_{12}B & \dots & a_{1n}B \\ a_{21}B & a_{22}B & \dots & a_{2n}B \\ \vdots & \vdots & & \vdots \\ a_{m1}B & a_{m2}B & \dots & a_{mn}B \end{vmatrix}$$

Prove that:

(a) With matrices A, B, C, D of appropriate dimensionality $(A \otimes B)(C \otimes D) = (AC) \otimes (BD)$.

(b) When A and B are non-singular $(A \otimes B)^{-1}$ exists and is given by $A^{-1} \otimes B^{-1}$.

(c) If A and B are square and of orders n and m respectively, $\det (A \otimes B) = (\det A)^m (\det B)^n$.

(d) The product of any eigenvalue of A with any eigenvalue of B is an eigenvalue of $A \otimes B$.

3 Let $N(s) = N_0 + N_1 s + \dots + N_r s^r$ be an $m \times n$ polynomial matrix with real coefficients. Let $V(\mathscr{R})$ be the space spanned by the columns of $N(s)$. Let $W(\mathscr{R})$ be the space spanned by the columns of $\begin{pmatrix} N_0 \\ N_1 \\ \vdots \\ N_r \end{pmatrix}$. Show that V and W are isomorphic.

4 Prove that any hermitian matrix H which satisfies $H^2 = I$ is unitary.

5 The matrix $B = (A + I)(A - I)^{-1}$ is of importance in the stability theory of linear systems. Show that:

(a) If A is skew-hermitian then B is unitary.

(b) If the eigenvalues of A all have negative real parts then the largest eigenvalue of B is less than one in modulus.

6 Prove that any normal matrix which is also triangular is necessarily diagonal.

7 Show that $A^{T}BA$ where A is a real non-singular matrix is positive definite if and only if the real matrix B is positive definite.

8 Prove that if A is any matrix over the field of complex numbers then A^*A is hermitian positive semi-definite and if, in addition, A is non-singular then A^*A is hermitian positive definite.

9 Investigate the definiteness of the quadratic forms

$$2x_1^2 - x_1x_2 + x_2^2$$

and

$$2x_1^2 + 2x_2^2 + 4x_3^2 + x_1x_2 - 6x_1x_3 + 3x_2x_3$$

10 Find all the eigenvalues and the eigenvectors of the matrix

$$\begin{pmatrix} 2 & -2 & 1 \\ 3 & -1 & -2 \\ 2 & 2 & -1 \end{pmatrix}$$

4 Limiting processes

1 Introduction

The preceding chapters have been concerned with essentially finite operations: operations which could be carried out completely by an actual computation. In this chapter the idea of a limit is introduced. This is a way of defining quantities, no longer by a finite set of operations which can be carried out, but rather by an infinite set of operations which cannot in practice be performed. The distinction is illustrated by the two following ways of generating the number 2

$$1 + 1 = 2 \tag{1.1}$$

$$\lim_{n \to \infty} \sum_{r=0}^{n} 2^{-r} = 2 \tag{1.2}$$

The operations in (1.1) can be carried out to give the result 2. The operations in (1.2) cannot be completed, and the sum on the left will never reach 2, though the reader will no doubt be already aware that it can be made as close to 2 as we wish.

The idea of a limit is one of the most fruitful in all mathematics. At the same time it is often a source of difficulty when mathematics is applied to physical problems. The usefulness of the idea lies in the simplification and completeness of the mathematical results to which it leads. The difficulties, on the other hand, spring from the fact that the limiting process invariably introduces an element of idealization or over-simplification into the physical situation. Usually we tolerate this for the sake of the mathematical simplicity which accompanies it, but sometimes there can be doubt whether the mathematics truly describes a physical process. The results of an analysis may then be suspect, or they may need some care in interpretation.

A simple example is provided by the definition of velocity v by

$$v = \lim_{\delta t \to 0} \frac{\delta x}{\delta t} \tag{1.3}$$

If we make measurements of δx and of δt, and take the ratio $\delta x/\delta t$, then as δt is reduced from large values, the ratio will usually come nearer and nearer to some definite value. Continued reduction of δt, however, will in the end lead to uncertainties in measurement, so that the ratio will diverge in a random way from the previous value. Increased accuracy of measurement will postpone this result, though not indefinitely. The difficulty here is not troublesome, but in other situations it can be. Caution is therefore needed when limiting processes occur in applied mathematics.

The results developed in this chapter are mainly for real variables. Consideration of complex variables is deferred in general to Chapter 7.

2 Point of accumulation

A *neighbourhood* $N(x, h)$ of a point x in E_n is the set of points y

$$N(x, h) = \{y \mid \|y - x\| < h\} \tag{2.1}$$

That is, $N(x, h)$ is the *open ball* of radius h centred on x. When h is unimportant for the purpose in hand it is omitted, so that $N(x)$ is a neighbourhood of x with some unspecified radius. The notation $N'(x, h)$ is used to denote $N(x, h)$ with the point x removed,

$$N'(x, h) = \{y \mid \|y - x\| < h, y \neq x\} \tag{2.2}$$

and is called a *deleted neighbourhood* of x.

Let S be a set of points in E_n, and let a be a point in E_n. Then a is called a *point of accumulation* of S if every neighbourhood of a contains a point of S distinct from a. Another term for a point of accumulation is *cluster point*.

Example 2.1 Let $n = 1$, and let $S \subset E_1$ be the set of points $1, \frac{1}{2}, \frac{1}{4}, \ldots$. Then $a = 0$ is a point of accumulation of S. Notice that a is not in S.

Example 2.2 Let $n = 2$, and let $S \subset E_2$ be the sphere $\|x\| = 1$ (the sphere in two dimensions is a circle). Then $a = \begin{pmatrix}1\\0\end{pmatrix}$ is a point of accumulation of S. Here $a \in S$.

Example 2.3 Let $n = 1$ and let $S \subset E_1$ be the set of points $\frac{1}{2}, 2, \frac{1}{3}, 3, \ldots$. Then $a = 0$ and $b = \infty$ are both points of accumulation of S. Note that in E_1 a neighbourhood $N(\infty, h)$ of $+\infty$ is defined by

$$N(\infty, h) = \left\{y \mid y > \frac{1}{h}\right\} \tag{2.3}$$

and $N(-\infty, h)$ is defined by

$$N(-\infty, h) = \left\{y \mid -y > \frac{1}{h}\right\} \tag{2.4}$$

In E_n, $n \geqslant 2$, there is only one neighbourhood of infinity with given h, defined by

$$N(\infty, h) = \left\{ y \,\middle|\, \|y\| > \frac{1}{h} \right\} \tag{2.5}$$

In the definition of a point of accumulation it is only required that every neighbourhood of a should contain (at least) one point of S distinct from a. In fact, if this is true, every such neighbourhood contains an infinite number of such points, leading to the following theorem.

Theorem 2.1 If a is a point of accumulation of the set S then every neighbourhood $N(a)$ contains an infinite number of points of S.

Proof Suppose that some neighbourhood $N(a, h)$ contains a finite number of points of S distinct from a: call these $y_1, y_2, \ldots, y_n$. Then the finite set of numbers

$$\|y_1 - a\|, \|y_2 - a\|, \ldots, \|y_n - a\|$$

contains a least member: call it 2ϵ. It follows that the neighbourhood $N(a, \epsilon)$ contains no points of S distinct from a, and so a cannot be a point of accumulation.

3 Open and closed sets

A set of points S in E_n is said to be *open* if every point $x \in S$ has a neighbourhood $N(x)$ lying entirely in S.

Example 3.1 The set of points S in E_1,

$$S = \{x \mid -1 < x < 1\} \tag{3.1}$$

is open. This *open interval* S is written $(-1, 1)$.

Example 3.2 The set of points S in E_2,

$$S = \left\{ x \,\middle|\, x = \begin{pmatrix} x_1 \\ 0 \end{pmatrix}, -1 < x_1 < 1 \right\} \tag{3.2}$$

is not open, because every neighbourhood $N(x, h)$ of $x = \begin{pmatrix} x_1 \\ 0 \end{pmatrix} \in S$ contains $\begin{pmatrix} x_1 \\ h/2 \end{pmatrix} \notin S$.

Example 3.3 The set of points S in E_2,

$$S = \left\{ x \,\middle|\, \|x\| < 1 \quad \text{or} \quad x = \begin{pmatrix} 1 \\ 0 \end{pmatrix} \right\} \tag{3.3}$$

is not open, because every neighbourhood $N(x, h)$ of the point $x = \begin{pmatrix} 1 \\ 0 \end{pmatrix}$ contains the point $\begin{pmatrix} 1 + h/2 \\ 0 \end{pmatrix}$ which is not in S.

A set of points S in E_n is said to be *closed* if S contains all its points of accumulation. If a set S_0 in E_n is not closed, then the set S obtained by adding to S_0 all its points of accumulation is known as the *closure* of S_0.

Example 3.4 The set of points S in E_1,

$$S = \{x \mid -1 \leqslant x \leqslant 1\} \tag{3.4}$$

is closed. This *closed interval* S is written $[-1, 1]$.

Example 3.5 The set of points in Example 3.3 is not closed.

Example 3.5 shows that in addition to open sets and closed sets there are sets which are neither open nor closed. There are also two special sets which are both open and closed: these are the whole space E_n and the null set $\varnothing$. The first of these satisfies the definitions both of an open and of a closed set. The null set contains no points and therefore has no neighbourhoods and no points of accumulation. It is conventional to agree that it therefore satisfies both definitions trivially.

Example 3.6 The plane $x = \begin{pmatrix} x_1 \\ x_2 \\ 0 \end{pmatrix}$, which is a subspace of E_3, is closed but not open. Note the distinction between this plane and the space E_2; compare Example 3.2.

EXERCISE 3.1 Show that the union of any (finite or infinite) collection of open sets is open, and that the intersection of any finite collection of open sets is open.

Example 3.7 The restriction of the second part of Exercise 3.1 to finite collections cannot be relaxed, as is shown by the fact that if in E_1

$$S_k = \left\{x \,\middle|\, -\frac{1}{k} < x < \frac{1}{k}\right\} \tag{3.5}$$

then $S_1 \cap S_2 \cap S_3 \cap \ldots$ is the single point $x = 0$, which constitutes a closed set.

EXERCISE 3.2 Show that if $S \subseteq E_n$ is open, then its complement $E_n - S$ is closed, and conversely.

EXERCISE 3.3 State and prove the result for closed sets analogous to the result given in Exercise 3.1 for open sets.

EXERCISE 3.4 Show that every point x of an open set $S \subseteq E_n$ is a point of accumulation of S.

4 Supremum and infimum

Let S be a set of real numbers. Then the *supremum* q of S, if it exists, is defined by the following property: no member of S exceeds q, but some member of S exceeds any number less than q. The *infimum* p of S, if it exists, is similarly defined: no member of S is less than p, but any number greater than p exceeds some member of S. An alternative name for supremum is *least upper bound* (l.u.b.). An infimum is alternatively known as the *greatest lower bound* (g.l.b.). For brevity it is usual to write $p = \inf S$ and $q = \sup S$.

If S is a set of real numbers, bounded above, then S has a supremum. This is either regarded as an axiom of the real number system or is derived from other axioms at about the same level. There is of course no way of physically testing such an assertion in any situation to which infinite sets of real numbers are applied. This is one of the points at which the demands of mathematical rigour and of physical justification part company, as mentioned in Section 1.

In the same way, if S is a set of real numbers, bounded below, it has an infimum.

Example 4.1 The set $\{0, \frac{1}{2}, \frac{2}{3}, \frac{3}{4}, \ldots\}$ has supremum 1, which is not a member of the set. It has infimum 0 which is a member of the set.

4.1 Maximum and minimum

Let S be a set of real numbers having supremum q. If q is a member of S it is called the *maximum* of S. Similarly, if S has an infimum $p \in S$, then p is the *minimum* of S. These statements are abbreviated to $p = \min S$, $q = \max S$. If S consists of a finite set then it has a maximum equal to its supremum and a minimum equal to its infimum.

Example 4.2 In Example 4.1 the set has minimum zero, but has no maximum.

5 Bounded sets

A set of points S in E_n is said to be *bounded* if there exists a neighbourhood $N(0, r)$ of the origin, with finite r, such that $S \subset N(0, r)$.

Theorem 5.1 (Bolzano–Weierstrass theorem) If a bounded set $S \subset E_n$ contains infinitely many points then S has at least one point of accumulation (which may or may not be a member of S).

Proof The set S lies in $N(0, r)$. It therefore lies in the region (a hypercube) defined by

$$x = \begin{pmatrix} x_1 \\ x_2 \\ \vdots \\ x_n \end{pmatrix}, \quad -r \leqslant x_i \leqslant r, \quad i = 1, 2, \ldots, n \tag{5.1}$$

Divide this region into two parts, one of which has $-r \leqslant x_1 \leqslant 0$, while the other has $0 \leqslant x_1 \leqslant r$. (The face common to these two parts is included in each of them, but this does not affect the argument.) Then at least one part contains an infinite set of points of S. For if both parts contained a finite number of points of S, their union would contain a finite number of such points; but this union includes the original hypercube (the common face being included twice) and so includes S.

Now select one of the two parts, containing an infinite number of points of S, and subdivide it into parts having respectively $-r \leqslant x_2 \leqslant 0$ and $0 \leqslant x_2 \leqslant r$. At least one of these parts contains an infinite number of points of S. Proceed thus for $x_1, x_2, \ldots, x_n$. The result will be a hypercube of side r which contains an infinite number of points of S. This can be further subdivided to give hypercubes of side $\frac{1}{2}r, \frac{1}{4}r, \ldots$, all containing an infinite number of points of S.

The projections of these successive hypercubes on the axis of x_1 define an infinite set of nested intervals. Let the left-hand ends of these successive intervals be $-r, p_1, p_2, \ldots$, and the right-hand ends $r, q_1, q_2, \ldots$. The sequence $p_1, p_2, \ldots$ is bounded above by r and so has a supremum p (the sequence is also non-decreasing). Similarly, $q_1, q_2, \ldots$ is bounded below by $-r$ and therefore has an infimum q (the sequence is also non-increasing). The difference $q - p \geqslant 0$, because if q were to the left of p there would be some p_i and q_i between q and p, from the definition of the infimum with $p_i > q_i$, which is a contradiction.

Now because the sequence $p_1, p_2, \ldots$ is non-decreasing, $p \geqslant p_k$, all k. Similarly $q \leqslant q_k$, all k. So $q - p \leqslant q_k - p_k = 2^{1-k}r$, regardless of the value of k. It follows that $q - p = 0$. Define $t_1 = p = q$, and define $t_2, t_3, \ldots, t_n$ in the same way for the axes of $x_2, x_3, \ldots, x_n$. Let t be the point $\begin{pmatrix} t_1 \\ t_2 \\ \vdots \\ t_n \end{pmatrix}$. Then for t_1 we have already shown that $p_k \leqslant t_1 \leqslant q_k$, all k, and similar results hold for $t_2, t_3, \ldots, t_n$. Consequently t is included in every one of the hypercubes generated by subdividing the original hypercube.

Consider a neighbourhood $N(t, h)$ of t. The point t lies in a hypercube with side $2^{1-k}r$, which has diameter $2^{1-k}r\sqrt{n}$. By a suitably large choice of k it can be ensured that $2^{1-k}r\sqrt{n} < h$, when the hypercube lies entirely within $N(t, h)$. Since the hypercube contains an infinite number of points of S, this shows that t is a point of accumulation of S.

EXERCISE 5.1 Assuming that π is irrational, show that the set of points S in E_1

$$S = \{x_k \mid x_k = \sin k, \quad k = 1, 2, \ldots\} \tag{5.2}$$

has a point of accumulation.

Theorem 5.2 The intersection $S = \bigcap_{i=1}^{\infty} R_i$ of a denumerable family of non-vacuous sets $R_i \subset E_n$ with each R_i closed, R_1 bounded, and $R_{i+1} \subseteq R_i$, $i = 1, 2, \ldots$, is itself closed and non-vacuous.

Proof That S is closed follows directly from Exercise 3.3. Either every set R_i contains infinitely many points, or there is some m such that R_i contains a finite number of points whenever $i > m$. In the second case the proof is immediate. In the first it is possible to select a point x_1 from R_1, x_2 (different from x_1) from R_2, x_3 (different from x_1 and x_2) from R_3, and so on. By Theorem 5.1 the infinite sequence $\{x_i\}$ has a point of accumulation x, say. Every neighbourhood of x contains infinitely many points of $\{x_i\}$ and therefore infinitely many points of R_k, because $x_j \in R_k$, $j \geqslant k$. It follows that x is a point of accumulation of every R_k and since R_k is closed $x \in R_k$, and consequently $x \in \bigcap R_k$ which cannot be vacuous.

A collection of sets $S_i \subseteq E_n$ is called a *covering* of a set $S \subseteq E_n$ if S is contained in the union $\bigcup_i S_i$. If the S_i are open then they form an *open covering* of S.

A *compact set* S in E_n is defined as a set such that every open covering of S contains a finite collection of sets which also cover S.

Theorem 5.3 (Heine–Borel covering theorem) Any closed and bounded set S in E_n is compact.

Proof The first step in the proof is to show that every covering C of $S \subset E_n$ contains a denumerable subfamily of sets which also covers S. Consider the denumerable collection of sets $C_1 = \{N_1, N_2, \ldots\}$ where the N_i are neighbourhoods in E_n having rational radii and their centres at the points with rational coordinates.

If $x \in G$ where G is an open set in E_n, then there is some neighbourhood $N(x, h) \subset G$. Suppose that $x = (x_1, x_2, \ldots, x_n)^T$ and $y = (y_1, y_2, \ldots, y_n)^T$ where each y_k is a rational number satisfying

$$|y_k - x_k| < \frac{h}{4n}, \quad k = 1, 2, \ldots, n$$

Then $\|y - x\| < h/4$, and if α is another rational number satisfying $h/4 < \alpha < h/2$ it follows that $x \in N(y, \alpha)$ and $N(y, \alpha) \subset N(x, h) \subset G$. This shows that at least one of the family N_i contains x and is itself contained in G.

If $x \in S$ then there is an open set G in C such that $x \in G$ and $x \in N_k \subset G$ for some k. The set of such neighbourhoods N_k as the point x takes on all values in S is a denumerable covering of S. If a $G \subset C$ is then associated with each N_k, such that $N_k \subset G$, a denumerable subset of C will be obtained which covers S.

Let $\{G_1, G_2, \ldots\}$ be such a denumerable subset of C which covers S, and let $I_m = \bigcup_{i=1}^{m} G_i$ for $m \geqslant 1$. Then certainly I_m is open (Exercise 3.1), and $E_n - I_m$ is closed (Exercise 3.2). Now put $R_1 = S$, $R_i = S \bigcap (E_n - I_i)$ and notice that R_i is closed, $R_{i+1} \subseteq R_i$ because $I_{i+1} \supseteq I_i$, and $S = R_1$ is bounded. If, therefore, all of the R_i are non-vacuous it follows from Theorem 5.2 that their intersection is non-empty. This is a contradiction, however, since $S \subset \bigcup_{i=1}^{\infty} G_i$ and so R_m must be empty for some m. Consequently S is covered by some I_m and the theorem is proved.

The converse of the Heine–Borel theorem is also important and will now be proved.

Theorem 5.4 Any compact set $S \subset E_n$ is closed and bounded.

Proof The family of open balls with centre at the origin and radii 1, 2, 3, ... form an open covering of S and, since S is compact, a finite subset of this family covers S. Thus by definition (Section 5) S is bounded.

It will now be shown that the assumption that S is open implies a contradiction. If S is open there must be an accumulation point, say y, of S which is not in S. Hence if $x \in S$, the distance $d_x = \|x - y\|$ cannot be zero and as x varies throughout S the family of neighbourhoods $N(x, \frac{1}{2}d_x)$ forms an open covering of S. Since S is compact by hypothesis, a finite subfamily of these neighbourhoods also covers S. Suppose that p is the smallest distance d_x in this finite subfamily. Then if $z \in N(y, \frac{1}{2}p)$ it follows that $\|z - y\| < \frac{1}{2}p \leqslant \frac{1}{2}d_k$ where $N(x_k, \frac{1}{2}d_k)$ is a general member of the finite subfamily. Also

$$\|z - x_k\| = \|y - x_k + z - y\| \geqslant \|y - x_k\| - \|z - y\| = d_k - \|z - y\| > \tfrac{1}{2}d_k$$

showing that $z \notin N(x_k, \frac{1}{2}d_k)$. This means that no point of $N(y, \frac{1}{2}p)$ lies in S, contradicting the assumption that y is an accumulation point of S. Therefore S is closed and the proof of the theorem is complete.

EXERCISE 5.2 Show that a set S in E_n is compact if and only if every infinite subset of S has a point of accumulation in S. [If S is closed and bounded the result follows directly from the Bolzano–Weierstrass theorem. If every infinite subset of S has an accumulation point in S show that the assumption that S is unbounded involves a contradiction, and show that S is closed by proving that every accumulation point x in S can be regarded as the only accumulation point of a sequence $\{x_k\} \subset S$ with $x_k \in N(x, 1/k)$, $k = 1, 2, \ldots$.]

6 Limits

Let f be a function, having domain $S \subseteq E_n$ and range $R \subseteq E_m$. That is, x is an n-vector, $y = f(x)$ is an m-vector, and to every $x \in S$ the function f ascribes a unique $y \in R$. Let x_0 be a point of accumulation of S. Then y is said to *tend to the limit y_0 as x tends to x_0* if for every neighbourhood $N(y_0) \subset E_m$ there exists a neighbourhood $N(x_0) \subset E_n$ such that

$$x \in N'(x_0) \cap S \Rightarrow y = f(x) \in N(y_0) \tag{6.1}$$

Of course, $y \in N(y_0) \cap R$, but this does not have to be stressed. It is, however, necessary to write $x \in N'(x_0) \cap S$ rather than $x \in N'(x_0)$, because if $x \in N'(x_0)$ but $x \notin S$, then $f(x)$ is undefined.

Alternative notations to express the fact that y tends to y_0 as x tends to x_0 are

$$y \to y_0 \quad \text{as } x \to x_0 \tag{6.2}$$

$$\lim_{x \to x_0} y = y_0 \tag{6.3}$$

$$\lim_{\substack{x \to x_0 \\ x \in S}} y = y_0 \tag{6.4}$$

Here (6.4) emphasizes the fact that $x \in E_n$ is restricted to the set S. If $S \subset E_1$ is an interval (a, b) or $[a, b]$, $\lim_{x \to a,\, x \in S}$ and $\lim_{x \to b,\, x \in S}$ are written respectively $\lim_{x \to a+}$, $\lim_{x \to b-}$.

Example 6.1 Let $S \subset E_1$ be the set of points $x = 1, 2, 3, \ldots$. Let $f(x) = 1/x$, so that $R \subset E_1$ is the set $y = 1/n$, $n = 1, 2, 3, \ldots$. Then $y \to 0$ as $x \to \infty$. For if $x \in N'(\infty, h) \cap S$, x is an integer greater than $1/h$. Then $y < h$ and so certainly $y \in N(0, h)$. Here $x_0 \notin S$ and $y_0 \notin R$, while y_0 is a point of accumulation of R.

Example 6.2 Let $S \subseteq E_1$ be the interval $(-\infty, \infty)$; that is $S = E_1$. Define f by

$$\left.\begin{aligned} f(x) &= 0, \quad x \text{ not a negative integer} \\ f(x) &= 1, \quad x \text{ a negative integer} \end{aligned}\right\} \tag{6.5}$$

so that R is the set $\{0, 1\}$. Then

$$\lim_{x \to \infty} f(x) = 0 \tag{6.6}$$

but $f(x)$ does not tend to a limit as $x \to -\infty$. Also neither 0 nor 1 is a point of accumulation of R, because $N(0, \frac{1}{2})$ contains no point of R distinct from 0, and similarly for $N(1, \frac{1}{2})$.

Example 6.3 Let $S \subseteq E_1$ be $(-\infty, \infty)$ and let $f(x)$ be

$$\left.\begin{aligned} f(x) &= 0, \quad x \text{ not an integer} \\ f(x) &= e^{-x}, \quad x \text{ an integer} \end{aligned}\right\} \tag{6.7}$$

Then

$$\lim_{x \to \infty} f(x) = 0 \tag{6.8}$$

while $f(x)$ does not tend to a limit as $x \to -\infty$. Here $y = 0$ and $y = \infty$ are both points of accumulation of

$$R = \{y \,|\, y = 0 \quad \text{or} \quad y = e^{\pm k},\ k = 0, 1, 2, \ldots\}$$

EXERCISE 6.1 Deduce from the preceding examples that y_0 may or may not be a point of accumulation of R. Also that if y_1 is a point of accumulation of R, it does not follow that $y \to y_1$ as x tends to some $x_1 \in R$. Notice, however, that '$f(x) \to y_0$ as $x \to x_0$' implies that x_0 must be a point of accumulation of S. So, for example, if

$$S = \{x \in E_1 \,|\, x = +1, -1, -\tfrac{1}{2}, -\tfrac{1}{4}, \ldots\}$$

$$f(x) = \begin{cases} x, & x < 0 \\ 0, & x \geqslant 0 \end{cases} \tag{6.9}$$

we do not say that $f(x) \to 0$ as $x \to +1$, even though for all $\epsilon > 0$ there exists a $\delta = 1 + \epsilon$ such that

$$x \in N'(1, \delta) \cap S \Rightarrow y \in N(0, \epsilon)$$

EXERCISE 6.2 Show that if S is a set in E_n possessing a point of accumulation x, then it is possible to choose a sequence of distinct points $x_1, x_2, \ldots$, all in S, such that $\lim_{k \to \infty} x_k = x$.

EXERCISE 6.3 The function f: $f(x) = x^2$ is said to *tend to infinity* as $x \to \pm\infty$. Reconcile this terminology with the notation of equation (2.3). Does $1/x$ tend to infinity as $x \to 0$? [No, because x may have either sign.]

6.1 The Cauchy condition

In the definition of a limit given above, the limiting value y_0 itself appears. Cauchy gave a criterion for the existence of a limit y_0 which does not presuppose knowledge of its value.

Theorem 6.1 Let the notation be the same as in the earlier part of this section and let x_0 be a point of accumulation of S. Then Cauchy's condition is the following. A necessary and sufficient condition for the existence of y_0 such that

$y = f(x) \to y_0$ as $x \to x_0$ is that for any $\epsilon > 0$ there exists a δ such that for every $x_1 \in N'(x_0, \delta) \cap S$ and $x_2 \in N'(x_0, \delta) \cap S$, $\|f(x_2) - f(x_1)\| < \epsilon$ (see Fig. 1).

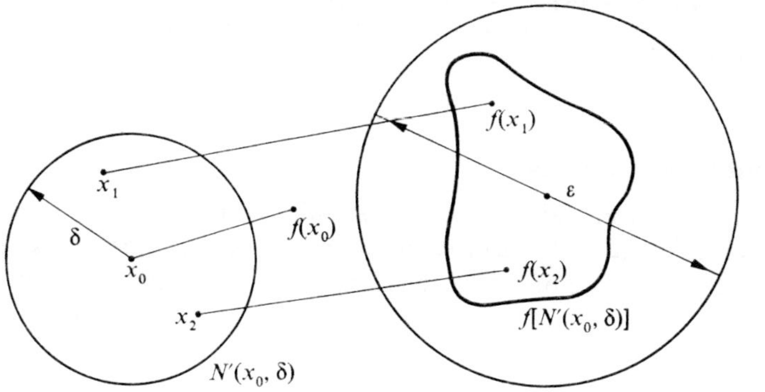

FIG. 1

Proof Suppose that y_0 exists such that $\lim_{x \to x_0} f(x) = y_0$. In the definition of Section 6 choose $N(y_0) = N(y_0, \epsilon/2)$. Then there exists an $N(x_0)$ such that $x \in N'(x_0) \cap S \Rightarrow f(x) \in N(y_0, \epsilon/2)$. Let the radius of this $N(x_0)$ be δ. Then if $x_1 \in N'(x_0, \delta) \cap S$, $x_2 \in N'(x_0, \delta) \cap S$, it follows that $f(x_1) \in N(y_0, \epsilon/2)$, $f(x_2) \in N(y_0, \epsilon/2)$, and so $\|f(x_2) - f(x_1)\| \leqslant \|f(x_2) - y_0\| + \|f(x_1) - y_0\| < \epsilon$. This proves necessity.

To prove sufficiency, notice first that given any δ_0 corresponding to some ϵ_0 according to the premises of the theorem, $f(N'(x_0, \delta_0) \cap S)$ is a bounded set, where (Fig. 1)

$$f(N'(x_0, \delta_0) \cap S) = \{y \mid y = f(x);\ x \in S,\ \|x - x_0\| < \delta_0,\ x \neq x_0\} \quad (6.10)$$

This follows because if $x_1 \in N'(x_0, \delta_0) \bigcap S$ is fixed, and x_2 is any other point of $N'(x_0, \delta_0) \cap S$ then

$$\|f(x_2)\| \leqslant \|f(x_2) - f(x_1)\| + \|f(x_1)\| < \epsilon_0 + \|f(x_1)\|$$

The proof of sufficiency now proceeds by selecting a sequence of points $p_k \in S$, $k = 1, 2, \ldots$, all distinct from x_0 but such that $\lim_{k \to \infty} p_k = x_0$. This is possible by Exercise 6.2 because x_0 is a point of accumulation of S. Then, from the definition of a limit, a k_0 will exist such that for all $k > k_0$, p_k lies in $N'(x_0, \delta_0)$, where δ_0 corresponds to ϵ_0. The sequence $f(p_k)$, $k > k_0$, therefore lies in $f(N'(x_0, \delta_0) \cap S)$, which is bounded. By the Bolzano–Weierstrass theorem, the sequence $f(p_k)$, $k > k_0$, therefore has a point of accumulation, which we call y_0.

Now choose some ϵ, and let δ correspond to ϵ as in the premises of the theorem. Since $p_k \in S$, $p_k \neq x_0$ and $\lim_{k \to \infty} p_k = x_0$, there is some l such that for all $k > l$, $p_k \in N'(x_0, \delta) \cap S$. Moreover, because y_0 is a point of accumulation of $f(p_k)$, $k > l$, there is some $m > l$ such that $f(p_m) \in N(y_0, \epsilon)$. Consider any

$x \in N'(x_0, \delta) \cap S$. By the hypotheses of the theorem, since p_m and x are both in $N'(x_0, \delta) \cap S$, $\|f(x) - f(p_m)\| < \epsilon$. Therefore

$$\|f(x) - y_0\| \leqslant \|f(x) - f(p_m)\| + \|f(p_m) - y_0\| < 2\epsilon \tag{6.11}$$

That is $f(x) \to y_0$ as $x \to x_0$.

6.2 Cauchy sequences

The previous theorem takes a specially simple form when S is the set $1, 2, 3, \ldots$ in E_1. Its statement then becomes the following. A necessary and sufficient condition for the existence of y such that $\lim_{k\to\infty} y_k = y$ is that for any $\epsilon > 0$ there should exist an l such that for all $p > l$, $q > l$, $\|y_p - y_q\| < \epsilon$.

EXERCISE 6.4 Derive the result in Section 6.2 from Theorem 6.1.

6.3 Real numbers

The real numbers can be defined in an axiomatic way by specifying their properties (compare Chapter 1). Historically they were introduced by a limiting process from the rational numbers. For example it is well known that there is no rational number p/q such that $(p/q)^2 = 2$. On the other hand it is possible to define a sequence in the following way. Let $q_k = 10^k$, and let p_k be the greatest integer such that $(p_k/q_k)^2 < 2$. Then the sequence

$$\frac{p_0}{q_0}, \frac{p_1}{q_1}, \frac{p_2}{q_2}, \ldots = \frac{1}{1}, \frac{14}{10}, \frac{141}{100}, \ldots \tag{6.12}$$

lies in $N(0, 2)$, and so by the Bolzano–Weierstrass theorem has a point of accumulation, say x. Moreover p_k/q_k is a non-decreasing sequence, from which it easily follows that $p_k/q_k \to x$ as $k \to \infty$. Now by definition

$$2 < \left(\frac{p_k + 1}{q_k}\right)^2 = \left(\frac{p_k}{q_k}\right)^2 + (2p_k + 1)10^{-2k} \tag{6.13}$$

and as $p_k < 2.10^k$

$$0 < 2 - \left(\frac{p_k}{q_k}\right)^2 < 5.10^{-k} \tag{6.14}$$

Also

$$0 < x^2 - (p_k/q_k)^2 = (x + p_k/q_k)(x - p_k/q_k) < 4(x - p_k/q_k) = 4\epsilon \tag{6.15}$$

where $\epsilon \to 0$ as $k \to \infty$. Hence

$$|2 - x^2| < 5.10^{-k} + 4\epsilon \tag{6.16}$$

and by suitably large choice of k the right-hand side of (6.16) can be made as small as we please. It follows that $x^2 = 2$, and x is therefore the real (irrational) number $\sqrt{2}$.

It will be noticed that the existence of x is inferred [after equation (6.12)] from the Bolzano–Weierstrass theorem, which depends on an axiom given in Section 4.

6.4 Operations with limits

If f_1 and f_2 are two functions, both with domain $S \subseteq E_n$ and range $R \subseteq E_m$, and if $f_1(x) \to y_1$ and $f_2(x) \to y_2$ as $x \to x_0$, then $f_1(x) + f_2(x) \to y_1 + y_2$ as $x \to x_0$. This is easily proved: for given $\epsilon_1 = \eta/2$ and $\epsilon_2 = \eta/2$ there exist δ_1 and δ_2 such that

$$x \in N'(x_0, \delta_1) \cap S \Rightarrow f_1(x) \in N(y_1, \epsilon_1) \tag{6.17}$$

$$x \in N'(x_0, \delta_2) \cap S \Rightarrow f_2(x) \in N(y_2, \epsilon_2) \tag{6.18}$$

Let $\delta = \min\{\delta_1, \delta_2\}$. Then for $x \in N'(x_0, \delta) \cap S$,

$$\|f_1(x) + f_2(x) - (y_1 + y_2)\| \leqslant \|f_1(x) - y_1\| + \|f_2(x) - y_2\| < \epsilon_1 + \epsilon_2 = \eta \tag{6.19}$$

That is, for any η, δ may be found such that

$$x \in N'(x_0, \delta) \cap S \Rightarrow f_1(x) + f_2(x) \in N(y_1 + y_2, \eta) \tag{6.20}$$

EXERCISE 6.5 If f_1 and f_2 are two functions with domain $S \subseteq E_n$ and range $R \subseteq E_1$, and $f_1(x) \to y_1, f_2(x) \to y_2$ as $x \to x_0$, show that the product $f_1(x)f_2(x) \to y_1y_2$ as $x \to x_0$.

EXERCISE 6.6 If f_1 and f_2 are two functions with domain $S \subseteq E_n$ and range $R \subseteq E_m$, and $f_1(x) \to y_1, f_2(x) \to y_2$ as $x \to x_0$, show that the inner product $f_1^{\mathrm{T}}(x)f_2(x) \to y_1^{\mathrm{T}}y_2$ as $x \to x_0$.

EXERCISE 6.7 If f_1, f_2, R, S, y_1 and y_2 are as in Exercise 6.5, show that $f_1(x)/f_2(x) \to y_1/y_2$ as $x \to x_0$, provided that $y_2 \neq 0$.

EXERCISE 6.8 Examine how the above results may be extended to sequences of operations.

EXERCISE 6.9 If f_1 and f_2 both have domain $S \subseteq E_n$ and range $R \subseteq E_m$, and if $f_1(x) \leqslant f_2(x)$ for all $x \in S$, show that $\lim_{x \to x_0} f_1(x) \leqslant \lim_{x \to x_0} f_2(x)$ provided that both limits exist.

7 Continuity

It will be noticed that in the definition of the limit of $f(x)$ as $x \to x_0$, the value of $f(x_0)$ plays no part. This value need not in fact be defined, or if it is defined it may be altered in any way without affecting the limit (if this exists).

Example 7.1 Let $S = E_1$ and consider the two functions defined by

$$f_1(x) = |x| \tag{7.1}$$

$$f_2(x) = \begin{cases} |x|, & x \neq 0 \\ 1, & x = 0 \end{cases} \tag{7.2}$$

Then $\lim_{x\to 0} f_1(x) = 0 = \lim_{x\to 0} f_2(x)$.

In Example 7.1 f_1 is a more 'natural' function than f_2, because $\lim_{x\to 0} f_1(x) = f_1(0)$, whereas for f_2 the corresponding statement is false. This is expressed by saying that f_1 is continuous at the origin, while f_2 is discontinuous at the origin. The definition is as follows: let a function f be defined on a domain $S \subseteq E_n$ and have range $R \subseteq E_m$. Let $x_0 \in S$ be a point of accumulation of S. Then if $\lim_{x\to x_0} f(x)$ exists and is equal to $f(x_0)$, f is said to be *continuous* at x_0.

EXERCISE 7.1 Show that the function defined by $f(x) = \sin(1/x)$ is not continuous at $x = 0$. [f has no limit as $x \to 0$.]

EXERCISE 7.2 Show that if $S = E_1$ and f is defined by

$$f(x) = 1/x \tag{7.3}$$

then f is not continuous at $x = 0$. [$f(0)$ is undefined.]

EXERCISE 7.3 Show that if $S = E_1$ and f is defined by

$$f(x) = \begin{cases} 0, & x \leqslant 0 \\ 1, & x > 0 \end{cases} \tag{7.4}$$

then f is not continuous at $x = 0$. The function f defined in (7.4) is usually called a *unit step* at the origin, and is written $U(x)$.

Example 7.2 Let $S = E_2$, $R \subset E_1$, and define f by

$$f(x) = \begin{cases} x_1x_2/(x_1^2 + x_2^2), & x_1^2 + x_2^2 \neq 0 \\ 0, & x_1^2 + x_2^2 = 0 \end{cases} \tag{7.5}$$

This function is zero along $x_1 = 0$ and along $x_2 = 0$, but it is discontinuous at $x = 0$. For if $x_2 = x_1$, $f(x) = \frac{1}{2}$, so that along the line $x_2 = x_1$ there are points, as close as we please to the origin, where $\|f(x) - f(x_0)\| = \frac{1}{2}$.

EXERCISE 7.4 Show that if two functions f_1 and f_2, with domains $S_1 \subseteq E_n$, $S_2 \subseteq E_n$ and ranges $R_1 \subseteq E_m$, $R_2 \subseteq E_m$, are continuous at a point of accumulation x_0 of $S_1 \cap S_2$, then the function $f = f_1 + f_2$ is continuous at x_0.

EXERCISE 7.5 State analogous results to that in Exercise 7.4 for the functions defined (when $m = 1$) by $f(x) = f_1(x)f_2(x)$ and $f(x) = f_1(x)/f_2(x)$.

When $S \subseteq E_1$ it is convenient to subdivide the notion of continuity. Thus, let f be a function with $S \subseteq E_1$ and $R \subseteq E_m$, and let x_0 be a point of accumulation of S. If $\lim_{x \to x_0,\, x < x_0} f(x)$ exists and is equal to $f(x_0)$, f is said to be *continuous on the left* at x_0. Similarly if $f(x) \to f(x_0)$ as $x \to x_0$ from above, f is said to be *continuous on the right* at x_0. A function (with $S \subseteq E_1$) which is continuous on the right, and also continuous on the left, at an interior point x_0 of S is clearly continuous at x_0 according to the previous definition.

Example 7.3 The following function with $S = E_1$ and $R \subset E_2$ is continuous on the left at every x

$$f(x) = \begin{pmatrix} |x| \\ U(x) \end{pmatrix} \tag{7.6}$$

When a function f has $S \subseteq E_1$ and $R \subseteq E_1$, its possible types of discontinuity can be classified. Particularly important is the simple discontinuity. A function f has a *simple discontinuity of magnitude* α *on the right* at x_0 if $\lim_{x \to x_0,\, x > x_0} f(x) - f(x_0) = \alpha$. Here, and in future definitions of the same kind, the statement about the limit is taken to imply that it exists. A simple discontinuity on the left is defined in the analogous way.

The following theorem regarding functions which are continuous on closed bounded sets will be useful later.

Theorem 7.1 Let f be a continuous function on a set $S \subseteq E_m$ and let $f(S) = R \subseteq E_n$. Then if S is compact so is R; that is, the image of a compact set under a continuous mapping is itself compact.

Proof If $x \in S$ and $C = \bigcup_i A_i$ is an open covering of R then $f(x) \in A_i$ for at least one value of i, say $i = r$. Since A_r is open there is a neighbourhood $N(f(x))$ of $f(x)$ such that $N(f(x)) \subset A_r$ and, since f is continuous, there exists a neighbourhood $N(x)$ such that $f(N(x) \cap S) \subset N(f(x)) \subset A_r$. As x takes on all values of S, the neighbourhoods $N(x)$ form an open covering of S from which a finite subset $N(x_p)$, $p = 1, 2, \ldots, q$ can be selected which also covers S (Theorem 5.3). Any $x \in S$ is contained in a member $N(x_p)$ of this finite family of sets. Then $f(x) \in f(N(x_p) \cap S) \subset N(f(x_p)) \subset A_p$ where A_p is the appropriate member of C. This proves that $f(S)$ is covered by a finite family of open sets A_p, $p = 1, 2, \ldots, q$ and is therefore compact.

EXERCISE 7.6 Show that if the function f in the previous Theorem is one-to-one then f^{-1} is also continuous. [Let y be a point of accumulation in R and $y = \lim_{n \to \infty} y_n$. Show that $x = f^{-1}(y)$ is the only accumulation point of $\{x_n\}$ where $x_n = f^{-1}(y_n)$.]

EXERCISE 7.7 Prove that a function which is continuous on a closed set is bounded.

7.1 Real-valued continuous functions

Some properties of continuous functions f with domain $S \subseteq E_n$ and range in E_1 are now derived. Such functions are known as *real-valued functions*. If there is a point $a \in S$ for which $f(a) \geqslant f(x)$ for all $x \in S$ then f is said to have a *weak absolute* (or *global*) *maximum* at a. If the inequality is strict, that is $f(a) > f(x)$ for all $x \in S$ other than a, then there is a *strong absolute* (*global*) *maximum* at a. Weak and strong absolute minima are defined in the same way with $\geqslant, >$ replaced by $\leqslant, <$. If the above conditions hold only in some neighbourhood of the point a then the maxima or minima are said to be *relative* (or *local*) rather than absolute. A point at which there occurs a relative maximum or minimum is termed an *extreme point*.

Theorem 7.2 A real-valued function f which is continuous on a compact set S in E_n attains an absolute maximum and an absolute minimum on S.

Proof By Theorem 7.1, $f(S)$ is compact and hence by Theorem 5.4 it is closed and bounded. If $f(S)$ has a finite number of elements the result is obvious (Section 4) so it is assumed that the set $f(S)$ is infinite. Because $f(S)$ is bounded $\sup f(S)$ exists. Given any $\epsilon > 0$ there is some point $x \in f(S)$ such that $\sup f(S) - \epsilon \leqslant x$. Hence $\sup f(S)$ must be a point of accumulation of $f(S)$, and therefore in $f(S)$ because $f(S)$ is closed. Since $\sup f(S) \geqslant f(x)$ for all $x \in S$, the absolute maximum of f on S is $\sup f(S)$. A similar argument proves that the absolute minimum of f on S is $\inf f(S)$. This absolute maximum and absolute minimum are at least weak, and may be strong.

Theorem 7.3 Let f be a real-valued function which is continuous on the closed interval $[a, b]$. Then if $f(a) < c < f(b)$ there is an $x_0 \in (a, b)$ such that $f(x_0) = c$.

Proof Denote the set $\{x \mid \xi \in [a, x] \Rightarrow f(\xi) < c\}$ by S. Then S is certainly non-vacuous since $a \in S$, and S is bounded above by b: let $x_0 = \sup S$ and suppose that $f(x_0) < c$. Then since f is continuous there is a $\delta > 0$ for which $\xi \in (x_0 - \delta, x_0 + \delta)$ implies $f(\xi) < c$. But since $x_0 = \sup S$ it follows that $x_0 - \delta \in S$ and the whole interval $[x_0 - \delta, x_0 + \delta) \subset S$. This implies that there is a point in S greater than x_0, contradicting the definition of x_0: therefore $f(x_0) \geqslant c$. Now assume that $f(x_0) > c$; then by continuity there is a $\delta > 0$ for which $\xi \in (x_0 - \delta, x_0 + \delta)$ implies $f(\xi) > c$. This means that $x_0 - \delta \notin S$, again contradicting $x_0 = \sup S$; hence $f(x_0) \leqslant c$. Thus $f(x_0) = c$ and because $f(a) \neq c, f(b) \neq c$, it follows that $x_0 \in (a, b)$ and the proof is complete.

EXERCISE 7.8 Prove that if f is real-valued and continuous on $[a, b]$ and $(a) f(b) < 0$ then there is at least one point x in (a, b) for which $f(x) = 0$.

EXERCISE 7.9 Prove that a real-valued, continuous function f on a closed interval $[a, b]$ takes on every value between its maximum and minimum.

EXERCISE 7.10 Show that if f is a continuous one-to-one mapping of the closed bounded interval $[c, d]$ on to the closed bounded interval $[a, b]$, with $f(c) = a$, then f^{-1} exists and is continuous on $[a, b]$, and both f and f^{-1} are strictly increasing. [It is implied in the notation $[a, b]$ that $b > a$. Use the result of Exercise 7.6.]

7.2 Uniform continuity

Consider the function f defined by $f(x) = 1/x$ on the half-open interval $I = (0, 1]$; then, as will be shown, f is continuous on I. Suppose that $\left|\frac{1}{x} - \frac{1}{a}\right| < \epsilon$, $x, a \in I$, which on multiplying both sides by $xa > 0$ gives $|x - a| < \epsilon xa$. This inequality is satisfied if and only if $|x - a| < \epsilon a^2/(1 - a\epsilon)$. This confirms directly that f is continuous at the point a [take $\delta = \epsilon a^2/(1 - a\epsilon)$] and shows that the value of the number δ in the definition of continuity depends on both ϵ and the point a under consideration.

It is natural to ask whether for a given ϵ it is possible to find a δ (depending, of course, on ϵ) which will be satisfactory for every point in I; that is, a δ which is independent of a. That this is not so for $f(x) = 1/x$ follows from the fact that $\epsilon a^2/(1 - a\epsilon) \to 0$ as $a \to 0+$.

This question leads to the following definition of uniform continuity. Let f be a function with domain $S \subseteq E_n$ and range $R \subseteq E_m$; then f is said to be *uniformly continuous on* S if, given $\epsilon > 0$, there exists a δ such that $\|x - y\| < \delta$ implies $\|f(x) - f(y)\| < \epsilon$ when $x, y \in S$. Notice that δ depends on ϵ only.

EXERCISE 7.11 Prove that f: $f(x) = 1/x$ is uniformly continuous on the closed interval $[a, 1]$ for any $0 < a < 1$.

EXERCISE 7.12 Prove that the function f: $f(x) = x$ is uniformly continuous on the closed interval $[0, 1]$.

EXERCISE 7.13 Prove that f: $f(x) = \sin(1/x)$ is continuous but not uniformly continuous on the open interval $(0, 1)$.

That a function f which is uniformly continuous on a set S is also continuous there is evident from the definition of uniform continuity. It can be shown that the converse of this statement is also true when the set S is compact. The proof depends on the Heine–Borel theorem (Theorem 5.3).

Theorem 7.4 Any function f continuous on a compact set S in E_n is also uniformly continuous on S.

Proof Since f is continuous it follows that given $\epsilon > 0$ and $x \in S$ there exists a $\delta_x > 0$ such that $y \in N(x, 2\delta_x) \cap S$ implies that $f(y) \in N(f(x), \epsilon)$. Now the family of neighbourhoods $N(x, \delta_x)$, as x takes on every value in S, must cover S. Hence by the Heine–Borel theorem it is possible to select a finite subset S_0 of S in such a way that $x \in S$ implies that $x \in N(a, \delta_a)$ for some $a \in S_0$.

Let $p = \min \delta_a$; then if $x \in S$ and $y \in N(x, p) \cap S$ it follows that there is an $a \in S_0$ such that $x \in N(a, \delta_a)$ and $y \in N(a, \delta_a + p) \subset N(a, 2\delta_a)$. This means that $f(x)$ and $f(y)$ are both in $N(f(a), \epsilon)$ and so $f(y) \in N(f(x), 2\epsilon)$ demonstrating the uniform continuity of f.

7.3 Continuity of a composite function

Theorem 7.5 Let f be a function with domain S in E_n and range $f(S) \subseteq E_m$. Let a second function g have domain $f(S)$ and range R in E_l. Let f be continuous at a point of accumulation $a \in S$ and let $f(a)$ be a point of accumulation of $f(S)$ at which g is continuous. Then the composite function gf is also continuous at the point a.

Proof Under the assumptions of the theorem if $f(a) = b$ and $g(b) = c$ then given a neighbourhood $N(c) \subset E_l$ there exists a neighbourhood $N(b) \subset E_m$ such that if $x \in N(b) \cap f(S)$ then $g(x) \in N(c)$. Similarly it is possible to find a neighbourhood $N(a) \subset E_n$ such that if $y \in N(a) \cap S$ then $f(y) \in N(b)$. Thus for every neighbourhood of $c = g(b) = g(f(a)) \in E_l$ there is a neighbourhood $N(a) \subset E_n$ such that if $y \in N(a) \cap S$ then $gf(y) \in N(c)$ and therefore the composite function gf is continuous at a.

7.4 Continuity of physical variables

A physical variable, such as a voltage or velocity, may be caused to change very suddenly. For example, in the circuit shown in Fig. 2 the switch may be closed,

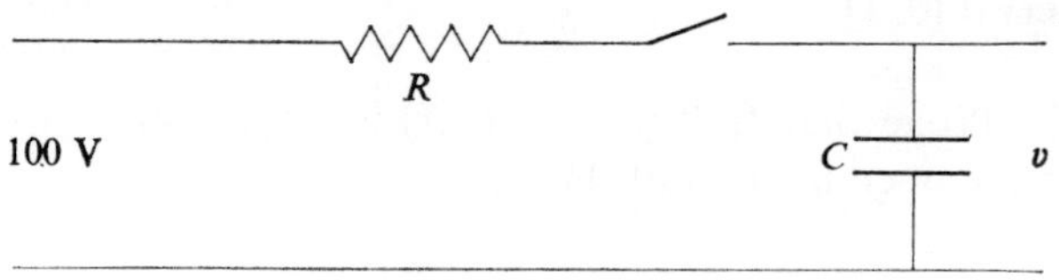

FIG. 2

or a golf ball may be struck by a club. In such cases, if sufficiently fast measuring equipment is used, the way in which the physical variable changes can be examined in detail. When this is done the variable will be found to change continuously. Even where the difficulties of measurement cannot be overcome, the supposition is that physical variables cannot change discontinuously.

On the other hand, there are many situations in which the details of the sudden change are of little interest. In the circuit of Fig. 2, for example, the switch may physically be a device which produces metal-to-metal contact, while C may represent a small stray capacitance. Then if closing of the switch is initiated at time $t = 0$, there will be a small delay until the contact is made. Then the voltage v will change in a way depending on the detailed (and random) way in which the contact resistance changes with time, and depending also on R and C. This is indicated in Fig. 3.

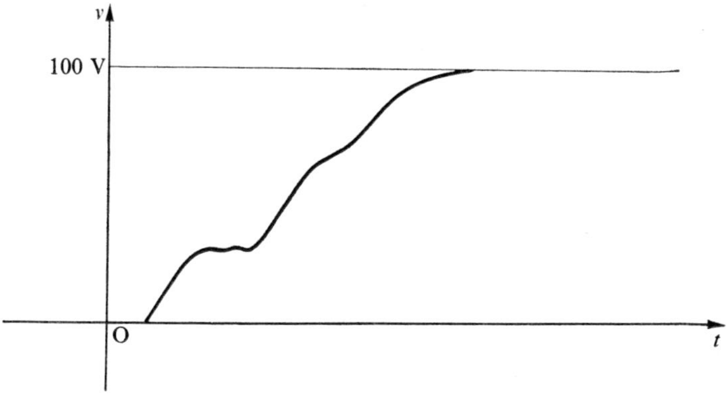

FIG. 3

The change in v shown in Fig. 3 might well be completed in a fraction of a millisecond. If v is then applied, for example, to the input of an amplifier driving a mechanical control system, the effect will be substantially the same as if v changed instantaneously from 0 to 100 V. That is to say, if, starting from the situation shown in Fig. 3, the behaviour of v were made to approximate more and more closely to $100\ U(t)$, then the behaviour of the mechanical system would be expected to show little alteration.

This is the justification for introducing discontinuous physical variables into the description of a system. The unit step $U(t)$ is easier to deal with mathematically than the actual time behaviour, and the latter may not even be known. In accordance with this discussion all physical variables will be assumed to be continuous on the left for all t but they will be allowed, in any finite interval, to have a finite number of simple discontinuities on the right.

Note that the unit step is sometimes defined in slightly different ways from the definition adopted here [equation (7.4)]. The alternatives are

Alternative 1 (not used here)
$$U(t) = \begin{cases} 0, & t < 0 \\ \text{undefined}, & t = 0 \\ 1, & t > 0 \end{cases} \tag{7.7}$$

Alternative 2 (not used here)
$$U(t) = \begin{cases} 0, & t < 0 \\ \frac{1}{2}, & t = 0 \\ 1, & t > 0 \end{cases} \tag{7.8}$$

It is also worth noticing that a simple discontinuity on the right belongs to one level of abstraction from reality. The discontinuity defined in equation (7.2) belongs to a quite different level of abstraction, and cannot be given any physical significance. Correspondingly, the various definitions of the unit step (which differ only at a single point) are equivalent for all physical purposes. The definition chosen above seems to be the one which arises most naturally.

8 The derivative

Let f be a function with domain $S \subseteq E_n$, where S is an open set, and with range $R \subseteq E_m$. Let x be a point of S, which is therefore (by Exercise 3.4) a point of accumulation in S; $f(x_0)$ will then exist and be in R. Consider the following linear function

$$g(x) = f(x_0) + J(x - x_0) \tag{8.1}$$

where J is an $m \times n$ constant (finite) matrix. The function g is so constructed that $g(x_0) = f(x_0)$, and J is at present undetermined. If

$$\lim_{\substack{x \to x_0 \\ x \in S}} \frac{\|f(x) - g(x)\|}{\|x - x_0\|} = 0 \tag{8.2}$$

we say that *f and g are tangent at x_0*.

For a given f and x_0, there can be at most one function g satisfying (8.2). For suppose g_1 and g_2 both satisfy (8.2), then

$$\frac{\|g_2(x) - g_1(x)\|}{\|x - x_0\|} \leqslant \frac{\|f(x) - g_2(x)\|}{\|x - x_0\|} + \frac{\|f(x) - g_1(x)\|}{\|x - x_0\|} \tag{8.3}$$

and so with the same limiting process as in (8.2), and using results from Section 6.4,

$$\lim \frac{\|g_2(x) - g_1(x)\|}{\|x - x_0\|} \leqslant \lim \frac{\|f(x) - g_2(x)\|}{\|x - x_0\|} + \lim \frac{\|f(x) - g_1(x)\|}{\|x - x_0\|} = 0 \tag{8.4}$$

But from Section 10 of Chapter 3, $x - x_0$ can be chosen so that

$$\frac{\|g_2(x) - g_1(x)\|}{\|x - x_0\|} = \frac{\|J_2(x - x_0) - J_1(x - x_0)\|}{\|x - x_0\|} \tag{8.5}$$

$$= \|J_2 - J_1\| \tag{8.6}$$

where the matrix norm is subordinate to the euclidean vector norm; and so $\|J_2 - J_1\| = 0$. This implies that $J_2 = J_1$, and so if J exists, satisfying (8.2) at x_0, it is unique.

The matrix J will be defined here to be the *derivative* of f. A more usual definition, which is essentially equivalent, is to call the function $g - f(x_0)$ the derivative of f. By Section 4 of Chapter 3 there is a one-to-one relation between this linear function (mapping) and the matrix J.

Common notations for the derivative J at x_0 are $f'(x_0)$ and $Df(x_0)$. These notations bring into evidence the dependence of J upon f and x_0. A useful informal notation is df/dx, but this needs to be used with care. If the dependence upon x_0 is to be stressed, this last notation becomes a little clumsy: possible notations are

$$\left.\frac{df}{dx}\right|_{x=x_0}, \qquad \left.\frac{df}{dx}\right|_{x_0}, \qquad \frac{df}{dx}(x_0), \text{ etc.}$$

Notice also that if the notation $f'(x_0)$ is used for J, the transpose of a vector u has to be denoted by u^{T}, or u^{t}, and not u'.

If f has a derivative at x_0, it is said to be *differentiable* at x_0. The essential feature of differentiability is that there exists a tangent plane to the function at the point considered.

Example 8.1 Let f: $E_2 \to E_2$ be the function defined by

$$\left.\begin{aligned} y_1 &= f_1(x) = a_1 + a_{11}x_1 + a_{12}x_2 \\ y_2 &= f_2(x) = a_2 + a_{21}x_1 + a_{22}x_2 \end{aligned}\right\} \tag{8.7}$$

Then f is differentiable (it is linear, and so defines its own tangent plane) and has derivative

$$J = \begin{pmatrix} a_{11} & a_{12} \\ a_{21} & a_{22} \end{pmatrix} \tag{8.8}$$

at every point $x \in E_2$.

If f is defined on a set T which is not open, then an open set S can be obtained by deleting the boundary points of T. At each point of S where (8.2) is fulfilled, f has a derivative. The derivative is not defined, however, at a boundary point of T.

8.1 Functions with domain $S \subseteq E_1$

For functions depending on a single variable (that is, functions with domain $S \subseteq E_1$) the existence of an ordering relation ($>$, $<$) allows the idea of differentiability to be subdivided (compare Section 7). Let f be a function with domain $S \subseteq E_1$ and range $R \subseteq E_m$. Let $x_0 \in S$ be a point of accumulation of S. Then if

$$\lim_{\substack{x \to x_0 \\ x \in S \\ x < x_0}} \frac{f(x) - f(x_0)}{x - x_0} \tag{8.9}$$

exists and is finite, f is said to be *differentiable on the left at* x_0 and the numerical value of the limit is called the *left-hand derivative* $D_-f(x_0)$ *of* f *at* x_0. The right-hand derivative, $D_+f(x_0)$, and differentiability on the right, are defined similarly. A function f which is differentiable both on the right and on the left at x_0 is differentiable at x_0, in the sense of Section 8, if $D_-f(x_0) = D_+f(x_0)$; in which case each of these one-sided derivatives is equal to $Df(x_0)$.

EXERCISE 8.1 Restate this definition by introducing a linear function g and using equation (8.2) in an appropriately specialized form. Reconcile the two definitions.

When $S \subseteq E_1$, it is no longer necessary to insist that S is open, as was done previously. It is still necessary for x_0 to be a point of accumulation of S, for otherwise the limiting operation is not defined (see Exercise 6.1). It is now possible, however, to allow x_0 to be (for example) the right-hand end point b of a closed interval $S = [a, b]$. The 'derivative at b' is then interpreted in a natural way as the left-hand derivative at b.

Because the derivatives at the end points a, b of the interval $S = [a, b]$ may be interpreted in this way, the phrase 'f has a derivative on $[a, b]$' becomes ambiguous. It may mean that f has a derivative in the ordinary sense at a and at b, which implies the existence of f on an open interval including $[a, b]$. Or it may mean that D_+f exists on $[a, b)$ and D_-f exists on $(a, b]$, and $D_+f = D_-f$ on (a, b). The second interpretation does not require f to be defined outside $[a, b]$. In this book the first of these two interpretations will be meant unless otherwise stated.

The above ideas can readily be extended to matrices $A(x)$ with elements $a_{ij}(x)$ depending on a real variable $x \in S \subseteq E_1$. If

$$\lim_{\substack{x \to x_0 \\ x \in S}} \frac{A(x) - A(x_0)}{x - x_0}$$

exists and is finite, A is said to be differentiable at x_0. Thus A is differentiable when its elements are differentiable, and $dA/dx = (da_{ij}/dx)$.

EXERCISE 8.2 Prove that if $A_1(x)$ and $A_2(x)$ are differentiable matrices and α is any scalar

(a) $\dfrac{d}{dx}(A_1 + A_2) = \dfrac{dA_1}{dx} + \dfrac{dA_2}{dx}$

(b) $\dfrac{d}{dx}(\alpha A_1) = \alpha\dfrac{dA_1}{dx}$

(c) $\dfrac{d}{dx}(A_1 A_2) = A_1\dfrac{dA_2}{dx} + \dfrac{dA_1}{dx}A_2$

Notice that

$$\frac{d}{dx}A^2 = A\frac{dA}{dx} + \frac{dA}{dx}A$$

which is only equal to $2A(dA/dx)$ if A and dA/dx commute.

8.2 Partial derivative, directional derivative

Let f be a function with domain an open set $S \subseteq E_n$, and range $R \subseteq E_m$. Restrict x to a subset T of S,

$$T = \{x \mid x = x_0 + ut,\ t \in E_1,\ x \in S\} \tag{8.10}$$

where u is a unit vector in E_n and t is a parameter. In other words, T is an open segment of the line through x_0 with direction u. Notice that because S is open, there is some neighbourhood $N(x, h)$ such that $N(x, h) \cap T$ is an open interval of length $2h$.

With the restriction $x = x_0 + ut$, f becomes a function h of $t \in E_1$. The derivative $h'(0)$ of h at $t = 0$, if it exists, is called the *directional derivative* of f at x_0, and is written $D_u f(x_0)$. Notice that $D_u f(x_0)$ is an m-vector.

The particular directional derivatives obtained by taking u parallel to the coordinate axes are called the *partial derivatives* with respect to $x_1, x_2, \ldots, x_n$, where $x_1, x_2, \ldots, x_n$ are the elements of the n-vector x. They are written $D_1 f(x_0)$, $D_2 f(x_0)$, etc., or $\partial f/\partial x_1$, $\partial f/\partial x_2$, etc. Written in full the m-vector $D_1 f(a)$ is

$$D_1 f(a) = \left(\frac{\partial f}{\partial x_1}\right)_{x=a} = \begin{pmatrix} \partial f_1/\partial x_1 \\ \partial f_2/\partial x_1 \\ \vdots \\ \partial f_m/\partial x_1 \end{pmatrix} \tag{8.11}$$

Example 8.2 Let f: $E_2 \to E_1$ be defined by $f(x) = x_1 x_2$. The partial derivative $D_1 f(a)$ at the point $a = \begin{pmatrix} a_1 \\ a_2 \end{pmatrix}$ is obtained by restricting

$$x = \begin{pmatrix} x_1 \\ x_2 \end{pmatrix} \text{ to } x = \begin{pmatrix} a_1 \\ a_2 \end{pmatrix} + t\begin{pmatrix} 1 \\ 0 \end{pmatrix}$$

So $f(x)$ becomes $h(t) = (a_1 + t)a_2$ and $h'(0) = a_2 = D_1 f(a)$. This clearly leads to the usual rule of 'treating x_2 as a constant'.

Theorem 8.1 Let f have a derivative $Df(x_0)$ at x_0. Then

$$D_u f(x_0) = Df(x_0)u = Ju \tag{8.12}$$

where u as before is a unit n-vector.

Proof Since the derivative exists,

$$\lim_{\substack{x\to x_0\\ x\in S}} \frac{\|f(x) - f(x_0) - J(x - x_0)\|}{\|x - x_0\|} = 0 \tag{8.13}$$

Let T be the set defined in (8.10), and write

$$f(x_0 + ut) = h(t) \tag{8.14}$$

Then it follows at once from (8.13) and the definition of a limit that when $x = x_0 + ut$

$$\lim_{\substack{t\to 0\\ t\in T\subset S}} \frac{\|h(t) - h(0) - Jut\|}{\|ut\|} = 0 \tag{8.15}$$

and since u is a unit vector, $\|ut\| = |t|$. So (8.15) gives

$$\lim_{\substack{t\to 0\\ t\in T\subset S}} \left\|\frac{h(t) - h(0)}{t} - Ju\right\| = 0 \tag{8.16}$$

$$\lim_{\substack{t\to 0\\ t\in T\subset S}} \frac{h(t) - h(0)}{t} = Ju \tag{8.17}$$

That is

$$D_u f(x_0) = h'(0) = Ju \tag{8.18}$$

In particular it follows that if $Df(x_0)$ exists then $D_1f(x_0)$ exists and is the first column of $J = Df(x_0)$; $D_2f(x_0)$ exists and is the second column; and so on. That is, the matrix J is given by

$$J = \begin{pmatrix} \partial f_1/\partial x_1 & \partial f_1/\partial x_2 & \dots & \partial f_1/\partial x_n \\ \partial f_2/\partial x_1 & \partial f_2/\partial x_2 & \dots & \partial f_2/\partial x_n \\ \vdots & \vdots & & \vdots \\ \partial f_m/\partial x_1 & \partial f_m/\partial x_2 & \dots & \partial f_m/\partial x_n \end{pmatrix} \tag{8.19}$$

This matrix is known as the *jacobian matrix* of f, and the determinant of J is the *jacobian* of f. On the other hand it is important to notice that the existence of $D_1f(x_0), D_2f(x_0), \dots, D_nf(x_0)$ in no way ensures that $Df(x_0)$ exists.

Example 8.3 The function f (compare Example 7.2) defined by

$$f(x) = \begin{cases} x_1x_2/(x_1^2 + x_2^2), & x_1^2 + x_2^2 \neq 0 \\ 0, & x_1^2 + x_2^2 = 0 \end{cases} \tag{8.20}$$

is zero along $x_1 = 0$ and along $x_2 = 0$. Therefore $D_1 f(0)$ and $D_2 f(0)$ both exist and are zero. Yet f is not differentiable at the origin, because along the line

$$\begin{pmatrix} x_1 \\ x_2 \end{pmatrix} = \frac{t}{\sqrt{2}} \begin{pmatrix} 1 \\ 1 \end{pmatrix} = tu \tag{8.21}$$

$f(x) = h(t) = \frac{1}{2}$, $t \neq 0$, but $h(0) = 0$. So

$$\lim_{t \to 0} \frac{h(t) - h(0)}{t} = \lim_{t \to 0} \frac{1}{2t} \tag{8.22}$$

The limit does not exist (because t may be positive or negative) and so $Df(0)$ cannot exist: for if $Df(0)$ existed, by (8.12) every $D_u f(0)$ would exist and be zero.

Example 8.4 Even the existence of (finite) directional derivatives $D_u f(x_0)$ for all u does not imply that $Df(x_0)$ will exist, as is shown by the function

$$f(x) = \begin{cases} x_1 x_2^2/(x_1^2 + x_2^4), & x_1^2 + x_2^2 \neq 0 \\ 0, & x_1^2 + x_2^2 = 0 \end{cases} \tag{8.23}$$

for which if $u = \begin{pmatrix} \alpha \\ \beta \end{pmatrix}$, with $\alpha^2 + \beta^2 = 1$,

$$D_u f(0) = \lim_{t \to 0} \frac{\alpha \beta^2}{\alpha^2 + \beta^4 t^2} \tag{8.24}$$

$$= \begin{cases} \left(\dfrac{\beta^2}{\alpha}\right), & \alpha \neq 0 \\ 0, & \alpha = 0 \end{cases} \tag{8.25}$$

Once again (8.12) is not satisfied, and so $Df(0)$ does not exist.

When the range R of f is E_1, the matrix J in (8.19) becomes the row vector

$$\left(\frac{\partial f}{\partial x_1}, \frac{\partial f}{\partial x_2}, \ldots, \frac{\partial f}{\partial x_n} \right) \tag{8.26}$$

This occurs so often in physical applications that it is given a special name: it (or sometimes its transpose) is called the *gradient* of f, and is usually written ∇f (pronounced del f) or grad f. In this special case equation (8.12) becomes

$$D_u f(x_0) = \nabla f(x_0) u \tag{8.27}$$

8.3 The chain rule

It was shown in Section 7.3 that the composite of two continuous functions is itself continuous. An important theorem concerning the derivative of a composite function is now given.

Theorem 8.2 (The chain rule) Let f be a continuous function with domain an open neighbourhood $N(x^0) \subseteq E_n$ and range in E_m. Let $y^0 = f(x^0)$ and let g be a continuous function with domain an open neighbourhood $N(y^0) \subseteq E_m$ and range in E_l. If f is differentiable at x^0 and g is differentiable at y^0 then the composite function $h = gf$ is differentiable at x^0 and

$$Dh = Dg\, Df \tag{8.28}$$

where $Dg\, Df$ is to be regarded as the product of two matrices.

Proof Since f is differentiable at x^0 and g is differentiable at y^0 it follows that

$$f(x^0 + x) = f(x^0) + f'(x^0)x + \epsilon_1 \tag{8.29}$$

and

$$g(y^0 + y) = g(y^0) + g'(y^0)y + \epsilon_2 \tag{8.30}$$

where $x \in N'(x^0)$, $y \in N'(y^0)$, and $\|\epsilon_1\| \to 0$, $\|\epsilon_2\| \to 0$ as $x \to x^0$ and $y \to y^0$ respectively. Therefore, using (8.29) and (8.30),

$$\begin{aligned} h(x^0 + x) &= g\big(f(x^0 + x)\big) \\ &= g\big(f(x^0) + f'(x^0)x + \epsilon_1\big) \\ &= g(y^0) + g'(y^0)\big(f'(x^0) + \epsilon_1\big) + \epsilon_2 \end{aligned}$$

that is,

$$h(x^0 + x) = g(y^0) + g'(y^0)f'(x^0) + \epsilon_3 \tag{8.31}$$

where $\epsilon_3 = g'(y^0)\epsilon_1 + \epsilon_2$.

The required result is obtained directly from (8.31) because $g'(y^0)$ is a finite matrix, $y \to y^0$ as $x \to x^0$ by continuity, and so $\|\epsilon_1\|$ and $\|\epsilon_2\|$ tend to zero as $x \to x^0$.

Suppose that

$$\begin{aligned} x^0 &= (x_1^0, x_2^0, \ldots, x_n^0) \\ y^0 &= (y_1^0, y_2^0, \ldots, y_m^0) \\ f &= (f_1, f_2, \ldots, f_m) \\ g &= (g_1, g_2, \ldots, g_l) \\ h &= (h_1, h_2, \ldots, h_l) \end{aligned}$$

then equation (8.28) becomes

$$\begin{pmatrix} \dfrac{\partial h_1}{\partial x_1} & \dfrac{\partial h_1}{\partial x_2} & \cdots & \dfrac{\partial h_1}{\partial x_n} \\ \dfrac{\partial h_2}{\partial x_1} & \dfrac{\partial h_2}{\partial x_2} & \cdots & \dfrac{\partial h_2}{\partial x_n} \\ \vdots & \vdots & & \vdots \\ \dfrac{\partial h_l}{\partial x_1} & \dfrac{\partial h_l}{\partial x_2} & \cdots & \dfrac{\partial h_l}{\partial x_n} \end{pmatrix} = \begin{pmatrix} \dfrac{\partial g_1}{\partial y_1} & \dfrac{\partial g_1}{\partial y_2} & \cdots & \dfrac{\partial g_1}{\partial y_m} \\ \dfrac{\partial g_2}{\partial y_1} & \dfrac{\partial g_2}{\partial y_2} & \cdots & \dfrac{\partial g_2}{\partial y_m} \\ \vdots & \vdots & & \vdots \\ \dfrac{\partial g_l}{\partial y_1} & \dfrac{\partial g_l}{\partial y_2} & \cdots & \dfrac{\partial g_l}{\partial y_m} \end{pmatrix} \begin{pmatrix} \dfrac{\partial f_1}{\partial x_1} & \dfrac{\partial f_1}{\partial x_2} & \cdots & \dfrac{\partial f_1}{\partial x_n} \\ \dfrac{\partial f_2}{\partial x_1} & \dfrac{\partial f_2}{\partial x_2} & \cdots & \dfrac{\partial f_2}{\partial x_n} \\ \vdots & \vdots & & \vdots \\ \dfrac{\partial f_m}{\partial x_1} & \dfrac{\partial f_m}{\partial x_2} & \cdots & \dfrac{\partial f_m}{\partial x_n} \end{pmatrix} \tag{8.32}$$

from which the formulae

$$\frac{\partial h_i}{\partial x_j} = \sum_{k=1}^{m} \frac{\partial g_i}{\partial y_k} \frac{\partial f_k}{\partial x_j} \tag{8.33}$$

for $i = 1, 2, \ldots, l$ and $j = 1, 2, \ldots, n$, can be obtained.

In the particular case when $h(x)$ is a scalar the gradient of h can be expressed as

$$\nabla h = Dg\, Df = \nabla g Df \tag{8.34}$$

or, in components

$$\frac{\partial h}{\partial x_i} = \frac{\partial g}{\partial y_1}\frac{\partial f_1}{\partial x_i} + \frac{\partial g}{\partial y_2}\frac{\partial f_2}{\partial x_i} + \cdots + \frac{\partial g}{\partial y_m}\frac{\partial f_m}{\partial x_i} \tag{8.35}$$

for $i = 1, 2, \ldots, n$.

EXERCISE 8.3 Give an independent proof of the chain rule for real-valued functions of a single variable using the definition of Section 8.1.

EXERCISE 8.4 Let f and g be continuous functions with a common domain S, which is an open subset of E_n, and with ranges in E_m. Prove that if f and g are differentiable at some point $x^0 \in S$, and if α and β are scalars, then $\alpha f + \beta g$ is also differentiable at x^0 and has derivative $\alpha Df(x^0) + \beta Dg(x^0)$.

Example 8.5 Let $g(r, \theta) = f(x, y)$ where $x = r\cos\theta$ and $y = r\sin\theta$. Then

$$\begin{aligned}\frac{\partial g}{\partial r} &= \frac{\partial f}{\partial x}\frac{\partial x}{\partial r} + \frac{\partial f}{\partial y}\frac{\partial y}{\partial r}\\ &= \cos\theta\frac{\partial f}{\partial x} + \sin\theta\frac{\partial f}{\partial y}\end{aligned} \tag{8.36}$$

and

$$\begin{aligned}\frac{\partial g}{\partial \theta} &= \frac{\partial f}{\partial x}\frac{\partial x}{\partial \theta} + \frac{\partial f}{\partial y}\frac{\partial y}{\partial \theta}\\ &= -r\sin\theta\frac{\partial f}{\partial x} + r\cos\theta\frac{\partial f}{\partial y}\end{aligned} \tag{8.37}$$

Equations (8.36) and (8.37) can be written

$$\begin{pmatrix}\dfrac{\partial g}{\partial r}\\[2ex] \dfrac{\partial g}{\partial \theta}\end{pmatrix} = \begin{pmatrix}\cos\theta & \sin\theta\\[2ex] -r\sin\theta & r\cos\theta\end{pmatrix}\begin{pmatrix}\dfrac{\partial f}{\partial x}\\[2ex] \dfrac{\partial f}{\partial y}\end{pmatrix} \tag{8.38}$$

which is a particular case of (8.32).

8.4 Mean-value theorem

The following theorem is often required in subsequent proofs.

Theorem 8.3 (Mean-value theorem) Let $I = [a, b]$ be a compact interval in E_1, f a continuous function with domain I and range in E_m, and g a continuous function with domain I and range in E_1. If f and g both have a derivative at each point ξ of (a, b) and $\|f'(\xi)\| \leqslant g'(\xi)$ on (a, b) then

$$\|f(b) - f(a)\| \leqslant g(b) - g(a) \tag{8.39}$$

Proof Let S be the subset of I for which if $\xi \in S$ and $\eta \in [a, \xi)$ then $\|f(\eta) - f(a)\| \leqslant g(\eta) - g(a) + \epsilon(\eta - a)$, where $\epsilon > 0$. S is non-vacuous since $a \in S$, and if $\xi \in S$ then any $\delta \in [a, \xi)$ is also in S by definition. If now γ is the supremum of S then S must either be $[a, \gamma]$ or $[a, \gamma)$, but from the definition it is $[a, \gamma]$.

Using the continuity of f and g it follows that

$$\|f(\gamma) - f(a)\| \leqslant g(\gamma) - g(a) + \epsilon(\gamma - a) \tag{8.40}$$

Assume, contrary to what is to be proved, that $\gamma < b$; then, from the definition of derivative, there must be an interval $[\gamma, \gamma + \lambda] \subset I$ such that if $\eta \in [\gamma, \gamma + \lambda]$ then

$$\|f(\eta) - f(\gamma) - f'(\gamma)(\eta - \gamma)\| \leqslant \frac{\epsilon}{2}(\eta - \gamma) \tag{8.41}$$

and

$$|g(\eta) - g(\gamma) - g'(\gamma)(\eta - \gamma)| \leqslant \frac{\epsilon}{2}(\eta - \gamma) \tag{8.42}$$

Hence it follows that

$$\begin{aligned}\|f(\eta) - f(\gamma)\| &\leqslant \|f'(\gamma)\|\,(\eta - \gamma) + \frac{\epsilon}{2}(\eta - \gamma)\\ &\leqslant g'(\gamma)(\eta - \gamma) + \frac{\epsilon}{2}(\eta - \gamma)\\ &\leqslant g(\eta) - g(\gamma) + \epsilon(\eta - \gamma)\end{aligned} \tag{8.43}$$

On using (8.40) and (8.43) there results

$$\|f(\eta) - f(a)\| \leqslant g(\eta) - g(a) + \epsilon(\eta - a) \tag{8.44}$$

which is a contradiction of the definition of γ. Thus $\gamma = b$ and the theorem follows from (8.40) and the fact that $\epsilon > 0$ is arbitrary.

Corollary 1 If $g(\xi)$ is taken as $M(\xi - a)$ with M positive and if $\|f'(\xi)\| \leqslant M$ on (a, b) then

$$\|f(b) - f(a)\| \leqslant M(b - a) \tag{8.45}$$

Corollary 2 Let f be a continuous function with domain S in E_n and range R in E_m and let a and b be two points of S. Then if the open line segment joining a to b lies entirely inside S, and f is differentiable at every point on this open line segment,

$$\|f(b) - f(a)\| \leqslant \|b - a\| \sup_{0<t<1} \|Df(a + t(b - a))\| \tag{8.46}$$

Proof The result is a direct consequence of the previous corollary since the function $h(t) = f(a + t(b - a))$ with domain $[0, 1]$ and range in E_m is differentiable at every point of $(0, 1)$ and has $Df(a + t(b - a))\,(b - a)$ as its derivative. The matrix norm in (8.46) must of course be consistent with the (euclidean) vector norm (Chapter 3, Section 10).

Note that Corollary 2 of Theorem 8.3 does not require that Df should be continuous on (a, b), nor that Df should exist at a or at b. If, in fact, Df exists and is continuous on $[a, b]$, $\sup_{0<t<1}$ in (8.46) can be replaced by $\max_{0\leqslant t\leqslant 1}$. Note also that if the open line segment joining a to b lies entirely within S whenever a and b are in S, then S is said to be *convex*. Consequently if S is known to be convex, the condition regarding the line segment in Corollary 2 is certainly satisfied.

EXERCISE 8.5 Let g be a continuous function with domain a closed finite interval $I = [a, b]$ in E_1 and range $R \subset E_1$. Prove that if g has a derivative at every point ξ of (a, b) and $m < g'(\xi) < M$ then $m(b - a) \leqslant g(b) - g(a) \leqslant M(b - a)$. Show, in addition, that $m(b - a) < g(b) - g(a) < M(b - a)$ except when $g(\xi) = g(a) + m(\xi - a)$, or $g(\xi) = g(a) + M(\xi - a)$ for all values of ξ in I.

EXERCISE 8.6 Prove that Theorem 8.3 still holds when there is a denumerable set $P = \{p_n\}$ of points in I at which f and g do not both possess derivatives. [Form a one-to-one correspondence $n \to p_n$ between the integers and the points $p_n \in S$ and add $\epsilon \sum_{p_n<\eta} 2^{-n}$ to the expression defining the set S in Theorem 8.3. Consider the cases $\gamma \notin P$ and $\gamma \in P$ separately.]

Consideration is now given to the important special case in which f is a real-valued function.

Theorem 8.4 (Rolle's theorem) Let f be real-valued, continuous on the closed interval $[a, b] \subset E_1$ and differentiable on the open interval (a, b). Let $f(a) = f(b) = 0$. Then for some point $x_0 \in (a, b)$ the derivative $f'(x_0) = 0$.

Proof Since f is continuous on $[a, b]$ it attains its maximum and minimum values on $[a, b]$ (Theorem 7.2). If both maximum and minimum values occur at end points of $[a, b]$ then the function is identically zero on $[a, b]$ and the theorem is trivial.

Suppose, then, that f has its maximum at $x_0 \in (a, b)$; then

$$f(x_0 + h) - f(x_0) \leqslant 0 \quad \text{for all } x_0 + h \in [a, b]$$

This means that $[f(x_0 + h) - f(x_0)]/h$ is negative when h is positive, and positive when h is negative. It follows, since the limit exists, that

$$f'(x_0) = \lim_{h \to 0} \frac{[f(x_0 + h) - f(x_0)]}{h} = 0$$

A similar argument holds if $f(x_0)$ is a minimum.

It is easily seen that Theorem 8.4 holds also when $f(a) = f(b) = c \neq 0$ by considering $f(x) - c$ instead of $f(x)$. The following theorem gives the form of the mean-value theorem which is most often quoted.

Theorem 8.5 (Mean-value theorem for real-valued functions of a single variable) If $f(x)$ is continuous on the closed interval $[a, b]$ and differentiable on the open interval (a, b) then for some point $x_0 \in (a, b)$

$$f(b) - f(a) = (b - a)f'(x_0) \tag{8.47}$$

Proof The proof follows directly from applying Rolle's theorem to the function

$$G(x) = f(b) - f(x) - \frac{b - x}{b - a}[f(b) - f(a)]$$

which satisfies the required conditions and has derivative

$$G'(x) = \frac{f(b) - f(a)}{b - a} - f'(x)$$

Notice that it is possible to write (8.47) in the equivalent form

$$f(b) - f(a) = (b - a)f'\big(a + \theta(b - a)\big) \tag{8.48}$$

where $0 < \theta < 1$.

Theorem 8.6 A real-valued function f which is continuous on $[a, b]$ and differentiable on (a, b) is constant if $f'(x) = 0$ for all $x \in (a, b)$, strictly decreasing if $f'(x) < 0$ on (a, b), and strictly increasing if $f'(x) > 0$ on (a, b).

Proof This follows very simply from equation (8.48) in the previous theorem and is left for the reader to verify.

Exercise 8.7 Prove that if two continuous functions f and g on $[a, b]$ have finite derivatives and $f'(x) = g'(x)$ for all $x \in (a, b)$ then $f - g$ is constant on $[a, b]$.

EXERCISE 8.8 Let the real-valued functions f and g be continuous on $[a, b]$ and differentiable on (a, b). Suppose that $g(b) \neq g(a)$, and that $f'(x)$ and $g'(x)$ cannot vanish simultaneously for any $x \in (a, b)$. Prove that there exists an $x_0 \in (a, b)$ for which

$$\frac{f(b) - f(a)}{g(b) - g(a)} = \frac{f'(x_0)}{g'(x_0)}$$

[Apply Rolle's theorem to a suitable function.]

EXERCISE 8.9 (Mean-value theorem for real-valued functions with domain in E_n) Let f be a real-valued function differentiable everywhere on an open set $S \subseteq E_n$ and let a and b be two points in S. Show that if the open line segment [the set of points $\theta a + (1 - \theta)b$, with $0 < \theta < 1$] joining a and b lies entirely within S there is a point c on this segment such that

$$f(b) - f(a) = \nabla f(c)(b - a)$$

[Apply Theorem 8.3 to the function of a single variable given by $g(t) = f(a + tu)$ with $0 < t < \|b - a\|$ and $u = (b - a)/\|b - a\|$.]

EXERCISE 8.10 Restate the result in Exercise 8.9 in a form suitable for a real-valued function f defined on a set $S \subseteq E_n$ which is not necessarily open.

8.5 Relation with continuity

If a function f has a derivative at x_0, then it is continuous at x_0. For by definition there exists a finite matrix J such that

$$\lim_{\substack{x \to x_0 \\ x \in S}} \frac{\|f(x) - f(x_0) - J(x - x_0)\|}{\|x - x_0\|} = 0 \tag{8.49}$$

whence

$$\lim_{\substack{x \to x_0 \\ x \in S}} \|f(x) - f(x_0) - J(x - x_0)\| = 0 \tag{8.50}$$

and so

$$\begin{aligned} \lim_{\substack{x \to x_0 \\ x \in S}} \|f(x) - f(x_0)\| &= \lim_{\substack{x \to x_0 \\ x \in S}} J(x - x_0) \\ &= 0 \end{aligned} \tag{8.51}$$

That is to say, f is continuous at x_0.

In particular, if $S \subseteq E_1$, the existence of finite $Df(x_0)$ implies the continuity of f at x_0. In all the definitions of a derivative given so far, it has been insisted that J [which when $S \subseteq E_1$ is just the vector $Df(x_0)$] should be finite. It is now convenient to relax this requirement when $S \subseteq E_1$. If $S \subseteq E_1$ and

$$\lim_{\substack{x \to x_0 \\ x \in S}} \frac{f(x) - f(x_0)}{x - x_0} \tag{8.52}$$

exists, then provided that f is continuous at x_0, $Df(x_0)$ will be said to exist whether the limit in (8.52) is finite or not. The requirement that f should be continuous at x_0 is superfluous if the limit (8.52) is finite, but otherwise is essential.

Example 8.6 The real function f: $f(x) = x^{1/3}$ is continuous at $x = 0$, and

$$\lim_{x\to 0} \frac{x^{1/3}}{x} = +\infty \tag{8.53}$$

The function will therefore be said to have derivative $+\infty$ at $x = 0$. On the other hand the function defined by

$$f(x) = \begin{cases} -1, & x < 0 \\ 0, & x = 0 \\ 1, & x > 0 \end{cases} \tag{8.54}$$

has

$$\lim_{x\to 0} \frac{f(x) - f(0)}{x} = \lim_{x\to 0} \frac{1}{|x|} = +\infty \tag{8.55}$$

yet because f is discontinuous at $x = 0$ there is no derivative there.

With this extended definition of a derivative when $S \subseteq E_1$, the existence of the derivative at x_0 still implies continuity at x_0.

In Section 8.2 it was seen that the existence of all partial derivatives $D_i f(x_0)$, or even the existence of all directional derivatives $D_u f(x_0)$, did not (when $S \subseteq E_n$, $n \geqslant 2$) imply that $Df(x_0)$ existed. These conditions do not even ensure continuity at x_0, as is shown by the following example.

Example 8.7 The function f of Example 8.4

$$f(x) = x_1 x_2^2/(x_1^2 + x_2^4) \tag{8.56}$$

is discontinuous at $x = 0$. For when $x_1 = x_2^2 \neq 0$, $f(x) = \frac{1}{2}$, yet when $x = 0$, $f(x) = 0$.

Of course, when $S \subseteq E_1$ the existence of $D_1 f(x_0)$ is the same as the existence of $Df(x_0)$, which implies continuity.

8.6 Partial derivatives in an open region

Though the existence of all partial derivatives at a point a does not imply differentiability, nor even continuity, it is highly desirable to have conditions based on the partial derivatives from which differentiability or continuity can be deduced. Otherwise it becomes tedious to verify in particular cases that a function is differentiable or continuous. The desired conditions can be found by requiring that the partial derivatives exist not only at a, but in an open region containing a. The two most useful results are the following.

Theorem 8.7 Let f be a function with $S \subseteq E_n$ and $R \subseteq E_m$. Let the partial derivatives $D_1f, D_2f, \ldots, D_nf$ exist in a neighbourhood $N(a, h) \subset S$ and be continuous at a. Then these conditions are sufficient to ensure that f has a derivative $Df(a)$ at a.

Proof If the derivative exists it is given by equation (8.19). What has to be proved is that with J given by this equation, equation (8.2) is true.

Let J_a, J_x denote the result of evaluating the partial derivatives in equation (8.19) at $a \in N(a, h) \subset S$, $x \in N(a, h) \subset S$ respectively. Apply Corollary 2 of Theorem 8.3 to the function

$$q(x) = f(x) - J_a x \tag{8.57}$$

Because the jacobian matrix of x is I_n, this gives

$$\|f(x) - f(a) - J_a(x - a)\| \leqslant \|x - a\| \sup_{0<t<1} \|J_{a+t(x-a)} - J_a\| \tag{8.58}$$

where the matrix norm (Chapter 3, Section 10) is

$$\|A\| = \sqrt{\{\lambda_{\max}(A^{T}A)\}} \tag{8.59}$$

Now by the continuity of the partial derivatives at a, it follows that for $z \in N(a, h) \subset S$, all the elements of $J_z - J_a$ tend to zero as $z \to a$. By Gershgorin's theorem (Chapter 3, Section 8.3) and (8.59) this implies that

$$\lim_{z \to a} \|J_z - J_a\| = 0 \tag{8.60}$$

That is, for any ε there exists $\delta < h$ such that

$$z \in N'(a, \delta) \subset S \Rightarrow \|J_z - J_a\| < \epsilon \tag{8.61}$$

which further implies that

$$\lim_{x \to a} \left\{ \sup_{0<t<1} \|J_{a+t(x-a)} - J_a\| \right\} = 0 \tag{8.62}$$

and so

$$\lim_{x \to a} \frac{\|f(x) - f(a) - J_a(x - a)\|}{\|x - a\|} = 0 \tag{8.63}$$

which is equation (8.2). This completes the proof of sufficiency.

Under the conditions of Theorem 8.7, $f(x)$ is differentiable at a, and continuity follows from Section 8.5. A weaker condition than the continuity of the partial derivatives is, however, sufficient to ensure continuity of f at a, as in the following theorem.

Theorem 8.8 Let f map $S \subseteq E_n$ onto $R \subseteq E_m$. Let all the partial derivatives $\partial f_i/\partial x_j$ exist and satisfy $|\partial f_i/\partial x_j| \leqslant M$ in a neighbourhood $N(a, h) \subset S$. Then f is continuous at a.

Proof Let $e_1, e_2, \ldots, e_n$ be the vectors

$$e_1 = \begin{pmatrix} 1 \\ 0 \\ \vdots \\ 0 \end{pmatrix}, \quad e_2 = \begin{pmatrix} 0 \\ 1 \\ \vdots \\ 0 \end{pmatrix}, \quad \ldots, \quad e_n = \begin{pmatrix} 0 \\ 0 \\ \vdots \\ 1 \end{pmatrix} \tag{8.64}$$

and define vectors $y_0, y_1, \ldots, y_n$ by

$$\begin{aligned} y_0 &= a \\ y_1 &= y_0 + e_1^{\mathrm{T}}(x - a)e_1 \\ y_2 &= y_1 + e_2^{\mathrm{T}}(x - a)e_2 \\ &\vdots \\ y_n &= y_{n-1} + e_n^{\mathrm{T}}(x - a)e_n = x \in N(a, \delta) \subset N(a, h) \end{aligned} \tag{8.65}$$

Then

$$\|f(x) - f(a)\| = \left\| \sum_{j=1}^{n} [f(y_j) - f(y_{j-1})] \right\| \leqslant \sum_{j=1}^{n} \|f(y_j) - f(y_{j-1})\| \tag{8.66}$$

and $y_j - y_{j-1} = e_j^{\mathrm{T}}(x - a)e_j$; that is to say, y_j and y_{j-1} differ only in their jth elements. By Corollary 1 of Theorem 8.3,

$$\|f(y_j) - f(y_{j-1})\| \leqslant M\delta$$

so that

$$\|f(x) - f(a)\| \leqslant nM\delta \tag{8.67}$$

and so $f(x) \to f(a)$ as $x \to a$. That is, f is continuous at a.

8.7 Higher derivatives

If f is a real-valued function with domain an open set $S \subseteq E_n$ and the partial derivatives $D_1 f, D_2 f, \ldots, D_n f$ exist everywhere in S they are themselves functions defined on S. It is possible, therefore, to consider the partial derivatives of the partial derivatives, when they exist. They are called the *second-order partial derivatives of f*. The process can obviously be repeated to give partial derivatives of third and higher orders. The partial derivative of $D_i f$ with respect to x_j, where $x = (x_1, x_2, \ldots, x_n)^T \subset S$, is written $D_{ji}f$ or $\partial^2 f/\partial x_j \partial x_i$. There are n^2 possible second partial derivatives of f and n^k partial derivatives of order k.

The following important theorem will be used in discussing maxima and minima later in this chapter.

Theorem 8.9 Let $f(x_1, x_2)$ be real-valued and defined on an open subset S of E_2 and let $D_1 f$, $D_2 f$, and $D_{2,1}f$ be continuous in some neighbourhood $N(a)$ of a point $a = \begin{pmatrix} a_1 \\ a_2 \end{pmatrix}$ in S. Then the mixed partial derivative $D_{1,2}f$ also exists at a and is equal to $D_{2,1}f(a)$.

Proof Let $x = \begin{pmatrix} x_1 \\ x_2 \end{pmatrix}$ be a point in $N(a)$; then

$$D_2 f(a_1 + h, a_2) - D_2 f(a_1, a_2) = \lim_{k \to 0} \frac{f(a_1 + h, a_2 + k) - f(a_1 + h, a_2)}{k} - \lim_{k \to 0} \frac{f(a_1, a_2 + k) - f(a_1, a_2)}{k} \quad (8.68)$$

For any fixed k, define

$$g_k(t) \equiv f(a_1 + t, a_2 + k) - f(a_1 + t, a_2) \quad (8.69)$$

Then (8.68) can be written

$$D_2 f(a_1 + h, a_2) - D_2 f(a_1, a_2) = \lim_{k \to 0} \frac{g_k(h) - g_k(0)}{k} \quad (8.70)$$

Using the mean-value theorem, the right-hand side of (8.70) can be written

$$\lim_{k \to 0} \frac{h}{k} g_k'(h^*) \quad (8.71)$$

where h^* lies between 0 and h and in general depends on k.

Now

$$g_k'(t) = D_1 f(a_1 + t, a_2 + k) - D_1 f(a_1 + t, a_2) \quad (8.72)$$

so from (8.70), (8.71), and (8.72) there follows

$$\frac{D_2 f(a_1 + h, a_2) - D_2 f(a_1, a_2)}{h} = \lim_{k \to 0} \frac{D_1 f(a_1 + h^*, a_2 + k) - D_1 f(a_1 + h^*, a_2)}{k} \quad (8.73)$$

Again, using the mean-value theorem the right-hand side of (8.73) can be written

$$\lim_{k \to 0} D_{2,1}f(a_1 + h^*, a_2 + k^*) \tag{8.74}$$

where k^* lies between 0 and k.

Since $D_{2,1}f$ is continuous there exists $N(a, \delta)$ such that

$$|D_{2,1}f(x_1, x_2) - D_{2,1}f(a_1, a_2)| < \epsilon \tag{8.75}$$

for any given $\epsilon > 0$, and all $(x_1, x_2) \in N(a, \delta)$. But $(a_1 + h^*, a_2 + k^*) \in N(a, \delta)$ if $|h| < \delta/2$ and $|k| < \delta/2$. Hence if h is kept fixed with $|h| < \delta/2$ then, if $|k| < \delta/2$,

$$|D_{2,1}f(a_1 + h^*, a_2 + k^*) - D_{2,1}f(a_1, a_2)| < \epsilon \tag{8.76}$$

Let $k \to 0$ in (8.76) to give

$$|\lim_{k \to 0} D_{2,1}f(a_1 + h^*, a_2 + k^*) - D_{2,1}f(a_1, a_2)| < \epsilon \tag{8.77}$$

for $h < \delta/2$. This last result simply means, however, that

$$\lim_{h \to 0} \{\lim_{k \to 0} D_{2,1}\, f(a_1 + h^*, a_2 + k^*)\} = D_{2,1}f(a_1, a_2) \tag{8.78}$$

The proof of the theorem now follows by noting, from (8.73) and (8.74), that the left-hand side of (8.78) is just $D_{1,2}f(a_1, a_2)$.

EXERCISE 8.11 Why is it not possible to shorten the proof of the theorem by writing

$$\lim_{k \to 0} D_{2,1}f(a_1 + h^*, a_2 + k^*) = D_{2,1}f(a_1 + h^*, a_2)$$

and then letting $h \to 0$? [h^* depends on k.]

EXERCISE 8.12 Show that if f is a real-valued function with domain an open set $S \subseteq E_n$ then $D_{ij}f(x) = D_{ji}f(x)$, for $i, j = 1, 2, \ldots, n$, provided that all the second-order partial derivatives $D_{ij}(x)$ are continuous in some neighbourhood of $x \in S$.

If the conditions of Exercise 8.12 are satisfied the matrix H with elements $h_{ij} = D_{ij}f$ is symmetric. The matrix H is called the *hessian matrix* of f.

Example 8.8 The following somewhat artificial example is often used to show that $\partial^2 f/\partial x_i \partial x_j$ is not always equal to $\partial^2 f/\partial x_j \partial x_i$. Let $f(x, y) = xy(x^2 - y^2) \div (x^2 + y^2)$ when x and y are not both zero and let $f(0, 0) = 0$. Then at the origin

$$D_1 f(0, 0) = \lim_{h \to 0} \frac{f(h, 0) - f(0, 0)}{h} = 0$$

and

$$D_2f(0, 0) = \lim_{k\to 0} \frac{f(0, k) - f(0, 0)}{k} = 0$$

On the other hand

$$D_1f(0, y) = -y, \quad y \neq 0$$
$$D_2f(x, 0) = x, \quad x \neq 0$$

Thus

$$D_{12}f(0, 0) = \lim_{h\to 0} \frac{D_2f(h, 0) - D_2f(0, 0)}{h}$$
$$= 1$$

and similarly

$$D_{21}f(0, 0) = -1$$

showing that the mixed second-order partials are not equal at (0, 0).

8.8 Taylor's theorem (General mean-value theorem)

The mean-value theorem in the form given in Theorem 8.5 shows that if f is continuous on $[a, b]$ and differentiable on (a, b) there is an $x_0 \in (a, b)$ such that $f(b) - f(a) = (b - a)f'(x_0)$ or, writing $x_0 = a + \theta(b - a)$ where $0 < \theta < 1$,

$$f(b) - f(a) = (b - a)f'\{a + \theta(b - a)\} \tag{8.79}$$

It is possible to extend this result in a fairly obvious way by assuming that f' is itself continuous on $[a, b]$ and differentiable on (a, b).

Consider the function g:

$$g(x) = f(b) - f(x) - (b - x)f'(x) - \left(\frac{b - x}{b - a}\right)^2 \{f(b) - f(a) - (b - a)f'(a)\} \tag{8.80}$$

It is easily seen that $g(a) = g(b) = 0$ and

$$g'(x) = \frac{2(b - x)}{(b - a)^2} \{f(b) - f(a) - (b - a)f'(a) - \tfrac{1}{2}(b - a)^2 f''(x)\} \tag{8.81}$$

Applying the mean-value theorem to g gives

$$f(b) = f(a) + (b - a)f'(a) + \tfrac{1}{2}(b - a)^2 f''(a + \theta_1(b - a)) \tag{8.82}$$

where $0 < \theta_1 < 1$.

This process can be repeated and leads to the following important theorem.

Theorem 8.10 (Taylor's theorem) Let f be a real-valued function with domain the closed interval $[a, b]$ and such that $f^{(n-1)}(x)$ is continuous on $[a, b]$ and differentiable on (a, b). Then

$$f(b) = f(a) + (b-a)f'(a) + \frac{1}{2!}(b-a)^2 f''(a) + \cdots$$
$$\cdots + \frac{1}{(n-1)!}(b-a)^{n-1}f^{(n-1)}(a) + \frac{1}{n!}(b-a)^n f^{(n)}(a + \theta_n(b-a)) \tag{8.83}$$

where $0 < \theta_n < 1$.

Proof Consider the function

$$g(x) = g_n(x) - \left(\frac{b-x}{b-a}\right)^n g_n(a) \tag{8.84}$$

where

$$g_n(x) = f(b) - f(x) - (b-x)f'(x) - \cdots - \frac{(b-x)^{n-1}}{(n-1)!} f^{(n-1)}(x) \tag{8.85}$$

Because $f^{(n-1)}$ is continuous on $[a, b]$, so are $f^{(n-2)}, f^{(n-3)}, \ldots, f$, by Section 8.3, and therefore so is g.

A direct evaluation shows that $g(a) = g(b) = 0$ and

$$g'(x) = \frac{n(b-x)^{n-1}}{(b-a)^n}\left\{g_n(a) - \frac{(b-a)^n}{n!} f^{(n)}(x)\right\} \tag{8.86}$$

Hence there is a θ_n, with $0 < \theta_n < 1$, such that $g'(a + \theta_n(b-a)) = 0$. This proves the theorem.

If in equation (8.83) the substitution $b = a + h$ is made it becomes

$$f(a+h) = f(a) + hf'(a) + \frac{1}{2!}h^2 f''(a) + \cdots$$
$$\cdots + \frac{1}{(n-1)!}h^{n-1}f^{(n-1)}(a) + \frac{1}{n!}h^n f^{(n)}(a + \theta_n h) \tag{8.87}$$

This is the form in which Taylor's theorem is usually quoted. If the conditions of the theorem hold on $[c, b]$, with a an interior point of $[c, b]$, h in (8.87) can be allowed to take negative as well as positive values.

Example 8.9 Let $f(x) = 1/x$ and suppose $0 < x < x + h$. Then the conditions of the above theorem are satisfied and $f^{(r)}(x) = (-1)^r r!/x^{r+1}$ giving

$$\frac{1}{x+h} = \frac{1}{x} - \frac{h}{x^2} + \frac{h^2}{x^3} - \cdots + \frac{(-1)^{n-1}h^{n-1}}{x^n} + \frac{(-1)^n h^n}{(x+\theta h)^{n+1}}, \quad 0 < \theta < 1 \tag{8.88}$$

EXERCISE 8.13 Show that

$$\cos(x+h) = \cos x - h\sin x - \frac{h^2}{2!}\cos x + \frac{h^3}{3!}\sin x + \cdots + (-1)^n \frac{h^{2n-1}}{(2n-1)!}\sin x + (-1)^n \frac{h^{2n}}{(2n)!}\cos(x+\theta h) \quad (8.89)$$

where

$$0 < \theta < 1. \quad (8.90)$$

It has been assumed so far in section 8.8 that f has derivatives of appropriate orders not only on (a, b) but also at a and at b. It is easy to see that for $b > a$ only the right-hand derivatives D_+f, $D_+^2 f = D_+(D_+f)$, etc., are important at a, and only the left-hand derivatives at b. So the conditions of Theorem 8.10 can be relaxed to read: let f be a real-valued function with domain the closed interval $[a, b]$ and such that $D_+^{n-1}f$ is continuous on $[a, b)$, $D_-^{n-1}f$ is continuous on $(a, b]$, and $D_+^{n-1}f = D_-^{n-1}f$ and is differentiable on (a, b). Then $f'(a)$, $f''(a)$ etc. in (8.83) can be replaced by $D_+f(a)$, $D_+^2 f(a)$, etc. Since the nth derivative is evaluated in the open interval (a, b) the only point requiring verification is the continuity of the (appropriately modified) function g. This is left as an exercise for the reader.

It is natural to extend Taylor's theorem to real-valued functions with domain S in E_n. The generalization of the preceding paragraph is then not available, and S is most conveniently taken to be open.

Theorem 8.11 Let f have continuous partial derivatives of mth order at each point of an open set $S \subseteq E_n$. If a and b are two distinct points of S such that the line segment joining a and b lies entirely in S then there is an interior point c on this line segment such that

$$f(b) = f(a) + \sum_{k=1}^{m-1} \frac{1}{k!} d^k f(a, b-a) + \frac{1}{m!} d^m f(c, b-a) \quad (8.91)$$

where if $y = (y_1, y_2, \ldots, y_n)^{\mathrm{T}}$,

$$d^k f(x, y) \equiv \sum_{j_1=1}^{n} \sum_{j_2=1}^{n} \cdots \sum_{j_k=1}^{n} D_{j_1 j_2 \ldots j_k} f(x) y_{j_1} y_{j_2} \ldots y_{j_k} \quad (8.92)$$

Proof To prove the theorem it is reduced to a problem in a single variable as follows. Let $g(\theta) = f(\theta b + (1-\theta)a)$ with $0 \leqslant \theta \leqslant 1$; then because f has continuous partial derivatives the chain rule (Theorem 8.2) gives

$$\begin{aligned} g'(\theta) &= \sum_{j_1=1}^{n} D_{j_1} f(\theta b + (1-\theta)a)(b_{j_1} - a_{j_1}) \\ &= d^1 f[\theta b + (1-\theta)a, b-a] \end{aligned} \quad (8.93)$$

$$\begin{aligned} g''(\theta) &= \sum_{j_2=1}^{n} \sum_{j_1=1}^{n} D_{j_2 j_1} f[\theta b + (1-\theta)a](b_{j_1} - a_{j_1})(b_{j_2} - a_{j_2}) \\ &= d^2 f[\theta b + (1-\theta)a, b-a] \end{aligned} \quad (8.94)$$

(because both j_1 and j_2 run through all values from 1 to n) and so on, giving generally

$$g^{(m)}(\theta) = d^m f[\theta b + (1 - \theta)a, b - a] \tag{8.95}$$

Using Theorem 8.10 on $g(1)$ gives

$$g(1) - g(0) = \sum_{k=1}^{m-1} \frac{g^k(0)}{k!} + \frac{1}{m!} g^{(m)}(\theta_1) \tag{8.96}$$

where $0 < \theta_1 < 1$. Since $g(1) = f(b)$ and $g(0) = f(a)$ the result follows directly by substituting from (8.95) into (8.96) and noticing that $\theta_1 b + (1 - \theta_1)a$ is some interior point c on the line segment joining a and b.

Example 8.10 If $b = (x_1 + \xi_1, x_2 + \xi_2)$ and $a = (x_1, x_2)$ then equation (8.91) becomes, for $m = 2$

$$f(x_1 + \xi_1, x_2 + \xi_2) = f(x_1, x_2) + \sum_{i=1}^{2} \xi_i \frac{\partial f(x_1, x_2)}{\partial x_i} + \frac{1}{2!} \sum_{i=1}^{2} \sum_{j=1}^{2} \xi_i \xi_j \frac{\partial^2 f(x_1 + \theta\xi_1, x_2 + \theta\xi_2)}{\partial x_i \partial x_j} \tag{8.97}$$

where $0 < \theta < 1$.

EXERCISE 8.13 Prove that when x and y are small, $y + xy$ is approximately equal to $e^x \sin y$.

EXERCISE 8.14 Show that in the neighbourhood of the origin $\sin x \sin y$ is approximately equal to xy.

8.9 Maxima and minima of continuous functions

Maxima and minima (absolute and relative) of continuous real-valued functions were defined in Section 7.1; their properties are discussed further in this section. First, an important necessary condition is proved.

Theorem 8.12 Let f be a real-valued function with domain the open interval $(a, b) \subset E_1$. If f has a relative maximum or minimum at a point $x_0 \in (a, b)$, and has a derivative there, then $f'(x_0) = 0$.

Proof Suppose that $f'(x_0) > 0$ and is finite; then from the definition of the derivative (Section 8.1) there is a neighbourhood $N(x_0)$ such that

$$\left| \frac{f(x) - f(x_0)}{x - x_0} - f'(x_0) \right| < \epsilon$$

for all x in the deleted neighbourhood $N'(x_0)$ and for any $\epsilon > 0$. This implies that $-\epsilon + f'(x_0) < [f(x) - f(x_0)]/(x - x_0) < \epsilon + f'(x_0)$ and by taking $\epsilon < f'(x_0)$ it follows that $f(x) - f(x_0)$ must have the same sign as $x - x_0$. Hence if $f'(x_0)$ is finite and positive there is some neighbourhood of x_0 in which $f(x)$ is increasing and so $f(x_0)$ cannot be a maximum or minimum.

If, on the other hand, $f'(x_0) = +\infty$ there must be some neighbourhood of x_0 for which $[f(x) - f(x_0)]/(x - x_0) > 1$ in the deleted neighbourhood $N'(x_0)$ and the same conclusion follows.

It can be shown in exactly the same way that f cannot have a maximum or a minimum value at $f(x_0)$ if $f'(x_0)$ is negative (finite or infinite). By hypothesis there is, however, a derivative at x_0 so that $f'(x_0) = 0$.

EXERCISE 8.15 Give an alternative proof of the above theorem using the mean-value theorem directly.

Notice that the condition $f'(x_0) = 0$ is only a necessary condition for a maximum or a minimum. The following simple example shows that it is by no means sufficient.

Example 8.11 Let $f(x) = x^3$, so that $f'(x) = 3x^2$ and $f'(0) = 0$. There is neither a maximum nor a minimum at $x = 0$, however, since x^3 is increasing in any open interval $(-h, h)$.

The following theorem gives sufficient conditions for a maximum or a minimum when the domain of f is in E_1.

Theorem 8.13 Let the domain of a real-valued function f be the open interval (a, b) and suppose that $f^{(n)}(x)$ is continuous on (a, b). If at a point $x_0 \in (a, b)$, $f'(x_0) = f''(x_0) = \cdots = f^{(n-1)}(x_0) = 0$ and $f^{(n)}(x_0) \neq 0$, then if n is an odd integer ($\geqslant 1$) there cannot be a maximum or a minimum at x_0; but if n is even there is a maximum at x_0 when $f^{(n)}(x_0) < 0$, and there is a minimum at x_0 when $f^{(n)}(x_0) > 0$.

Proof This follows directly from Taylor's theorem which under the above conditions reduces to

$$f(x) - f(x_0) = \frac{f^{(n)}(y)}{n!}(x - x_0)^n$$

where x lies in a suitable neighbourhood of x_0 and $y \in N'(x_0, |x - x_0|)$. The details are left as an exercise for the reader.

Example 8.12 Let $f(x) = (x - 1)^3$ so that $f'(x) = 3(x - 1)^2, f''(x) = 6(x - 1)$ and $f^{(3)}(x) = 6$. The first non-vanishing derivative at $x = 1$ is the third so there is no maximum or minimum there (Example 8.10). On the other hand, if $f(x) = (x - 1)^4$ it follows that $f'(1) = f''(1) = f^{(3)}(1) = 0, f^{(4)}(1) = 24 > 0$ and $(x - 1)^4$ has a minimum at $x = 1$.

An extension of the necessary condition of Theorem 8.12 to real-valued functions defined over a set in E_n is now given.

Theorem 8.14 Let the real-valued function f have first partial derivatives $\partial f/\partial x_i$, $i = 1, 2, \ldots, n$, at each point x of an open set $S \subset E_n$. Then for a relative maximum or minimum at the point $a \in S$ a necessary condition is that $\partial f/\partial x_i = 0$ at a, $i = 1, 2, \ldots, n$.

Proof Let $a = (a_1, a_2, \ldots, a_n)^T$; then if $f(a)$ is a maximum (or a minimum) the function $f(x_1, a_2, \ldots, a_n)$ must be a maximum (a minimum) with respect to the variable x_1. Therefore $\partial f/\partial x_1 = 0$ at a, and a similar result applies to the other variables.

EXERCISE 8.16 Give an alternative proof of this theorem based on Taylor's theorem.

A useful sufficient condition for the existence of maxima or minima of functions of more than one variable will now be proved.

Theorem 8.15 Let f have continuous second-order partial derivatives $D_{ij}f$, $i, j = 1, 2, \ldots, n$, on an open set S in E_n. Suppose that all the first-order partial derivatives of f vanish at a point $a \in S$ and that $\det(D_{ij}f)$ is not zero at a. Then f has a relative minimum at a if

$$d_i > 0, \quad i = 1, 2, \ldots, n \tag{8.98}$$

and a relative maximum at a if

$$\left.\begin{array}{ll} d_i > 0, & i = 2, 4, 6, \ldots \\ d_i < 0, & i = 1, 3, 5, \ldots \end{array}\right\} i \leqslant n \tag{8.99}$$

where the d_i are the leading minors

$$d_i = \begin{vmatrix} D_{11}f(a) & D_{12}f(a) & \ldots & D_{1i}f(a) \\ D_{21}f(a) & D_{22}f(a) & \ldots & D_{2i}f(a) \\ \vdots & \vdots & & \vdots \\ D_{i1}f(a) & D_{i2}f(a) & \ldots & D_{ii}f(a) \end{vmatrix} \tag{8.100}$$

Proof Under the conditions of the theorem (using a Taylor expansion of second degree)

$$f(a + \xi) - f(a) = \xi^T H(a + \theta\xi)\xi \tag{8.101}$$

where H is the hessian matrix, $0 < \theta < 1$, and $a + \xi$ is a point in some neighbourhood $N(a)$ which lies entirely in S.

If (8.98) is satisfied the matrix $H(a)$, which is symmetric by Exercise 8.12, is positive definite (Chapter 3, Section 9.5) and since $H(y)$ is continuous for $y \in S$ it follows that there must be some neighbourhood $N_1(a) \subset S$ for which $H(y) > 0$, $y \in N_1(a)$. If now $N(a)$ is assumed to be contained in $N_1(a)$ then $a + \theta\xi$ lies in $N_1(a)$ and, from (8.101), $f(a + \xi) - f(a) > 0$ in $N_1'(a)$, showing that a is a local minimum of f.

The conditions for a maximum follow in exactly the same way (using the conditions for negative definiteness in Chapter 3, Section 9.5). In case $H(a)$ is semi-definite (positive or negative) then the Taylor expansion (8.101) is not sufficient to determine whether a is an extreme point of f. Finally when $H(a)$ is indefinite $f(a + \xi) - f(a)$ can be both positive and negative in an arbitrarily small neighbourhood of a and so a cannot be an extreme point of f.

Example 8.13 Let f have continuous second-order partial derivatives at all points in an open set $S \subset E_2$ and let $D_1f(a) = D_2f(a) = 0$ at a point $a \in S$.

Then if

$$\Delta = D_{11}f(a)\, D_{22}f(a) - [D_{12}f(a)]^2 > 0 \tag{8.102}$$

a is a maximum of f if

$$D_{11}f(a) < 0 \tag{8.103}$$

and a minimum of f if

$$D_{11}f(a) > 0 \tag{8.104}$$

If $\Delta < 0$ there can be neither a maximum nor a minimum at a and finally, if $\Delta = 0$, the test is inconclusive and higher order derivatives must be investigated.

For a treatment of maxima and minima where the variables cannot change freely but are subjected to equality constraints the reader is referred to Apostol (1957).

EXERCISE 8.17 Consider the ratio $r(x) = x^*Hx/x^*x$, $x \neq 0$, where H is a hermitian matrix and x a vector with complex components. ($r(x)$ is called a *Rayleigh quotient*.) Prove that $r(x)$ attains maximum and minimum values on the set $\|x\| = 1$ and that these values are respectively the maximum and minimum eigenvalues of H. There may, of course, be more than one greatest or least eigenvalue. [Diagonalize the quadratic form in the numerator as in Section 9.2 of Chapter 3.]

9 Differentiability of physical variables

In the study of dynamical systems we are particularly concerned with vector-valued functions of time. Reasons were given in Section 7.4 for regarding these as continuous except perhaps at a finite number of points in any finite interval, where they have simple discontinuities on the right. For example, the unit step is defined by

$$U(t) = \begin{cases} 0, & t \leqslant 0 \\ 1, & t > 0 \end{cases} \tag{9.1}$$

It is convenient to go further, and to attribute properties of differentiability to such physical variables. The following definitions will be used.

(a) A function f, with domain a (finite or infinite) open interval $I \subseteq E_1$ and range $R \subseteq E_m$, belongs to the class D_R on I if on any bounded open interval $J \subset I$ it can be written

$$f(t) = \sum_{i=1}^{p} f_i U(t - p_i) + \phi(t) \tag{9.2}$$

where the $p_i \in J$ and the f_i are constants, and ϕ is continuous and bounded on J.

(b) A function f, with domain a (finite or infinite) open interval $I \subseteq E_1$ and range $R \subseteq E_m$, belongs to the class D_R^1 on I if on any bounded open interval $J \subset I$ it can bc written

$$f(t) = \sum_{i=1}^{p} f_i U(t - p_i) + \sum_{i=1}^{q} (t - q_i)\phi_i U(t - q_i) + \psi(t) \tag{9.3}$$

where the $p_i \in J$, $q_i \in J$, f_i and ϕ_i are constants and ψ is continuous and bounded on J and has continuous and bounded derivative $D\psi$ on J.

If the interval I is not specified, it will be understood that $I = E_1$. The fact that a function is in D_R or D_R^1 will be signified by $f \in D_R$ or $f \in D_R^1$. The classes D_L and D_L^1 are defined in a similar way, with $U(t - p_i)$, $U(t - q_i)$ replaced by $U(p_i - t)$, $U(q_i - t)$. From the properties of $U(t)$ it follows that $f \in D_R$ has the properties described in Section 7.4. A function $f \in D_L$ has its discontinuities on the left rather than on the right.

If $f \in D_R^1$ on I then it has a left-hand derivative everywhere on $J \subset I$ given by

$$D_- f(t) = \sum_{i=1}^{q} \phi_i U(t - q_i) + D\psi(t) \tag{9.4}$$

as follows at once from the properties of $U(t)$. From equation (9.4) it is clear that $D_- f \in D_R$. At each point in J, except the points $t = q_i$, $i = 1, 2, \ldots, q$, f has a derivative Df equal to $D_- f$. At the points q_i, $D_+ f$ does not exist, and therefore Df does not exist.

Functions which are continuous, or differentiable, everywhere except at a finite set of points in any finite interval, are called *piecewise continuous* or *piecewise differentiable*. The classes D_R and D_L define subsets of the piecewise continuous functions, and the classes D_R^1, D_L^1 define subsets of the piecewise differentiable functions.

9.1 Bounded variation

Functions of bounded variation will be important in later chapters. Consider first a function f with domain $S \subseteq E_1$, range $R \subseteq E_1$. Let $[a, b]$ be an interval in S and let a set of points in the interval be defined by

$$a = x_0 < x_1 < \cdots < x_{n-1} < x_n = b$$

Such a set of points is called a *partition* P of $[a, b]$ (or sometimes a *dissection* of $[a, b]$). Let

$$\Delta f_k = f(x_k) - f(x_{k-1}), \quad k = 1, 2, \ldots, n$$

If there exists a bound M such that $V_P(f) = \sum_{k=1}^{n} |\Delta f_k| < M$ for all choices of P (that is, for all choices of n and of the points $x_1, x_2, \ldots, x_{n-1}$) then f is said to have *bounded variation* on $[a, b]$.

If now a function g has domain $S \subseteq E_1$ and range $R \subseteq E_m$, and if each element of g has bounded variation on an interval $[a, b]$ as just defined, g is said to have bounded variation on $[a, b]$.

An important result which will be used later is the following.

Theorem 9.1 If a function g with domain $S \subseteq E_1$ and range $R \subseteq E_m$ belongs to D_R^1 [or to D_L^1] in an open interval I, then it has bounded variation on $[a, b] \subset I$.

Proof Let f be an element of the vector g. Write, as in (9.2)

$$f(t) = \sum_{i=1}^{p} f_i U(t - p_i) + \phi(t) \tag{9.5}$$

where, by (9.3),

$$\phi(t) = \sum_{i=1}^{q} (t - q_i)\phi_i U(t - q_i) + \psi(t) \tag{9.6}$$

For any given partition P,

$$V_P(f) = \sum_{k=1}^{n} \left| \sum_{i=1}^{p} f_i[U(x_k - p_i) - U(x_{k-1} - p_i)] + \phi(x_k) - \phi(x_{k-1}) \right| \tag{9.7}$$

$$\leqslant \sum_{i=1}^{p} |f_i| + V_P(\phi) \tag{9.8}$$

The introduction of a new point in P cannot reduce the sum $\sum |\Delta\phi|$, because if $x_{i-1} < x_\alpha < x_i$,

$$|\phi(x_i) - \phi(x_\alpha)| + |\phi(x_\alpha) - \phi(x_{i-1})| \geqslant |\phi(x_i) - \phi(x_{i-1})| \tag{9.9}$$

Introduce into P each of those points q_i, at which $D\phi$ is discontinuous, which was not already included; call the result P_1. Within each interval (x_{i-1}, x_i) of P_1, $D\phi$

exists and $|D\phi| < M$, by (9.6) and the boundedness of $D\psi$. Apply Theorem 8.5 to give

$$V_P(\phi) \leqslant V_{P1}(\phi) = \sum |\phi(x_i) - \phi(x_{i-1})| \tag{9.10}$$
$$= \sum |x_i - x_{i-1}|\, |D\phi(y_i)| \tag{9.11}$$
$$\leqslant M\, |b - a| \tag{9.12}$$

where in (9.11), $x_{i-1} < y_i < x_i$, all i.

Inequalities (9.8) and (9.12) show that for any P whatever

$$V_P(f) \leqslant \sum_{i=1}^{p} |f_i| + M\,|b - a| \tag{9.13}$$

so that f has bounded variation on $[a, b]$. With the necessary changes, the proof applies also for $g \in D_L^1$.

In later work the following theorem will be useful because it allows proofs to be very much simplified.

Theorem 9.2 If a function $f\colon S \subseteq E_1 \to R \subseteq E_1$ has bounded variation on $[a, b] \subset S$, then $f = g - h$ where g and h are both non-decreasing on $[a, b]$.

Proof The outline of the proof is as follows. Let P be a partition of $[a, b]$ and let

$$p(x) = \sup_P \sum_a^x |\Delta f_k|$$

where the sum is taken over the intervals of P which lie in $[a, x]$ and for which $\Delta f_k > 0$. The supremum is with respect to all partitions of $[a, b]$. Similarly, let

$$n(x) = \sup_P \sum_a^x |\Delta f_k|$$

where now the $\Delta f_k < 0$.

It is plain from their definitions that p and n are non-decreasing functions of x. The proof of the theorem follows now from the fact that $f(x) = f(a) + p(x) - n(x)$.

Exercise 9.1 Fill in the details of the preceding proof. [For any partition P including x, $f(x) = f(a) + \sum_p |\Delta f_k| - \sum_n |\Delta f_k|$ where $\sum_p$ includes all $\Delta f_k > 0$, $\sum_n$ includes all $\Delta f_k < 0$. Use the properties of limits.]

Note that if $f \in D_R^1$ it has bounded variation on any finite interval. The functions p and n can then be defined for all $x \geqslant a$.

10 Infinite series

Let E be any euclidean space and $\{x_n\}$, for $n \geqslant 0$, a set of elements of E. Now consider the related sequence $\{X_n\}$, $n \geqslant 0$, in which

$$X_n = x_0 + x_1 + \cdots + x_n$$
$$= \sum_{k=0}^{n} x_n$$

The expression $x_0 + x_1 + \cdots$ is identified with the sequence $\{X_n\}$ and is called the *series* with *general term* x_n; X_n is called the *nth partial sum* of the series. Notice that $x_0 = X_0$ and $x_n = X_n - X_{n-1}$ for $n \geqslant 1$.

If $\lim_{n\to\infty} X_n$ exists and is finite, say $\lim_{n\to\infty} X_n = X$, then the sequence X_n is said to *converge* to the sum X. When the sequence is not convergent it is said to *diverge* and when X_n for the divergent sequence does not tend to a limit (that is, to $+\infty$ or $-\infty$) the sequence is said to *oscillate*. When the sequence X_n converges, the series $x_0 + x_1 + \cdots$ is said to *converge*, and we write $X = \sum_{n=0}^{\infty} x_n$ or, where no confusion can arise, $X = \sum x_n$. The series is said to *diverge* when the sequence diverges and to *oscillate* when the sequence oscillates.

The Cauchy condition of Section 6.1 can be easily adapted to infinite series and results in the following.

Theorem 10.1 (Cauchy condition for series) A necessary and sufficient condition for the convergence of the series $\sum_{n=0}^{\infty} x_n$ is that, for every $\epsilon > 0$, there exists an integer N such that $n > N$ implies

$$\|x_{n+1} + x_{n+2} + \cdots + x_{n+r}\| < \epsilon \quad \text{for } r = 1, 2, \ldots$$

Proof This is immediate on applying Theorem 6.1 to the sequence $\{X_n\}$.

Corollary It follows directly from the theorem that the general term $x_n = X_n - X_{n-1}$ of a convergent series must tend to zero as $n \to \infty$.

The following counter-example shows that $\lim_{n\to\infty} x_n = 0$ is not, however, a sufficient condition for convergence. If $x_n = 1/n$ then certainly $\lim_{n\to\infty} x_n = 0$ but taking $n = 2^q$ and $r = 2^q$ gives

$$\sum_{i=n+1}^{n+r} x_i = \frac{1}{2^q + 1} + \frac{1}{2^q + 2} + \cdots + \frac{1}{2^q + 2^q} > \frac{2^q}{2.2^q} = \frac{1}{2}$$

so that the series must diverge by Theorem 10.1.

Theorem 10.2 If $\sum_{n=0}^{\infty} x_n$ and $\sum_{n=0}^{\infty} y_n$ are two convergent series, with sums s and r respectively, then the series $\sum_{n=0}^{\infty} (\alpha x_n + \beta y_n)$ is also convergent and has sum $\alpha s + \beta r$.

Proof

$$\sum_{n=0}^{r} (\alpha x_n + \beta y_n) = \alpha \sum_{n=0}^{r} x_n + \beta \sum_{n=0}^{r} y_n$$

and the result follows directly from Section 6.4.

EXERCISE 10.1 Prove that any two series which have all but a finite number of their terms equal must either both converge or both diverge.

EXERCISE 10.2 The sequence $\{s_n\}$ is said to be *monotonic increasing* (resp. *monotonic decreasing*) if $s_1 \leqslant s_2 \leqslant s_3 \leqslant \cdots \leqslant s_n \leqslant \cdots$ (resp. $s_1 \geqslant s_2 \geqslant s_3 \geqslant \ldots \geqslant s_n \geqslant \cdots$). When the equality signs are not allowed the sequences are said to be *strictly monotonic* (increasing or decreasing). Prove that any monotonic sequence is convergent if and only if it is bounded.

10.1 Absolute and conditional convergence

A series $\sum_{n=0}^{\infty} x_n$ is said to *converge absolutely* when $\sum_{n=0}^{\infty} \|x_n\|$ converges. If, however, $\sum_{n=0}^{\infty} x_n$ converges but $\sum_{n=0}^{\infty} \|x_n\|$ diverges, the series is said to *converge conditionally*.

Theorem 10.3 A necessary and sufficient condition for a series $\sum_{n=0}^{\infty} x_n$ of non-negative real numbers to converge is that its sequence of partial sums be bounded.

Proof Since $x_n \geqslant 0$ it follows that $X_n \geqslant X_{n-1}$ so that the sequence of partial sums is monotonic increasing. The result now follows from Exercise 10.2.

Theorem 10.4 An absolutely convergent series $\sum_{n=0}^{\infty} x_n$ is convergent.

Proof Since $\sum_{n=0}^{\infty} \|x_n\|$ is convergent it follows from the Cauchy condition that for any $\epsilon > 0$ there exists an N for which if $n > N$

$$\|x_n\| + \|x_{n+1}\| + \cdots + \|x_{n+r}\| < \epsilon \quad \text{for } r = 1, 2, \ldots$$

Hence from Chapter 3, Exercise 2.5,

$$\|x_n + x_{n+1} + \cdots + x_{n+r}\| < \epsilon$$

and so $\sum_{n=0}^{\infty} x_n$ is convergent by the Cauchy condition.

EXERCISE 10.3 Show by counter-example that the converse of Theorem 10.4 does not hold.

Theorem 10.5 Any rearrangement of an absolutely convergent series is absolutely convergent and has the same sum as the original series.

Proof Suppose that the series $\sum_{k=0}^{\infty} x_k$ is absolutely convergent and $\sum_{k=0}^{\infty} \|x_k\|$ has the sum X. Let $\sigma(n)$ be a one-to-one mapping of the set of integers $\{0, 1, 2,\ldots\}$ onto itself, and let the rearrangement of $\{x_n\}$ be written $\{y_n\}$ where $y_n = x_{\sigma(n)}$. Suppose that $\sum_{k=0}^{n} x_k = X_n$ and $\sum_{k=0}^{n} y_k = Y_n$. If for each n the largest integer of $\{\sigma(0), \sigma(1), \ldots, \sigma(n)\}$ is denoted by m, then

$$\sum_{k=0}^{n} \|y_k\| \leqslant \sum_{k=0}^{m} \|x_k\| \leqslant \sum_{k=0}^{\infty} \|x_k\| = X$$

and $\sum_{k=0}^{\infty} \|y_k\|$ is therefore convergent by Theorem 10.3. Now by Cauchy's condition it is possible, for any $\epsilon > 0$, to find an integer N such that if $n > N$ and $r > 0$ then

$$\|x_{n+1}\| + \|x_{n+2}\| + \cdots + \|x_{n+r}\| < \epsilon$$

If M is the largest integer in the set $\{\sigma^{-1}(0), \sigma^{-1}(1), \ldots, \sigma^{-1}(N)\}$ then

$$\|y_{n+1}\| + \|y_{n+2}\| + \cdots + \|y_{n+r}\| < \epsilon \quad \text{for } n > M \quad \text{and} \quad r > 0$$

and the difference $\|Y_M - X_N\|$ consists of a sum of terms x_j with $j > N$. This shows that $\|Y_M - X_N\| < \epsilon$ and hence for $n > \max(N, M)$,

$$\|Y_n - X_n\| < 3\epsilon$$

showing that $\sum_{n=0}^{\infty} x_n = \sum_{n=0}^{\infty} y_n$.

It should be emphasized at this stage that the preceding results in this section are true (except where stated otherwise) for series with elements in a general euclidean space, and therefore hold for series of vectors. For matrices the definitions of limit, convergence, and absolute convergence are as follows.

The sequence $\{A_k\}$ of $n \times n$ matrices $A_k = (a_{ij}^k)$ is said to *tend to the limit* $A = (a_{ij})$ if

$$\lim_{k\to\infty} a_{ij}^k = a_{ij}, \quad i, j = 1, 2, \ldots, n$$

The infinite matrix series $\sum_{k=0}^{\infty} A_k$ *converges to the sum* A provided that the sequence of partial sums

$$Y_m = \sum_{k=0}^{m} A_k$$

tends to A as n tends to ∞. The series $\sum_{k=0}^{\infty} A_k$ is *absolutely convergent* if $\sum \|A_k\|$ is convergent. The norm is subject to choice: a convenient norm is $\|A\|_c = n \max_{i,j} |a_{ij}|$ which (Chapter 3, Exercise 10.1) is consistent with the euclidean vector norm.

Theorem 10.6 An absolutely convergent matrix series is convergent.

Proof We have $|a_{ij}^k| \leqslant \|A_k\|_c$ so that the convergence of $\sum_{k=0}^{\infty} \|A_k\|_c$ implies that of $\sum_{k=0}^{\infty} |a_{ij}^k|$, and so (Theorem 10.4) that of $\sum_{k=0}^{\infty} a_{ij}^k$, $i, j = 1, 2, \ldots, n$.

Example 10.1 If $\|A\|_c = r < 1$, the matrix series $I_n + A + A^2 + \cdots$ converges absolutely to $(I_n - A)^{-1}$. For $\|I_n\|_c = n$, $\|A\|_c = r$, $\|A^2\|_c \leqslant r^2, \ldots$, and $n + r + r^2 + \cdots$ is convergent: therefore the matrix series converges absolutely. Also

$$\|I_n - (I_n - A)(I_n + A + A^2 + \cdots + A^k)\|_c = \|A^{k+1}\|_c \leqslant r^{k+1} \to 0$$

so that the matrix series converges to $(I_n - A)^{-1}$.

EXERCISE 10.3 Assuming the properties of e^x (Section 12) where x is real, show that $\sum_{k=0}^{\infty} A^k/k!$ is absolutely convergent for all A. This series defines the matrix function e^A.

10.2 Tests for convergence

It is useful to be able to test an infinite series for convergence and two such tests are given here. For fuller details of convergence tests the reader may refer to Apostol (1957), Rankin (1963), or Bromwich (1926).

Theorem 10.7 (The comparison test) If the infinite series of positive real numbers $\sum b_n$ is convergent to the sum b and a positive number k can be found such that $\|a_n\| \leqslant kb_n$ for $n \geqslant n_0$, then $\sum a_n$ is absolutely convergent. Furthermore, if the above inequality holds for all values of n, and a is the sum of $\sum a_n$, then $\|a\| \leqslant kb$.

Proof Since the partial sums of $\sum b_n$ are bounded (Theorem 10.3) so are those of $\sum \|a_n\|$; hence $\sum a_n$ is absolutely convergent. Suppose now that $\|a_n\| \leqslant kb_n$ for all values of n and denote the partial sums of $\sum a_n$ and $\sum b_n$ by A_n and B_n. Then

$$\begin{aligned}\|A_n\| &= \left\|\sum_{i=0}^{n} a_i\right\| \\ &\leqslant \sum_{i=0}^{n} \|a_i\| \\ &\leqslant k\sum_{i=0}^{n} b_i = kB_n\end{aligned}$$

That $\|a\| \leqslant kb$ now follows on taking the limit as $n \to \infty$.

Before stating the next theorem the definitions of upper and lower limits of a sequence $\{a_n\}$ are required:

U is said to be the *limit superior* (or *upper limit*) of $\{a_n\}$ if, given any $\epsilon > 0$,

(a) $a_n < U + \epsilon$, for all sufficiently large n
(b) $a_n > U - \epsilon$, for an infinite number of values of n.

The upper limit of a sequence is written $\limsup_{n\to\infty} a_n$. Similarly the *limit inferior* (or *lower limit*) L is defined by

(*c*) $a_n > L - \epsilon$, for all sufficiently large n
(*d*) $a_n < L + \epsilon$, for an infinity of values of n

and written $\liminf_{n\to\infty} a_n$. Notice that $\liminf_{n\to\infty} a_n = -\limsup_{n\to\infty} b_n$, where $b_n = -a_n$. In case the sequence $\{a_n\}$ does not have an upper bound then $\limsup_{n\to\infty} a_n$ is defined to be $+\infty$. If $\{a_n\}$ is bounded above but not below and does not have a finite upper limit then $\limsup_{n\to\infty} a_n = -\infty$.

Example 10.2 The sequence $\frac{1}{2}, -\frac{1}{2}, \frac{2}{3}, -\frac{2}{3}, \frac{3}{4}, -\frac{3}{4}, \ldots$ has limit superior 1, limit inferior -1. The sequence $1, 2, 3, \ldots$ has limit superior ∞ and also limit inferior ∞. The sequence $-1, -2, -3, \ldots$ has limit superior $-\infty$, and also limit inferior $-\infty$. The real numbers in the intervals (a, b), $(a, b]$, $[a, b)$, $[a, b]$ all have limit inferior a, limit superior b.

EXERCISE 10.4 Prove that a necessary and sufficient condition for the convergence of the sequence $\{a_n\}$ is that $\limsup_{n\to\infty} a_n$ and $\liminf_{n\to\infty} a_n$ are equal and finite.

Series with complex terms will be treated in detail in Chapter 7. It is convenient to include such series in the following tests, and for this purpose we define convergence of complex series. Let the complex terms be $a_n = x_n + iy_n$. Then if $b_n = \begin{pmatrix} x_n \\ y_n \end{pmatrix}$ is considered as a vector in E_2, $\sum a_n$ converges when $\sum b_n$ converges; $\sum a_n$ converges absolutely when $\sum b_n$ converges absolutely; and if $\sum b_n = b = \begin{pmatrix} x \\ y \end{pmatrix}$, then $\sum a_n = a = x + iy$. Notice that $\|b_n\| = |a_n|$.

Theorem 10.8 (The ratio test) Let $\sum a_n$ be an infinite series of non-zero real or complex numbers and set

$$r = \liminf_{n\to\infty} \left|\frac{a_{n+1}}{a_n}\right|, \quad R = \limsup_{n\to\infty} \left|\frac{a_{n+1}}{a_n}\right|$$

Then if $R < 1$, $\sum a_n$ is absolutely convergent; if $r > 1$, $\sum a_n$ is divergent. This result holds for (real or complex) scalar series only.

Proof By definition of the upper limit there exists an N for which $|a_{n+1}/a_n| < x$ if $n > N$, where $R < x < 1$. When $n > N$ it follows that

$$\frac{|a_{n+1}|}{x^{n+1}} < \frac{|a_n|}{x^n} < \cdots < \frac{|a_N|}{x^N}$$

and so

$$|a_n| < \frac{|a_N|}{x^N} x^n$$

This shows that $\sum |a_n|$ is convergent by the comparison test, because the geometric series is convergent for $|x| < 1$.

If $r > 1$ then there exists an N such that $|a_{n+1}| > |a_n|$ for $n > N$ implying $\lim_{n \to \infty} a_n \neq 0$ and hence that $\sum a_n$ cannot converge.

EXERCISE 10.5 Use the series $\sum 1/n^\alpha$ with (a) $\alpha > 1$ and (b) $\alpha \leqslant 1$ to show that the ratio test is inconclusive when $1 \in [r, R]$.

10.3 Product of absolutely convergent series

Theorem 10.9 (Cauchy product theorem) If the absolutely convergent series, with real or complex (scalar) terms, $\sum_{n=0}^{\infty} a_n$, $\sum_{n=0}^{\infty} b_n$ converge to the sums a and b respectively, then the series $\sum_{n=0}^{\infty} c_n$, where $c_n = \sum_{i=0}^{n} a_i b_{n-i}$, is absolutely convergent to the sum $c = ab$.

Proof Suppose first that $\sum a_n$ and $\sum b_n$ are series of positive real numbers and let A_n and B_n denote their respective partial sums. Let C_n be the partial sums of $\sum c_n$; clearly C_n is non-decreasing. Now consider the square array of numbers

$$\begin{array}{cccc|ccc}
a_0b_0 & a_1b_0 & \cdots & a_nb_0 & a_{n+1}b_0 & \cdots & a_{2n}b_0 \\
a_0b_1 & a_1b_1 & \cdots & a_nb_1 & a_{n+1}b_1 & \cdots & a_{2n}b_1 \\
\vdots & \vdots & & \vdots & \vdots & & \vdots \\
a_0b_n & a_1b_n & \cdots & a_nb_n & a_{n+1}b_n & \cdots & a_{2n}b_n \\
\hline
\vdots & \vdots & & \vdots & \vdots & & \vdots \\
a_0b_{2n} & a_1b_{2n} & & a_nb_{2n} & a_{n+1}b_{2n} & \cdots & a_{2n}b_{2n}
\end{array} \tag{10.1}$$

It is clear that A_nB_n consists of the sum of all those elements enclosed in the solid lines in the top left-hand corner of the array, whereas C_n is the sum of elements on and above the marked diagonal. This proves that

$$C_n \leqslant A_nB_n \tag{10.2}$$

and because $A_n \leqslant a$, $B_n \leqslant b$, C_n has a limit satisfying

$$c = \lim_{n \to \infty} C_n \leqslant ab \tag{10.3}$$

It is also apparent from the array that

$$C_{2n} \geqslant A_n B_n \tag{10.4}$$

From (10.4) and Section 6.4,

$$c = \lim_{n \to \infty} C_n \geqslant \lim A_n B_n = \lim A_n \lim B_n$$

so that

$$c \geqslant ab \tag{10.5}$$

Combining (10.3) and (10.5) gives $c = ab$.

Now return to the original assumption of the theorem that $\sum a_n$ and $\sum b_n$ are absolutely convergent. Let $d_n = \sum_{i=0}^{n} |a_i|\,|b_{n-i}|$ and let D_n be the partial sums of the series d_n. Because $|c_n| \leqslant |d_n|$ and $\sum d_n$ is convergent by the first part of the proof, it follows that $\sum c_n$ is absolutely convergent.

To complete the proof it is necessary to show that $c = ab$. Let E_n and F_n denote the partial sums of $\sum |a_n|$ and $\sum |b_n|$ and e and f their sums. Then

$$|A_n B_n - C_n| \leqslant |E_n F_n - D_n| \tag{10.6}$$

and since $D_n \to ef$ it is seen from equation (10.6) that $c = ab$.

It can be shown that it is only necessary for one of the series in the preceding theorem to be absolutely convergent (the other must be convergent) for the result to hold. This stronger result will not be needed in this book, however, and the reader is referred to Apostol (1957) for a proof.

Theorem 10.9 extends immediately to two (real or complex) matrix series provided that they are conformable and the order of the products is preserved. The changes in the proof consist only of substituting norms for moduli, and are left as an exercise.

EXERCISE 10.6 Show that $e^A e^A = e^{2A}$, where the matrix exponential e^A is defined in Section 10.1. Give sufficient conditions for $e^A e^B = e^{A+B}$. [A and B commute, so that we can write, for example, $A^2 + 2AB + B^2 = A^2 + AB + BA + B^2 = (A + B)^2$.]

10.4 Double sequences and series

Let f be a function, with domain the cartesian product of the set of positive integers with itself, and range in $\mathscr{R}$ or $\mathscr{C}$. If m, n are positive integers then $f(m, n)$ is written $a_{m,n}$ and $\{a_{m,n}\}$ is called a *double sequence*. The double sequence $\{a_{m,n}\}$ is said to *converge to the value a* if a positive integer N exists such that $|a_{m,n} - a| < \epsilon$, for any given $\epsilon > 0$, whenever $m > N$ and $n > N$. In this case a is said to be the *limit* of $\{a_{m,n}\}$ and written $\lim_{m,n\to\infty} \{a_{m,n}\} = a$.

Let $\{a_{m,n}\}$ be a double sequence and define a new double sequence $\{A_{m,n}\}$ by putting

$$A_{m,n} = \sum_{p=1}^{m} \sum_{q=1}^{n} a_{m,n} \tag{10.7}$$

Then $\{A_{m,n}\}$ is said to be a *double series* and to *converge to the sum a* if

$$\lim_{m,n\to\infty} \{A_{m,n}\} = a \tag{10.8}$$

Every $A_{m,n}$ is called a *partial sum* of the double series: the latter will be written $\sum_{m,n} a_{m,n}$. As for the single series, $\sum_{m,n} a_{m,n}$ is said to be *absolutely convergent* if $\sum_{m,n} |a_{m,n}|$ is convergent.

EXERCISE 10.7 Generalize Theorems 10.3 and 10.4 to double series.

EXERCISE 10.8 Let $\lim_{m,n\to\infty} \{a_{m,n}\} = a$ and suppose that $\lim_{n\to\infty} \{a_{m,n}\}$ exists for each fixed value of m. Show that $\lim_{m\to\infty} \lim_{n\to\infty} \{a_{m,n}\}$ also exists and is equal to A. [$\lim_{m\to\infty} \lim_{n\to\infty} \{a_{m,n}\}$ is called a *repeated limit* to distinguish it from the double limit $\lim_{m,n\to\infty} \{a_{m,n}\}$.]

In summing a double series $\sum_{m,n} a_{m,n}$ it is obviously possible to form the partial sums $A_{m,n}$ in many different ways. In particular, if the sums $\sum_{n=1}^{\infty} a_{m,n}$ are evaluated first for fixed values of m and then the sum of these sums found, the result is called a *repeated series* and is written

$$\sum_{m=1}^{\infty} \sum_{n=1}^{\infty} a_{m,n} \tag{10.9}$$

(Notice that the inner sum is evaluated first.) If the summation with respect to m is carried out first and then the summation with respect to n, the result is the repeated series

$$\sum_{n=1}^{\infty} \sum_{m=1}^{\infty} a_{m,n} \tag{10.10}$$

Expression (10.9) is called the *sum by rows* of $\sum_{m,n} a_{m,n}$ and expression (10.10) the *sum by columns*.

An important theorem concerning these two repeated series will now be proved.

Theorem 10.10 Let all terms of the sequence $\{a_{m,n}\}$ be real and non-negative. Suppose that $\sum_{n=1}^{\infty} a_{m,n}$ converges for each fixed value of m and that the repeated sum $\sum_{m=1}^{\infty} \sum_{n=1}^{\infty} a_{m,n}$ converges to the sum a. Then $\sum_{m=1}^{\infty} a_{m,n}$ converges for each fixed value of n, and $\sum_{n=1}^{\infty} \sum_{m=1}^{\infty} a_{m,n}$ also converges and is equal to a.

Proof Since $a_{m,n} \geqslant 0$ it follows that

$$a_{m,n} \leqslant \sum_{k=1}^{\infty} a_{m,k} \tag{10.11}$$

for each fixed value of n, and on summing with respect to m

$$\sum_{m=1}^{M} a_{m,n} \leqslant \sum_{m=1}^{M} \sum_{k=1}^{\infty} a_{m,k} \leqslant a \tag{10.12}$$

If now $M \to \infty$ in (10.12) it is seen that $\sum_{m=1}^{\infty} a_{m,n}$ converges for each fixed value of n.

Denote the sum $\sum_{m=1}^{\infty} a_{m,n}$ by A_n for any fixed n. Then summing with respect to n gives

$$\begin{aligned}\sum_{n=1}^{N} A_n &= \sum_{n=1}^{N} \sum_{m=1}^{\infty} a_{m,n} \\ &= \sum_{m=1}^{\infty} \sum_{n=1}^{N} a_{m,n} \\ &\leqslant \sum_{m=1}^{\infty} \sum_{n=1}^{\infty} a_{m,n} \\ &= a \end{aligned} \tag{10.13}$$

because $a_{m,n} \geqslant 0$. This shows that $\sum A_n$ converges to a sum $A \leqslant a$. It can be shown in exactly the same way that $a \leqslant A$ so that $a = A$ and the proof of the theorem is complete.

EXERCISE 10.9 Show by counter-example that if both the repeated series $\sum_{m=1}^{\infty} \sum_{n=1}^{\infty} a_{m,n}$ and $\sum_{n=1}^{\infty} \sum_{m=1}^{\infty} a_{m,n}$ converge for a general sequence $\{a_{m,n}\}$ they are not necessarily equal.

11 Uniform convergence

Let $\sum_{n=0}^{\infty} a_n(z)$ be a series in which each term is a function of a complex variable z with domain S and range R, both in $\mathscr{C}$. The series $\sum_{n=0}^{\infty} a_n(z)$ is said to be *uniformly convergent* on S to the sum $T(z)$ if, given $\epsilon > 0$, there exists an integer $N(\epsilon)$ such that $|T(z) - T_n(z)| < \epsilon$ for all $n \geqslant N$ and all $z \in S$, where $T_n(z)$ is the nth partial sum $\sum_{i=0}^{n} a_i(z)$. This definition should be compared with that of uniform continuity in Section 7.1.

Theorem 11.1 The sum $T(z)$ of a uniformly convergent series $\sum_{n=0}^{\infty} a_n(z)$ is continuous at each point $z_0 \in S$ at which every $a_n(z_0)$ is continuous.

Proof Let $T_n = \sum_{i=0}^{n} a_i(z)$: then by Section 7, T_n is continuous at z_0. By definition, z_0 is a point of accumulation. Since the series is uniformly convergent to $T(z)$ an integer p exists such that $|T_n(z) - T(z)| < \epsilon/3$ when $n \geqslant p$, for all $z \in S$ and $\epsilon > 0$. Also since $T_n(z)$ is continuous at z_0 there exists a neighbourhood $N(z_0)$ with the property

$$z \in N(z_0) \cap S \Rightarrow |T_n(z) - T_n(z_0)| < \frac{\epsilon}{3}$$

Now since

$$|T(z) - T(z_0)| \leqslant |T(z) - T_n(z)| + |T_n(z) - T_n(z_0)| + |T_n(z_0) - T(z_0)|$$

it follows that for $z \in N(z_0) \cap S$, $|T(z) - T(z_0)| < \epsilon$, which proves the theorem.

Example 11.1 Consider the series with general term $a_n = x^2/(1 + x^2)^n$, $n = 0, 1, \ldots$; here x is a real number which, to agree with the preceding discussion, is regarded as the complex number $x + i0$. This is a geometric series with common ratio $1/(1 + x^2)$, which is always less than 1 except for $x = 0$. Hence the series is absolutely convergent for all real values of x. Notice, however, that

$$\begin{aligned}\lim_{n\to\infty} T_n &= \lim_{n\to\infty}\left\{1 + x^2 - \frac{1}{(1 + x^2)^n}\right\}\\ &= 1 + x^2 \quad \text{when } x \neq 0\end{aligned}$$

A direct inspection of the series shows that its sum is zero when x is zero, thus demonstrating that the sum has a discontinuity at $x = 0$. The reason for this is, of course, that $\sum a_n$ is not uniformly convergent in any interval containing the origin.

EXERCISE 11.1 (Cauchy condition for uniform convergence) Prove that a necessary and sufficient condition for $\sum_{n=0}^{\infty} a_n(z)$ to converge uniformly on S is that for every $\epsilon > 0$ there exists an N such that if $n > N$ then $|a_{n+1}(z) + a_{n+2}(z) + \cdots + a_{n+p}(z)| < \epsilon$ for all $p > 0$ and all $z \in S$.

Theorem 11.2 (Weierstrass M-test) The series $\sum_{n=0}^{\infty} a_n(z)$ converges uniformly on S if $|a_n(z)| \leqslant M_n$ for $n \geqslant 0$ and all $z \in S$, where $\sum_{n=0}^{\infty} M_n$ is a convergent series of non-negative numbers.

Proof By Cauchy's condition (Theorem 10.1) an N can be found such that for $\epsilon > 0$ and $n > N$, $\sum_{i=n+1}^{n+r} M_i < \epsilon$ for all $r > 0$. But by the hypothesis $|\sum_{i=n+1}^{n+r} a_i(z)| \leqslant \sum_{i=n+1}^{n+r} M_i < \epsilon$, for all $z \in S$, and the result now follows from Exercise 11.1.

EXERCISE 11.2 Let $\{a_{m,n}\}$ be a double sequence with complex values and suppose that $\sum_{n=1}^{\infty} a_{m,n}$ is absolutely convergent when m is held fixed and that the repeated series $\sum_{m=1}^{\infty}\sum_{n=1}^{\infty}|a_{m,n}|$ is convergent. Prove that $\sum_{m=1}^{\infty} a_{m,n}$ is absolutely convergent if n is held fixed and that $\sum_{n=1}^{\infty}\sum_{m=1}^{\infty} a_{m,n}$ converges absolutely and is equal to $\sum_{m=1}^{\infty}\sum_{n=1}^{\infty} a_{m,n}$. [The absolute convergence of $\sum_{m=1}^{\infty} a_{m,n}$ for fixed n, and of $\sum_{n=1}^{\infty}\sum_{m=1}^{\infty} a_{m,n}$, follows directly from Theorem 10.10. Now study the series $\sum_{m=1}^{\infty} g_m(x)$, where $x \in \{0, 1, 1/2, \ldots, 1/q, \ldots\}$ and $g_m(0) = \sum_{n=1}^{\infty} a_{m,n}$, $g_m(1/q) = \sum_{n=1}^{q} a_{m,n}$, using Theorems 11.1 and 11.2 to show that $\sum_{m=1}^{\infty} g_m(x)$ is continuous at $x = 0$. The result now follows from the fact that $\sum_{m=1}^{\infty} g_m(0) = \lim_{q\to\infty}\sum_{m=1}^{\infty} g_m(1/q)$ and the definition of $g_m(x)$.]

EXERCISE 11.3 Generalize Theorems 11.1 and 11.2 to vector and matrix series.

12 Power series

Series of the form

$$\sum_{n=0}^{\infty} a_n(z - z_0)^n \tag{12.1}$$

where z, z_0, and the a_n are complex numbers, are known as *power series* and are of special importance. We conclude this chapter with a brief discussion of some of their properties. The complex number z_0 is called the *centre* of the power series and when $z_0 = 0$, equation (12.1) simplifies to

$$a_0 + a_1 z + a_2 z^2 + \cdots + a_n z^n + \cdots \tag{12.2}$$

Example 12.1 The series $1 + z + z^2 + \cdots + z^n + \cdots$ is known as the geometric series. Notice that

$$T_n = \sum_{k=0}^{n} z^k = \frac{1 - z^{n+1}}{1 - z}$$

and since $\lim_{n\to\infty} z^{n+1} = 0$ if $|z| < 1$, the series converges to $1/(1 - z)$ when $|z| < 1$. If, however, $|z| \geqslant 1$ the series must diverge because then $|z^n| \geqslant 1$.

As will be shown, power series converge inside and diverge outside a certain circle which is known as the *circle of convergence*: they may or may not converge on the circle itself. The radius of the circle of convergence is known as *the radius of convergence* and may in particular cases be 0 or ∞. When the radius of convergence is infinite the series converges in the whole complex plane.

Theorem 12.1 Every power series (12.2) has a radius of convergence R, with $0 \leqslant R \leqslant \infty$. The series is absolutely convergent for $|z| < R$, is uniformly convergent for $|z| \leqslant r$, where r is any positive number less than R, and is divergent for $|z| > R$. These results extend in an obvious way to series such as (12.1).

Proof Let

$$R = \frac{1}{\limsup_{n\to\infty} \sqrt[n]{|a_n|}}$$

Then if $R = 0$ the series is clearly divergent for all $|z| > 0$. If $R > 0$

$$\limsup_{n\to\infty} \sqrt[n]{|a_n z^n|} = \frac{|z|}{R} \tag{12.3}$$

By the definition of limit superior it follows that $|a_n z^n| < x^n$ for all sufficiently large n, provided that $|z|/R < x < 1$. By the comparison test, the power series converges absolutely if $|z| < R$, for then x can be taken as $\frac{1}{2}(1 + |z|/R)$.

If $|z| \leqslant r$ where $0 < r < R$, $|a_n z^n| < |a_n r^n|$, and $\sum |a_n r^n|$ is a series of non-negative numbers, convergent by the previous paragraph. So by the Weierstrass M-test, $\sum a_n z^n$ is uniformly convergent for $|z| \leqslant r$.

Finally equation (12.3) with $|z| > R$ implies $|a_n z^n| > 1$ for infinitely many values of n so that $\lim_{n\to\infty} a_n z^n \neq 0$ and the series is divergent.

That the sum of a power series is continuous within its circle of convergence follows directly from Theorem 11.1. Also, given two power series $f(z) = \sum a_n z^n$ and $g(z) = \sum b_n z^n$ with radii of convergence r and R respectively, their product is given by $f(z)g(z) = \sum c_n z^n$ with radius of convergence $\min\{r, R\}$, where $c_n = \sum_{i=0}^{n} a_i b_{n-i}$ (Theorem 10.9).

Example 12.2 (The exponential series) If $f(z) = \sum_{n=0}^{\infty} z^n/n!$ then, by the ratio test, the radius of convergence of $f(z)$ is $+\infty$. The sum function of this series is defined as the *exponential function* and written

$$\exp(z) = \sum_{n=0}^{\infty} \frac{z^n}{n!} \tag{12.4}$$

(another notation is to write e^z for $\exp(z)$). For all z, $\exp(z)$ is continuous by Theorems 12.1 and 11.1.

EXERCISE 12.1 Use Theorem 10.9 to show that $e^{z_1}e^{z_2} = e^{z_1+z_2}$. Also show that for real t and all positive r and β, $\lim_{t\to+\infty} t^r e^{-\beta t} = 0$. [Write $e^{-\beta t} = 1/e^{\beta t}$ and use the series.]

An important property of power series is that they may be differentiated term by term in any neighbourhood where they converge. As a preparation for Theorem 12.2, in which this property is more accurately stated, the following Lemma is required.

Lemma Let $\sum a_n(z - z_0)^n$ converge in an open ball B with centre z_0 and radius H and suppose that $f(z) = \sum_{n=0}^{\infty} a_n(z - z_0)^n$ holds for all $z \in S$, where S is an open subset of B. For each point $z_1 \in S$ there is a neighbourhood $N(z_1, h) \subset S$ in which

$$f(z) = \sum_{k=0}^{\infty} b_k(z - z_1)^k \tag{12.5}$$

where

$$b_k = \sum_{n=k}^{\infty} \binom{n}{k} a_n(z_1 - z_0)^{n-k} \quad \text{for } k = 0, 1, 2, \ldots \tag{12.6}$$

and

$$\binom{n}{k} = \frac{n!}{k!(n-k)!}$$

Proof For any point $z \in S$

$$f(z) = \sum_{n=0}^{\infty} a_n(z - z_0)^n \tag{12.7}$$

$$= \sum_{n=0}^{\infty} a_n\{(z - z_1) + (z_1 - z_0)\}^n \tag{12.8}$$

On expanding the expression in braces there follows

$$f(z) = \sum_{n=0}^{\infty} a_n \sum_{k=0}^{n} \binom{n}{k}(z - z_1)^k(z_1 - z_0)^{n-k} \tag{12.9}$$

which can be written as a double series $\sum_{n=0}^{\infty} \sum_{k=0}^{\infty} c_n(k)$ with $c_n(k) = \binom{n}{k}a_n(z - z_1)^k(z_1 - z_0)^{n-k}$ for $k \leqslant n$ and $c_n(k) = 0$ otherwise.

Consider the absolute convergence of this double series; if z is any point in the neighbourhood $N(z_1, h) \subset S$ then

$$\sum_{n=0}^{\infty} \sum_{k=0}^{\infty} |c_n(k)| \leqslant \sum_{n=0}^{\infty} |a_n|\{|z - z_1| + |z_1 - z_0|\}^n$$

This proves the absolute convergence of the double series since $|z - z_1| + |z_1 - z_0| < h + |z_1 - z_0| \leqslant H$.

By Exercise 11.2 it is now possible to change the order of summation in (12.9) to give

$$f(z) = \sum_{k=0}^{\infty} \sum_{n=k}^{\infty} \binom{n}{k}a_n(z - z_1)^k(z_1 - z_0)^{n-k}$$

which proves the Lemma.

The derivative $Df(z)$ of a complex function of a complex variable will be considered in Chapter 7. It is convenient to anticipate the definition of $Df(z)$ for the purposes of this section: if $f(z)$ exists on $N(z_1)$ and the limit

$$\lim_{z\to z_1}\frac{f(z)-f(z_1)}{z-z_1}$$

exists, it is defined to be $Df(z_1)$.

Theorem 12.2 If the series $\sum a_n(z-z_0)^n$ converges in a neighbourhood $N(z_0, h)$ then the sum function $f(z) = \sum_{n=0}^{\infty} a_n(z-z_0)^n$ has a derivative, in the same neighbourhood, given by

$$Df(z) = \sum_{n=1}^{\infty} na_n(z-z_0)^{n-1} \tag{12.10}$$

Proof From equations (12.5) and (12.6)

$$f(z) = b_0 + b_1(z-z_1) + \sum_{k=1}^{\infty} b_{k+1}(z-z_1)^{k+1}$$

where $b_0 = f(z_1)$, $b_1 = \sum_{n=1}^{\infty} na_n(z_1-z_0)^{n-1}$, and $z_1 \in N(z_0, h)$. Thus

$$\frac{f(z)-f(z_1)}{z-z_1} = \sum_{n=1}^{\infty} na_n(z_1-z_0)^{n-1} + \sum_{k=1}^{\infty} b_{k+1}(z-z_1)^k$$

and the result follows on taking the limit as $z \to z_1$.

Exercise 12.2 By repeated application of the previous theorem show that the power series in fact possesses derivatives of all orders inside its circle of convergence and that

$$D^k f(z) = \sum_{n=k}^{\infty} \frac{n!}{(n-k)!} a_n(z-z_0)^{n-k} \tag{12.11}$$

Deduce that if a function $f(z)$ can be represented by a power series about a point z_0 then the representation is unique and

$$f(z) = \sum_{n=0}^{\infty} \frac{D^n f(z_0)}{n!}(z-z_0)^n \tag{12.12}$$

Example 12.3 Applying equation (12.10) to the exponential series

$$\exp(z) = \sum_{n=0}^{\infty} \frac{z^n}{n!}$$

gives

$$D\exp(z) = \sum_{n=1}^{\infty} \frac{nz^{n-1}}{n!} = \exp(z) \tag{12.13}$$

Example 12.4 By Taylor's theorem the real-valued function $f(x)$ can be written

$$f(x) = \sum_{k=0}^{n-1} \frac{f^{(k)}(x_0)}{k!} (x - x_0)^k + \frac{f^{(n)}(x_1)}{n!} (x - x_0)^n \qquad (12.14)$$

where $f(x)$ has been assumed to possess derivatives of all orders on the closed interval $[a, b]$; x, $x_0 \in [a, b]$; and x_1 is a point between x and x_0. The series

$$\sum_{n=0}^{\infty} \frac{f^{(n)}(x_0)}{n!} (x - x_0)^n$$

is called the *Taylor series* for $f(x)$ and a necessary and sufficient condition for convergence of the Taylor series is

$$\lim_{n \to \infty} \frac{f^{(n)}(x_1)}{n!} (x - x_0)^n = 0$$

[Note that $f(x)$ is finite on $[a, b]$ by Section 8.5 and Theorem 7.2.]

13 Functions of a matrix

The *spectral radius* $\rho(A)$ of a real $q \times q$ matrix A is defined by

$$\rho(A) = \max_i \{|\lambda_i(A)|\}$$

It is easily seen that $\rho(A) \leqslant \|A\|$ for any norm. For if x is a non-zero eigenvector and λ the corresponding eigenvalue of A then

$$Ax = \lambda x$$

and

$$\|\lambda x\| = |\lambda| \, \|x\| = \|Ax\| \leqslant \|A\| \, \|x\|$$

giving

$$|\lambda| \leqslant \|A\|$$

from which the result follows directly.

Exercise 13.1 Prove that if

$$\lim_{n \to \infty} \sum_{i=1}^{n} A_i = A$$

then

$$\lim_{n \to \infty} \sum_{i=1}^{n} BA_iC = BAC$$

Exercise 13.2 Show that a necessary and sufficient condition for

$$\lim_{n \to \infty} A^n = 0$$

is $\rho(A) < 1$. [Use Exercise 13.1 with $C = B^{-1}$, and Chapter 3, Exercise 7.19.]

Theorem 13.1 Consider the geometric series

$$I + A + A^2 + A^3 + \cdots \tag{13.1}$$

where A is a real $q \times q$ matrix. If $\rho(A) < 1$ then the series (13.1) converges to the limit $(I - A)^{-1}$.

Proof Since $\rho(A) < 1$ it follows that $|I - A| \neq 0$ and so $I - A$ is non-singular. Furthermore if

$$S_n = I + A + A^2 + \cdots + A^n$$

then

$$S_n(I - A) = I - A^{n+1}$$

and by Exercise 13.2 it follows that

$$\lim_{n \to \infty} S_n(I - A) = I$$

proving the result.

A weaker form of Theorem 13.1 was given in Example 10.1.

Theorem 13.2 More generally, consider the power series

$$\alpha_0 I + \alpha_1 A + \alpha_2 A^2 + \cdots \tag{13.2}$$

where the α_i are scalars. If the associated scalar power series

$$f(\beta) = \alpha_0 + \alpha_1 \beta + \alpha_2 \beta^2 + \cdots \tag{13.3}$$

has a radius of convergence $r > \rho(A)$ then the matrix series (13.2) converges; its limit is written $f(A)$.

Proof To prove this result on the convergence of (13.2) the Jordan canonical form J of A is used. Write

$$A = SJS^{-1} \tag{13.4}$$

Then J is the quasi-diagonal matrix

$$J = \begin{pmatrix} J_1 & & & & \\ & J_2 & & 0 & \\ & & \cdot & & \\ & 0 & & \cdot & \\ & & & & J_r \end{pmatrix} \tag{13.5}$$

with J_i the ith Jordan block. (Chapter 3, Section 7.4.) Consider the partial sum

$$
\begin{aligned}
f_n(A) &= \sum_{p=0}^{n} \alpha_p A^p \\
&= \sum_{p=0}^{n} \alpha_p (SJS^{-1})^p \\
&= S\left(\sum_{p=0}^{n} \alpha_p J^p\right) S^{-1}
\end{aligned}
\tag{13.6}
$$

But

$$
J^p = \begin{pmatrix} J_1^p & & & & \\ & J_2^p & & 0 & \\ & & \cdot & & \\ & 0 & & \cdot & \\ & & & & J_r^p \end{pmatrix}
\tag{13.7}
$$

so that on substituting (13.7) into (13.6) and using the result of Chapter 3, Exercise 7.19, all the relevant series converge as $n \to \infty$ provided $r > \rho(A)$. This completes the proof.

Notice, in particular, that if the series (13.3) converges in the whole complex plane, then so does the series (13.2). Notice also that in the special case when

$$J = \operatorname{diag}(\lambda_1, \lambda_2, \ldots, \lambda_k)$$

then

$$J^p = \operatorname{diag}(\lambda_1^p, \lambda_2^p, \ldots, \lambda_k^p)$$

and

$$f_n(A) = S \operatorname{diag}\left(\sum_{p=0}^{n} \alpha_p \lambda_1^p, \sum_{p=0}^{n} \alpha_p \lambda_2^p, \ldots, \sum_{p=0}^{n} \alpha_p \lambda_k^p\right) S^{-1}$$

The three important functions $\cos \beta$, $\sin \beta$ and $\exp \beta$ have convergent power series expansions for all values of β and therefore the matrix functions

$$\exp A = I + A + \frac{1}{2!} A^2 + \frac{1}{3!} A^3 + \cdots \tag{13.8}$$

$$\cos A = I - \frac{1}{2!} A^2 + \frac{1}{4!} A^4 - \cdots \tag{13.9}$$

$$\sin A = A - \frac{1}{3!} A^3 + \frac{1}{5!} A^5 - \cdots \tag{13.10}$$

converge for all A. Similarly the power series expansion for the function $\log(1 + \beta)$ converges for all β having $|\beta| < 1$ so that

$$\log(1 + A) = A - \tfrac{1}{2}A^2 + \tfrac{1}{3}A^3 - \cdots$$

converges for all matrices A with $\rho(A) < 1$.

EXERCISE 13.3 Prove that

$$\lim_{h\to 0} \frac{1}{h}\{\exp(Ah) - I\} = A$$

13.1 Matrix exponential

It has already been seen that the matrix exponential e^A is convergent. Now consider e^{At} where t is a real number. From Example 12.2, $e^{\|A\|t}$ is convergent for all real t and hence e^{At} is absolutely convergent for all real t. The following theorems will be required later.

Theorem 13.3 The derivative of e^{At} is given by $De^{At} = Ae^{At}$.

Proof By the Cauchy product theorem (Theorem 10.9; compare Exercise 10.6)

$$e^{At_1}e^{At_2} = e^{A(t_1+t_2)} \tag{13.11}$$

Now

$$De^{At} = \lim_{h\to 0} \frac{e^{A(t+h)} - e^{At}}{h} \tag{13.12}$$

which on using (13.11) becomes

$$De^{At} = \lim_{h\to 0} e^{At}\frac{e^{Ah} - I}{h} \tag{13.13}$$

$$= e^{At}\lim_{h\to 0} \frac{e^{Ah} - I}{h} = Ae^{At} \tag{13.14}$$

by Exercise 13.3.

Theorem 13.4 The determinant of e^{At} cannot vanish for any real number t.

Proof Let $A = S^{-1}JS$, where J is the Jordan form of A. Define a function ϕ by

$$\phi(t) = |e^{At}| = \det(e^{At}) \tag{13.15}$$

Then

$$\phi(t) = |S^{-1}e^{Jt}S|$$

$$= |e^{Jt}|$$

$$= e^{\lambda_1 t}e^{\lambda_2 t}\ldots e^{\lambda_q t} \tag{13.16}$$

so that $\phi(t) \neq 0$ unless $e^{\lambda_i t} = 0$ for some λ_i and some t. But this is impossible, because $e^{-\lambda_i t}$ is finite by Example 12.2, and $e^{-\lambda_i t}e^{\lambda_i t} = e^0 = 1$ by Exercise 12.1.

Theorem 13.5 If c is a real n-vector and α is a real number exceeding the greatest real part of any eigenvalue of A, there exists an M such that $\|e^{At}c\| < Me^{\alpha t}$ for all $t \geqslant 0$.

Proof As in Theorem 13.4, let $A = S^{-1}JS$, and consider

$$e^{-\alpha t}\,\|e^{At}c\| \leqslant \|S\|_c \|S^{-1}\|_c \|c\|\;\|e^{-\alpha t}e^{Jt}\|_c \tag{13.17}$$

where the vector norm is as usual the euclidean norm, with which the cubic matrix norm is consistent by Exercise 10.1 of Chapter 3. On using the result of Exercise 7.19 of Chapter 3 it is easily seen that the non-zero elements of e^{Jt} have the form

$$\frac{t^r}{r!}\,e^{\lambda_q t}, \qquad r = 0, 1, \ldots, k_q < n \tag{13.18}$$

where $\lambda_q = \sigma_q + i\omega_q$ is an eigenvalue of A. By Exercise 12.1, for $t \geqslant 0$,

$$\phi_{qr}(t) \triangleq \left|e^{-\alpha t}\frac{t^r}{r!}\,e^{\lambda_q t}\right| = \frac{t^r}{r!}\,e^{-(\alpha-\sigma_q)t} \to 0 \text{ as } t \to \infty \tag{13.19}$$

For given λ_q and r there is therefore a $T_{qr} \geqslant 0$ such that $\phi_{qr}(t) < 1$ for all $t > T_{qr}$. Let T be the greatest of the T_{qr}. On the interval $0 \leqslant t \leqslant T$, each of the functions ϕ_{qr} is continuous and therefore has a maximum M_{qr}. Let M_0 be the greatest of the M_{qr}. Then for all $t \geqslant 0$,

$$\|e^{-\alpha t}e^{Jt}\|_c \leqslant \max\{M_0, 1\} \tag{13.20}$$

which with (13.17) proves the theorem with

$$M > \|S\|_c\,\|S^{-1}\|_c\,\|c\|\,\max\{M_0, 1\} \tag{13.21}$$

Further reading

An excellent general account of analysis is given by Apostol (1957). More advanced texts, which deal with analysis at a deeper and more abstract level, are Dieudonné (1960) and Graves (1956). For a full treatment of infinite series Bromwich (1926) may be consulted.

A selection of undergraduate books on analysis are Phillips (1950), Munroe (1965), Scott and Tims (1966), Flett (1966), and Rankin (1963).

References

APOSTOL, T. M., 1957, *Mathematical analysis*, Addison-Wesley, Reading, Mass.
BROMWICH, T. J. I'A., 1926, *An introduction to the theory of infinite series*, Macmillan, London.
DIEUDONNÉ, J., 1960, *Foundations of modern analysis*, Academic Press, New York.
FLETT, T. M., 1966, *Mathematical analysis*, McGraw-Hill, New York.

GRAVES, L. M., 1956, *The theory of functions of a real variable*, McGraw-Hill, New York.
MUNROE, M. E., 1965, *Introductory real analysis*, Addison-Wesley, Reading, Mass.
PHILLIPS, E. G., 1950, *A course of analysis*, Cambridge University Press, London.
RANKIN, R. A., 1963, *An introduction to mathematical analysis*, Pergamon Press, Oxford.
SCOTT, D. B. and S. R. TIMS, 1966, *Mathematical analysis: an introduction*, Cambridge University Press, London.

Problems

1 Prove that a neighbourhood $N(a, h)$ of a point $a \in E_n$ is a convex set.

2 Suppose that the functions f and g have finite derivatives on the open interval $a < x < b$ and $f(c) = g(c) = 0$ where $c \in (a, b)$. Prove that

$$\lim_{x \to c} \frac{f(x)}{g(x)} = \frac{f'(c)}{g'(c)}$$

provided that $g'(x)$ does not vanish at any point of (a, b).

3 Discuss the continuity of the function f defined by $f(x) = x \sin(1/x)$, $x \neq 0$, $f(0) = 0$. Does $Df(0)$ exist?

4 Find all the maxima and minima of the functions $\cos x/(1 + \cos^2 x)^{2/3}$ and $xe^{a/x}$.

5 Show how a number N can be divided into three parts a, b, c in such a way that the product a^3b^2c is maximized.

6 Expand $\cos^{-1}(1 - 2x)$ as a power series in x and give an expression for the nth term of the series. Use the result to show that $\cos^{-1}(0{\cdot}98) \doteqdot 0{\cdot}2003$.

7 Show that a double series $\sum_{m,n} a_{m,n}$ of non-negative terms has the same sum for any method of summation whatever.

8 Show that the positive (resp. negative) direction of the gradient vector $(\partial f/\partial x_1, \partial f/\partial x_2, \ldots, \partial f/\partial x_n)$ of a function $f(x_1, x_2, \ldots, x_n)$ is the direction in which f is increasing (resp. decreasing) most rapidly.

9 Prove that the quadratic curve passing through the three points (x_1, y_1), (x_2, y_2) and (x_3, y_3) has a minimum value at the point x_4 given by

$$x_4 = \frac{1}{2}\frac{(x_2^2 - x_3^2)y_1 + (x_3^2 - x_1^2)y_2 + (x_1^2 - x_2^2)y_3}{(x_2 - x_3)y_1 + (x_3 - x_1)y_2 + (x_1 - x_2)y_3}$$

provided that

$$\frac{(x_2 - x_3)y_1 + (x_3 - x_1)y_2 + (x_1 - x_2)y_3}{(x_1 - x_2)(x_2 - x_3)(x_3 - x_1)} < 0$$

10 Find for what real values of x the infinite series $\sum_{k=1}^{\infty} (\sin kx)/k^2$ is convergent and investigate its uniform convergence.

5 Riemann-Stieltjes integration

1 Introduction

It will be assumed that the reader is already familiar with the idea of an integral as the area under a curve. The purpose of this chapter will be to give a mathematical formulation of this idea, and at the same time to extend it in a natural way.

The area of a rectangle of length a and breadth b is defined as ab. Physically, it is clear that the areas of rectangles cut from a uniform sheet are proportional to their mass. Then, again physically, if we wish to define the area under an arbitrary curve such as c_1g in Fig. 1 we may say that it is proportional to the mass when

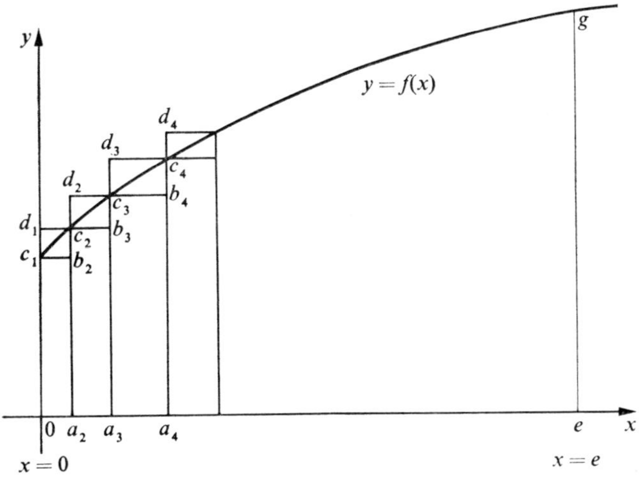

FIG. 1

the area oc_1ge is cut from a uniform sheet. In this way the area under the arbitrary curve can be related to the area, already defined, of a rectangle.

To give a mathematical definition of the area under a curve we have to find again some way of relating it to the area of a rectangle. The Riemann integral does this in a familiar way by a limiting process. Ordinates oc_1d_1, $a_2b_2c_2d_2, \ldots$ are erected, not necessarily at equal intervals. The sum of the areas of the rectangles $od_1c_2a_2$, $a_2d_2c_3a_3, \ldots$ is well defined, and so is the sum of the areas of $oc_1b_2a_2$, $a_2c_2b_3a_3, \ldots$. Also, the first sum is not less than the second. If, as the number of points a_i in the fixed length oe is increased in a suitable way, the two sums tend to a common limit, then this is called the area under the curve. This definition is clearly in agreement with our physical ideas.

The Riemann–Stieltjes integral generalizes the Riemann integral by allowing us to weight the areas of the elementary rectangles according to their position on the x axis. Suppose, for example, that the sheet from which the area oc_1eg in Fig. 1 is cut has mass $\rho(x)$ per unit area. Then the weight of the piece will be

$$w = \int_0^e f(x)\rho(x)\,dx \tag{1.1}$$

Write

$$\alpha(x) = \int_0^x \rho(x)\,dx \tag{1.2}$$

Then equation (1.1) can be written (formally at this stage)

$$w = \int_0^e f(x)\frac{d\alpha(x)}{dx}\,dx \tag{1.3}$$

$$= \int_0^e f(x)\,d\alpha(x) \tag{1.4}$$

So far nothing has been gained, but the theory of the Riemann–Stieltjes integral starts by defining (1.4) rather than (1.1), and when this is suitably done $\alpha(x)$ may be allowed to have simple discontinuities. If $\alpha(x)$ jumps from 1 to 1·5 at $x = 3$, for example, $\rho(3)$ is undefined. It will be shown below, however, that subject to appropriate conditions (1.4) remains well defined, and there is a contribution to the integral at $x = 3$ which is equal to $0{\cdot}5f(3)$. This corresponds to an idealized physical situation in which there is a concentration of mass on the line $x = 3$ of intensity $\frac{1}{2}$ per unit length. The density $\rho(x)$ cannot be defined at $x = 3$ but the physical situation is meaningful, and the freedom which the Riemann–Stieltjes integral gives us to deal with it is therefore valuable.

There are other generalizations of the Riemann integral, notably the Lebesgue integral. This is aimed at extending the class of functions $f(x)$ which have an integral. For example the 'Dirichlet function' is defined by

$$\left.\begin{aligned} f(x) &= 0, \quad x \text{ rational} \\ f(x) &= 1, \quad x \text{ irrational} \end{aligned}\right\} \tag{1.5}$$

In any finite interval $f(x)$ takes both the value 0 and the value 1, and the Riemann integral does not exist. The Lebesgue integral on the other hand exists and on the interval $[a, b]$ has the value $b - a$.

This example illustrates the type of function for which Lebesgue integration is needed. The difficulties for which it provides a solution arise generally from the mathematical formulation of a physical problem, rather than from the nature of the problem itself. That is to say, they can usually be avoided by changing the mathematical formulation, and this will be the course adopted here.

2 Definition of the integral

Attention will be restricted first to functions f with domain $S \subset E_1$, and range $R_f \subset E_1$; more explicitly, S will be an interval $[a, b]$. A second function α, which will occur in the definition, is assumed also to have domain $[a, b]$ and range $R_\alpha \subset E_1$.

It will be necessary on many occasions to consider sets of points x_i subdividing $[a, b]$:

$$a = x_0 < x_1 < \cdots < x_{n-1} < x_n = b \tag{2.1}$$

A notation for such subdivision is needed, and the following will be used. The set of points x_i in (2.1) will be called a partition (or dissection) $f[a, b]$, and denoted by P (cf. Section 9.1 of Chapter 4). The *norm* $|P|$ of a partition is the largest interval $x_i - x_{i-1}$. A *refinement* P_1 of a partition P is a new partition obtained from P by intercalating fresh points in P. Clearly $|P_1| \leqslant |P|$.

Example 2.1 If $[a, b] = [0, 1]$, a partition P of $[a, b]$ is

$$P = \{0, 0{\cdot}5, 1\}$$

and $|P| = 0{\cdot}5$. The partition

$$P_1 = \{0, 0{\cdot}1, 0{\cdot}5, 1\}$$

is a refinement of P, and

$$|P_1| = 0{\cdot}5 = |P|$$

The partition P_2,

$$P_2 = \{0, 0{\cdot}3, 0{\cdot}6, 1\}$$

which has $|P_2| = 0{\cdot}4 < |P|$, is nevertheless not a refinement of P because the point $x_i = 0{\cdot}5$ which occurs in P is absent from P_2.

Suppose that P is a partition of $[a, b]$, and let the set of points t_k, $k = 1, 2, \ldots, n$, satisfy

$$x_{k-1} \leqslant t_k \leqslant x_k \tag{2.2}$$

By a *Riemann–Stieltjes* sum of f with respect to α will be meant a sum

$$\sum_{k=1}^{n} f(t_k)[\alpha(x_k) - \alpha(x_{k-1})] \triangleq S(P, f, \alpha) \tag{2.3}$$

EXERCISE 2.1 Let f and α be defined by

$$f(t) = U(t - a), \qquad \text{that is,} \quad f(t) = \begin{cases} 0, & t \leqslant a \\ 1, & t > a \end{cases}$$

$$\alpha(t) = U(b - t), \qquad \text{that is,} \quad \alpha(t) = \begin{cases} 1, & t < b \\ 0, & t \geqslant b \end{cases}$$

and let $P = \{a, \frac{1}{2}(a + b), b\}$. Show that $S(P, f, \alpha) = -1$, and that refinement of P does not change this sum.

The function f is said to be *integrable with respect to* α (in the Riemann–Stieltjes sense), and the *Riemann–Stieltjes integral* of f with respect to α on $[a, b]$ is said to be I, if I exists with the following properties:

Given any $\epsilon > 0$, there exists a partition P_ϵ such that $|S(P, f, \alpha) - I| < \epsilon$, where P is any refinement of the partition P_ϵ, and where in $S(P, f, \alpha)$ the points t_k may be chosen anywhere in $[x_{k-1}, x_k]$.

Example 2.2 In Exercise 2.1, $I = -1$ satisfies this definition, with $P_\epsilon = \{a, \frac{1}{2}(a + b), b\}$. Therefore f is integrable with respect to α, where f and α are the functions defined in Exercise 2.1.

An alternative notation is $f \in R(\alpha)$ on $[a, b]$, which means that f is integrable with respect to α on $[a, b]$, in the Riemann–Stieltjes sense. The integral I is written

$$I = \int_a^b f(x)\, d\alpha(x) \tag{2.4}$$

or

$$I = \int_a^b f\, d\alpha \tag{2.5}$$

Here f is called the *integrand* and α the *integrator*. The ordinary Riemann integral corresponds to the special case $\alpha(x) = x$, and the notation for the integral is then

$$I = \int_a^b f(x)\, dx$$

Now if f is a vector-valued function of x, with domain $S \subset E_1$, range $R_f \subset E_m$, and if α as before has domain $S \subset E_1$, range $R_\alpha \subset E_1$, the integral

$$I = \int_a^b f(x)\, d\alpha(x) \tag{2.6}$$

is defined in the natural way. That is to say, I is an m-vector having as its ith component

$$I_i = \int_a^b f_i(x)\, d\alpha(x) \tag{2.7}$$

provided that each of the I_i, $i = 1, 2, \ldots, m$, exists according to the earlier definition.

Similarly if f and α are both m-vectors,

$$I = \int_a^b f^{\mathrm{T}}(x)\, d\alpha(x) \tag{2.8}$$

is defined by

$$I = \int_a^b \sum_{i=1}^{m} f_i(x)\, d\alpha_i(x) \tag{2.9}$$

The definition can be further extended to the case in which $A(x) = (a_{ij}(x))$ is a $p \times q$ matrix, and $B(x) = (b_{jk}(x))$ is a $q \times r$ matrix. The integral

$$I = \int_a^b A(x)\, dB(x) \tag{2.10}$$

is then a $p \times r$ matrix having as its components

$$I_{ik} = \int_a^b \sum_{j=1}^{q} a_{ij}(x)\, db_{jk}(x) \tag{2.11}$$

3 Properties of the integral

It will be shown in the next section that sufficient (though not necessary) conditions for the existence of the Riemann–Stieltjes integral (2.4) are $f \in D_L^1$ on I, $\alpha \in D_R^1$ on I, where the open interval I contains $[a, b]$. As a preliminary to the proof of this result, certain properties of the integral are developed which the integral must possess whenever it exists. The first of these is the formula for *integration by parts*, which follows.

Theorem 3.1 Let f and α be real-valued functions defined on $[a, b]$. Then if $f \in R(\alpha)$ on $[a, \mathrm{b}]$ it follows that $\alpha \in R(f)$ on $[a, b]$, and the two integrals which are thus asserted to exist satisfy

$$\int_a^b f(x)\, d\alpha(x) + \int_a^b \alpha(x)\, df(x) = f(b)\alpha(b) - f(a)\alpha(a) \tag{3.1}$$

Proof Let $S(P, \alpha, f)$ be a Riemann–Stieltjes sum for the integral $\int_a^b \alpha\, df$. Then

$$S(P, \alpha, f) = \sum_{k=1}^{n} \alpha(t_k)\{f(x_k) - f(x_{k-1})\} \tag{3.2}$$

$$= f(x_n)\alpha(x_n) - f(x_0)\alpha(x_0) - \sum_{k=1}^{n} \{f(x_k)\alpha(x_k) - f(x_{k-1})\alpha(x_{k-1})\}$$

$$+ \sum_{k=1}^{n} \alpha(t_k)\{f(x_k) - f(x_{k-1})\} \tag{3.3}$$

because the terms introduced in (3.3) sum to zero. So, using $x_0 = a$, $x_n = b$,

$$S(P, \alpha, f) = f(b)\alpha(b) - f(a)\alpha(a) - \sum_{k=1}^{n} f(x_k)\{\alpha(x_k) - \alpha(t_k)\}$$

$$- \sum_{k=1}^{n} f(x_{k-1})\{\alpha(t_k) - \alpha(x_{k-1})\} \tag{3.4}$$

Now let P_1 be a refinement of P consisting of the original points x_k and the points t_k. The sum

$$\sum_{k=1}^{n} f(x_k)\{\alpha(x_k) - \alpha(t_k)\} + \sum_{k=1}^{n} f(x_{k-1})\{\alpha(t_k) - \alpha(x_{k-1})\} \tag{3.5}$$

will then be recognized as a Riemann–Stieltjes sum $S(P_1, f, \alpha)$ in which a particular choice has been made of the points where f is evaluated, namely at the left- or right-hand end of each of the constituent intervals of P_1.

Because $\int_a^b f\, d\alpha$ exists, there is an I and a P_ϵ such that for any refinement P of P_ϵ,

$$|S(P, f, \alpha) - I| < \epsilon \tag{3.6}$$

But with this P

$$S(P, \alpha, f) = f(b)\alpha(b) - f(a)\alpha(a) - S(P_1, f, \alpha) \tag{3.7}$$

where P_1 is a refinement of P, and therefore of P_ϵ. So by another application of (3.6),

$$|S(P_1, f, \alpha) - I| < \epsilon \tag{3.8}$$

which by (3.7) gives

$$|f(b)\alpha(b) - f(a)\alpha(a) - S(P, \alpha, f) - I| < \epsilon \tag{3.9}$$

where in $S(P, \alpha, f)$, P is any refinement of P_ϵ and the points t_k were arbitrarily chosen. Inequality (3.9) therefore shows, by the definition of a Riemann–Stieltjes integral, that $\alpha \in R(f)$ and

$$\int_a^b \alpha\, df = f(b)\alpha(b) - f(a)\alpha(a) - I \tag{3.10}$$

where

$$I = \int_a^b f\,d\alpha \tag{3.11}$$

This completes the proof.

Some other properties of the Riemann–Stieltjes integral, which follow in a simple way from the definition and the properties of limits, are given below as exercises.

EXERCISE 3.1 If $f \in R(\alpha)$ on $[a, b]$, show that $cf \in R(\alpha)$ on $[a, b]$, where c is any real constant, and

$$\int_a^b cf\,d\alpha = c\int_a^b f\,d\alpha \tag{3.12}$$

EXERCISE 3.2 If $h = f + g$ and $f, g \in R(\alpha)$ on $[a, b]$, show that $h \in R(\alpha)$ on $[a, b]$, and

$$\int_a^b h\,d\alpha = \int_a^b f\,d\alpha + \int_a^b g\,d\alpha \tag{3.13}$$

EXERCISE 3.3 If $f \in R(\alpha)$ on $[a, b]$, and $f \in R(\beta)$ on $[a, b]$, show that $f \in R(c_1\alpha + c_2\beta)$ on $[a, b]$, where c_1, c_2 are any real constants, and

$$\int_a^b f\,d(c_1\alpha + c_2\beta) = c_1\int_a^b f\,d\alpha + c_2\int_a^b f\,d\beta \tag{3.14}$$

EXERCISE 3.4 Show that equations (2.9) and (2.11) can be written

$$I = \sum_{i=1}^{m}\int_a^b f_i(x)\,d\alpha_i(x) \tag{3.15}$$

$$I_{ik} = \sum_{j=1}^{q}\int_a^b a_{ij}(x)\,db_{jk}(x) \tag{3.16}$$

provided that each integral in the sums in (3.15) and (3.16) exists. Extend the results in Exercises 3.1, 3.2, and 3.3 to integrals of the type of (2.8) and (2.10).

EXERCISE 3.5 Extend the formula for integration by parts to show that if f and α are m-vectors

$$\int_a^b f^{\mathrm{T}}\,d\alpha + \int_a^b \alpha^{\mathrm{T}}\,df = f^{\mathrm{T}}(b)\alpha(b) - f^{\mathrm{T}}(a)\alpha(a) \tag{3.17}$$

and if A is a $p \times q$ matrix, B a $q \times r$ matrix,

$$\int_a^b A\,dB + \left\{\int_a^b B^{\mathrm{T}}\,dA^{\mathrm{T}}\right\}^{\mathrm{T}} = A(b)B(b) - A(a)B(a) \tag{3.18}$$

[Where no confusion can arise, (3.18) may be written

$$\int_a^b A\,dB + \int_a^b dA\,.\,B = A(b)B(b) - A(a)B(a)]$$

EXERCISE 3.6 If $f \in R(\alpha)$ on $[a, b]$, show that $f \in R(\alpha)$ on $[b, a]$, and

$$\int_b^a f\,d\alpha = -\int_a^b f\,d\alpha \tag{3.19}$$

EXERCISE 3.7 If $f \in R(\alpha)$ on $[a, b]$, and $\beta(b - s) = \alpha(s)$, show that

$$\int_a^b f(s)\,d\alpha(s) = \int_a^b f(s)\,d\beta(b - s) = -\int_a^b f(a + b - s)\,d\beta(s - a) \tag{3.20}$$

EXERCISE 3.8 If $f \in R(\alpha)$ on $[a, c]$ and $f \in R(\alpha)$ on $[a, b]$, $c > b$, show that $f \in R(\alpha)$ on $[b, c]$, and

$$\int_a^c f\,d\alpha = \int_a^b f\,d\alpha + \int_b^c f\,d\alpha \tag{3.21}$$

Show that this result remains true if $b > c$. Extend the result to a finite collection of contiguous intervals.

EXERCISE 3.9 Show that when the integral $\int_a^b d\alpha$ exists, its value is $\alpha(b) - \alpha(a)$.

4 Existence of the integral

It will be recalled that, in Chapter 4, physical variables were assumed to be continuous on the left, and to belong in fact to D_R or to D_R^1. It will be seen later that such variables often occur as integrands or integrators, and that they usually occur as functions of time t, in which case they are as stated in D_R or D_R^1. Sometimes, however, they occur as functions of $-t$, and in such a case they belong to D_L or to D_L^1.

Example 4.1 The function defined by

$$U(t) = \begin{cases} 0, & t \leqslant 0 \\ 1, & t > 0 \end{cases}$$

is continuous on the left and is in D_R: it is also in D_R^1. The function defined by

$$U(-t) = \begin{cases} 1, & t < 0 \\ 0, & t \geqslant 0 \end{cases}$$

is continuous on the right and is in D_L: it is also in D_L^1. The function represented by $U(-t)$ does not represent a physical variable but it nevertheless occurs frequently in later results.

Consider first the special case where the integrator $\alpha(x)$ is a unit step $U(x - c)$, $a \leqslant c \leqslant b$, and where $f \in D_L$ on an open interval $I \supset [a, b]$. Then we have:

Theorem 4.1 If $f \in D_L$ on $I \supset [a, b]$, the integral

$$\int_a^b f(x)\, dU(x - c), \quad a \leqslant c \leqslant b \tag{4.1}$$

exists and has the value $f(c)$ when $a \leqslant c < b$. When $c = b$, the integral is zero.

Proof Let P be a partition of $[a, b]$ containing the point $x_{r-1} = c$, and suppose first that $a \leqslant c < b$. Then because $U(x - c)$ is constant except at c, the only contribution to the sum $S[P, f(x), U(x - c)]$ is a single term

$$S[P, f(x), U(x - c)] = f(t_r)[U(x_r - c) - U(x_{r-1} - c)] \tag{4.2}$$

$$= f(t_r) \tag{4.3}$$

where $c = x_{r-1} \leqslant t_r \leqslant x_r$. Now $f(x) \in D_L$, and so is continuous on the right. So given ϵ we may choose δ such that

$$|f(t_r) - f(c)| < \epsilon \quad \text{if } 0 < t_r - c < \delta \tag{4.4}$$

Choose for P_ϵ any partition of $[a, b]$ including c and having $|P_\epsilon| < \delta$. Then by (4.3) and (4.4), if P is any refinement of P_ϵ,

$$|S[P, f(x), U(x - c)] - f(c)| < \epsilon \tag{4.5}$$

That is, $f(x) \in R[U(x - c)]$ and the integral has the value $f(c)$. When $c = b$, $U(x - c) = 0$ everywhere in $[a, b]$, and so the integral exists and is zero.

From the method of proof it is clear that

$$\int_a^b f(x)\, dU(c - x) \tag{4.6}$$

exists if $f(x) \in D_R$ on $I \supset [a, b]$ and $a \leqslant c \leqslant b$, and that the integral has the value $-f(c)$ if $a < c \leqslant b$, and 0 if $c = a$. In (4.1) and (4.6) it will be noticed that not both of f and α are discontinuous on the right at any point, nor are they both discontinuous on the left at any point. Apostol (1957), p. 212, shows that this is a necessary condition for the existence of the integrals.

Example 4.2 A function $f \in D_R$ on $I \supset [0, T]$ can be written

$$f(t - c) = \int_c^t f(t - s)\, dU(s - c) \tag{4.7}$$

if $0 \leqslant c \leqslant t \leqslant T$. Note that if $f(t - s) = \phi(s)$, $\phi(s) \in D_L$ on $[c, t]$.

Example 4.3 The partial sums S_n of a series

$$S_n = f(0) + f(1) + \cdots + f(n) \tag{4.8}$$

where f is continuous on any finite interval, can be written

$$S_n = \int_0^{n+1/2} f(x)\, d\alpha(x) \tag{4.9}$$

where

$$\alpha(x) = U(x) + U(x-1) + \cdots + U(x-n) \tag{4.10}$$

As a second special case we have:

Theorem 4.2 If α is continuous, and f has bounded variation, on $[a, b]$, then $f \in R(\alpha)$.

Proof By Theorem 9.2 of Chapter 4 it is sufficient to assume that f is non-decreasing on $[a, b]$. What will be proved is that $\alpha \in R(f)$. The required theorem then follows from Theorem 3.1. Let P be a partition of $[a, b]$ and let $S(P, \alpha, f)$ be a Riemann–Stieltjes sum corresponding to P. Then $S(P, \alpha, f)$ is a function of all the t_k, which are subject to choice in the range $x_{k-1} \leqslant t_k \leqslant x_k$. The contribution to $S(P, \alpha, f)$ depending on one particular t_i is

$$\alpha(t_i)[f(x_i) - f(x_{i-1})] \tag{4.11}$$

Because α is continuous on the closed interval $[x_{i-1}, x_i]$, this expression takes its maximum value for some t_i in $[x_{i-1}, x_i]$. When all the t_k are adjusted in this way, let the value of $S(P, \alpha, f)$ be $S_{\max}(P, \alpha, f)$. Similarly let $S_{\min}(P, \alpha, f)$ be the result of choosing every t_k to minimize $S(P, \alpha, f)$. Clearly $S_{\max}(P, \alpha, f) \geqslant S_{\min}(P, \alpha, f)$ for any P.

Any refinement P_1 of P can be obtained by the successive addition of single points. Let y be intercalated in this way between x_{i-1} and x_i. Then the contribution to $S_{\max}(P_1, \alpha, f)$ from the intervals $[x_{i-1}, y]$ and $[y, x_i]$ is

$$\max_{s_1 \in [x_{i-1}, y]} \alpha(s_1)[f(y) - f(x_{i-1})] + \max_{s_2 \in [y, x_i]} \alpha(s_2)[f(x_i) - f(y)]$$

$$\leqslant \max_{s_1 \in [x_{i-1}, x_i]} \alpha(s_1)[f(y) - f(x_{i-1})] + \max_{s_2 \in [x_{i-1}, x_i]} \alpha(s_2)[f(x_i) - f(y)] \tag{4.12}$$

$$= \max_{t_i \in [x_{i-1}, x_i]} \alpha(t_i)[f(y) - f(x_{i-1}) + f(x_i) - f(y)] \tag{4.13}$$

$$= \max_{t_i \in [x_{i-1}, x_i]} \alpha(t_i)[f(x_i) - f(x_{i-1})] \tag{4.14}$$

because in the right-hand side of (4.12) the range of variation of s_1 and s_2 is increased, and in (4.13), $f(y) - f(x_{i-1})$ and $f(x_i) - f(y)$ are both non-negative. The expression on the right-hand side of (4.14) is the contribution to $S_{\max}(P, \alpha, f)$

before the new point s was added. Therefore $S_{max}(P, \alpha, f)$ cannot increase as P is refined. Similarly $S_{min}(P, \alpha, f)$ cannot decrease as P is refined.

Now consider successive refinements $P_2, P_3, \ldots$ of a given P_1, such that $|P_r| \to 0$ as $r \to \infty$. The sequence

$$S_{max}(P_1, \alpha, f), S_{max}(P_2, \alpha, f), S_{max}(P_3, \alpha, f), \ldots \tag{4.15}$$

is non-increasing, and is bounded below by $S_{min}(P_1, \alpha, f)$, because

$$S_{max}(P_r, \alpha, f) \geqslant S_{min}(P_r, \alpha, f) \geqslant S_{min}(P_1, \alpha, f) \tag{4.16}$$

Consequently the sequence (4.15) has a limit L. Similarly the sequence

$$S_{min}(P_1, \alpha, f), S_{min}(P_2, \alpha, f), S_{min}(P_3, \alpha, f), \ldots \tag{4.17}$$

has a limit l. It will now be shown that $l = L$.

The variation of the monotonic function f on $[a, b]$ has the upper bound $M = f(b) - f(a)$. Now α is continuous, and therefore uniformly continuous, on $[a, b]$, and consequently, for any $\epsilon > 0$, there exists a δ such that if r_0 is chosen to make $|P_r| < \delta$ whenever $r > r_0$,

$$|\alpha(s) - \alpha(t)| < \frac{\epsilon}{M}, \quad x_{i-1} \leqslant s, t \leqslant x_i \tag{4.18}$$

where x_{i-1}, x_i are points of P_r with $r > r_0$, and so $|s - t| < \delta$. The difference

$$S_{max}(P_r, \alpha, f) - S_{min}(P_r, \alpha, f)$$
$$= \sum_{i=1}^{n} [\alpha(t_i) - \alpha(s_i)][f(x_i) - f(x_{i-1})] \geqslant 0 \tag{4.19}$$

where t_i, s_i are respectively the maximizing and minimizing values. So for all $r > r_0$

$$S_{max}(P_r, \alpha, f) - S_{min}(P_r, \alpha, f) < \frac{\epsilon}{M} \sum_{i=1}^{n} [f(x_i) - f(x_{i-1})] \tag{4.20}$$
$$= \epsilon \tag{4.21}$$

This means that

$$0 \leqslant L - l \leqslant S_{max}(P_r, \alpha, f) - S_{min}(P_r, \alpha, f) < \epsilon \tag{4.22}$$

where ϵ is arbitrary, whence $L - l$ cannot be other than zero. Let the common value of L and l be I. It will be shown that I fulfils the requirements of a Riemann–Stieltjes integral.

Let P_r be a partition satisfying (4.21), and let $S(P, \alpha, f)$ be any Riemann–Stieltjes sum corresponding to a refinement P of P_r: that is, the t_k may be chosen arbitrarily in their respective intervals. Clearly

$$S_{min}(P, \alpha, f) \leqslant S(P, \alpha, f) \leqslant S_{max}(P, \alpha, f) \tag{4.23}$$

and by this and what has been proved above

$$S_{\min}(P_r, \alpha, f) \leqslant S_{\min}(P, \alpha, f) \leqslant S(P, \alpha, f)$$
$$\leqslant S_{\max}(P, \alpha, f) \leqslant S_{\max}(P_r, \alpha, f) \tag{4.24}$$

$$S_{\min}(P_r, \alpha, f) \leqslant I \leqslant S_{\max}(P_r, \alpha, f) \tag{4.25}$$

$$S_{\max}(P_r, \alpha, f) - S_{\min}(P_r, \alpha, f) < \epsilon \tag{4.26}$$

It follows that

$$|S(P, \alpha, f) - I| < \epsilon \tag{4.27}$$

where $S(P, \alpha, f)$ is any Riemann–Stieltjes sum corresponding to a refinement P of P_r, where P_r depends on ϵ and ϵ can be chosen at will. This completes the proof.

EXERCISE 4.1 If $f \in R(\alpha)$ on $[a, b]$ and $|f| < M_1$ on $[a, b]$, and if the variation of α is bounded by M_2 on $[a, b]$, show that

$$\left|\int_a^b f\, d\alpha\right| < M_1 M_2 \tag{4.28}$$

EXERCISE 4.2 Show that the Riemann integral $\int_a^b f(x)\, dx$ exists if either f is continuous on $[a, b]$, or f has bounded variation on $[a, b]$.

On the basis of Theorems 4.1 and 4.2 it is now easy to obtain the following theorem.

Theorem 4.3 Let f be a function belonging to D_L^1 on an open interval $I \supset [a, b]$, and let α be a function belonging to D_R^1 on I. Then $f \in R(\alpha)$ on $[a, b]$.

Proof The function α, because it belongs to D_R^1, has at most a finite number of simple discontinuities on the right in $[a, b]$. Let these be at $p_1 < p_2 < \cdots < p_n$, and let the corresponding jumps of α be $\alpha_1, \alpha_2, \ldots, \alpha_n$, these being reckoned positive when $\alpha(x)$ increases with x. Then the function

$$\gamma(x) = \alpha(x) - \sum_{i=1}^{n} \alpha_i U(x - p_i) = \alpha(x) - \beta(x) \tag{4.29}$$

is continuous on $[a, b]$.

The function f belongs to D_L^1, and therefore to D_L, on I. So by Theorem (4.1), $f \in R(\beta)$ on $[a, b] \subset I$, and in fact

$$\int_a^b f(x)\, d\beta(x) = \begin{cases} \sum_{i=1}^{n} \alpha_i f(p_i), & p_n < b \\ \sum_{i=1}^{n-1} \alpha_i f(p_i), & p_n = b \end{cases} \tag{4.30}$$

Because f belongs to D_L^1 it has bounded variation on $[a, b]$ by Theorem 9.1 of Chapter 4. Then, by Theorem 4.2, because γ is continuous on $[a, b]$, $f \in R(\gamma)$. Finally

$$\int_a^b f(x)\, d\alpha(x) = \int_a^b f(x)\, d[\beta(x) + \gamma(x)] \tag{4.31}$$

$$= \int_a^b f(x)\, d\beta(x) + \int_a^b f(x)\, d\gamma(x) \tag{4.32}$$

by Exercise 3.3. Both of these last integrals have been shown to exist, and so $f \in R(\alpha)$ on $[a, b]$.

EXERCISE 4.3 Show that if $f \in D_R$ on $I \supset [a, b]$, the Riemann integral $\int_a^b f(x)\, dx$ exists. [Let the discontinuity of $f(x)$ at $x = p_i$ be f_i and write

$$f(x) = \phi(x) + \sum_{i=1}^{n} f_i U(x - p_i) = \phi(x) + \xi(x) \tag{4.33}$$

so that $\phi(x)$ is continuous and $\xi(x)$ has bounded variation.]

EXERCISE 4.4 If $f \in D_R$ on $I \supset [a, b]$, show that the Riemann integral $\int_a^b |f(x)|\, dx$ exists, and

$$\left| \int_a^b f(x)\, dx \right| \leqslant \int_a^b |f(x)|\, dx \tag{4.34}$$

[Show that $|f| \in D_R$, so that the integral in question exists. For any partition of $[a, b]$, compare the Riemann sums to obtain (4.34).]

EXERCISE 4.5 Let the function f be defined by

$$f(x) = \begin{cases} 0, & x \neq c \\ 1, & x = c, \quad a \leqslant c \leqslant b \end{cases} \tag{4.35}$$

Show that $\int_a^b f(x)\, dx$ exists and is zero. Hence show that if $\int_a^b g(x)\, dx$ exists, its value is unchanged by alteration of the value of g at any finite set of points in $[a, b]$. In particular

$$\int_0^t g(x) U(x)\, dx = \int_0^t g(x)\, dx \tag{4.36}$$

provided that the second integral in (4.36) exists.

The classes of functions used in Theorem 4.3, namely D_L^1 and D_R^1, can certainly be widened; indeed in Theorem 4.1, α belonged to the wider class D_R and in Theorem 4.2, f belonged to the wider class of functions having bounded variation. The aim here, however, is not to make the results as general as possible, but rather to give results which are adequate for the applications envisaged.

4.1 A mean-value theorem for integrals

The following result will be needed later.

Theorem 4.4 Let α be non-decreasing, and therefore of bounded variation, on $[a, b]$. Let f be continuous on $[a, b]$. Then

$$\int_a^b f\,d\alpha = f(c)[\alpha(b) - \alpha(a)], \quad a \leqslant c \leqslant b \tag{4.37}$$

Proof If $\alpha(b) = \alpha(a)$, equation (4.37) is true. Assume $\alpha(b) > \alpha(a)$. Because f is continuous on $[a, b]$ it has a maximum M and a minimum m on $[a, b]$. Also by the continuity of f, if any y is given, $m \leqslant y \leqslant M$, then there is a $c \in [a, b]$ such that $f(c) = y$. Every Riemann–Stieltjes sum satisfies

$$m[\alpha(b) - \alpha(a)] \leqslant S(P, \alpha, f) \leqslant M[\alpha(b) - \alpha(a)] \tag{4.38}$$

whence (4.25) gives

$$m \leqslant \frac{S_{\min}(P_r, \alpha, f)}{\alpha(b) - \alpha(a)} \leqslant y = \frac{I}{\alpha(b) - \alpha(a)} \leqslant \frac{S_{\max}(P_r, \alpha, f)}{\alpha(b) - \alpha(a)} \leqslant M \tag{4.39}$$

and so $y = f(c)$, which completes the proof.

EXERCISE 4.6 Use the theorem just proved to show that if f is continuous on $[a, b]$ then the function g defined by

$$g(t) = \int_a^t f(\tau)\,d\tau \tag{4.40}$$

is also continuous on $[a, b]$.

5 Step response of a system

In this section a particularly important result will be developed on the basis of the preceding sections. It will be remembered from Chapter 1 that a space can be constructed having as its elements functions of some particular type—continuous functions, differentiable functions, etc. A mapping of such a function space into itself is known as an *operator*. Mathematically, the theorem which will be proved shows that a certain class of operators on a particular function space can be represented by *Riemann–Stieltjes convolution integrals*, that is, integrals of the type $\int_0^t f(t - \tau)\,dg(\tau)$.

For the sake of concreteness, however, and because it represents an important application of the theorem, the result will be developed for *physical systems*. By these we understand any object to which an *input* (a function of time) may be applied and which responds by producing an *output* (another function of time). Examples are: electrical circuits with input voltage v and output current i; mechanical systems with input force f and output displacement x; and a heated metal bar with input heat flow w and output temperature θ at a given point. Control systems are another example. Though we choose to speak of physical systems it will be clear that systems could be defined abstractly, and could then be identified with suitable operators.

Suppose then that a *linear system* is given, with one input variable u and one output variable y; both u and y are real-valued functions of time. By *linearity* of the system we mean that if y_1, y_2 are the outputs corresponding to u_1, u_2 respectively, then the output corresponding to $c_1u_1 + c_2u_2$ is $c_1y_1 + c_2y_2$, for any real constants c_1 and c_2. The system will be assumed to have the following further properties:

(i) It is *non-anticipative* [sometimes called *physically realizable* or *causal*]: that is to say, if $u(t) = 0$, $t \leqslant 0$, then $y(t) = 0$, $t \leqslant 0$.

(ii) If $u(t) = 0$, $t \leqslant 0$, and $|u(t)| < M_1$, $0 < t \leqslant T$, then $|y(t)| < M_2$, $0 < t \leqslant T$ (where M_2 depends only on M_1 and T, and not on u).

(iii) The system is *time-invariant*: that is, if $y(t)$ is the response to $u(t)$, then $y(t - \tau)$ is the response to $u(t - \tau)$, $\tau > 0$.

A particularly important quantity associated with the system is its *step response* $H(t)$: this is the particular output obtained when the input is a unit step at $t = 0$. It will be assumed that $H \in D_R^1$. This and the earlier properties will be shown in Chapter 6 to hold when the system can be described mathematically in certain common ways.

By Example 4.1, if $0 < t \leqslant T$,

$$H(t) = \int_0^t H(t - s)\, dU(s) \tag{5.1}$$

This equation is true whether or not $H(t)$ is the step response. When $H(t)$ is the step response, however, the equation may be regarded as giving the output $H(t)$ in terms of the input $U(t)$. Moreover, if the input is $U(t - c)$, where $0 \leqslant c \leqslant T$, then the output, by assumption (iii), is $H(t - c)$ which can be written, for $0 \leqslant c < t \leqslant T$,

$$H(t - c) = \int_c^t H(t - s)\, dU(s - c) \tag{5.2}$$

by another application of equation (4.7). Because $H(0) = 0$, equation (5.2) holds also for $0 \leqslant c \leqslant t \leqslant T$, with the convention $\int_c^c = 0$. Also because

$$U(x) = 0, \quad x \leqslant 0$$

$$0 = \int_0^c H(t - s)\, dU(s - c), \quad 0 \leqslant c \leqslant t \tag{5.3}$$

So for $0 \leqslant c \leqslant t \leqslant T$, on adding (5.2) and (5.3),

$$H(t - c) = \int_0^t H(t - s)\, dU(s - c) \tag{5.4}$$

which is also true for $0 \leqslant t < c \leqslant T$ because $U(x) = 0$, $x < 0$. Therefore (5.4) is true for all c, t satisfying $0 \leqslant c \leqslant T$, $0 \leqslant t \leqslant T$. If now the input is a sum of step functions,

$$v_n(t) = \sum_{i=1}^{r_n} v_{ni} U(t - c_i) \tag{5.5}$$

where $0 \leqslant c_i \leqslant T$, $i = 1, 2, \ldots, r_n$, then the corresponding output $y_n(t)$, by the linearity of the system, will be the sum of the responses to the individual $v_{ni}U(t - c_i)$. It follows from (5.4) that $y_n(t)$ can be written

$$y_n(t) = \sum_{i=1}^{r_n} v_{ni} \int_0^t H(t - s)\, dU(s - c_i) \tag{5.6}$$

$$= \int_0^t H(t - s)\, d\left[\sum_{i=1}^{r_n} v_{ni} U(s - c_i)\right] \tag{5.7}$$

$$= \int_0^t H(t - s)\, dv_n(s) \tag{5.8}$$

On using the formula for integration by parts, and the fact that $H(0) = v_n(0) = 0$, equation (5.8) becomes

$$y_n(t) = -\int_0^t v_n(s)\, dH(t - s) \tag{5.9}$$

Now suppose that $u \in D_R^{\frac{1}{R}}$ is some given input to the system with $u(t) = 0$ for $t \leqslant 0$ and that a sequence $v_1(t), v_2(t), \ldots$ can be found, each given by (5.5), such that given any δ there exists an n_0 with the property that

$$|u(t) - v_n(t)| < \delta, \quad 0 \leqslant t \leqslant T \tag{5.10}$$

whenever $n > n_0$. The existence of such a sequence will be established below. Because $H \in D_R^{\frac{1}{R}}$ it has variation bounded by M on $[0, T]$, and by Exercise 4.1

$$\left| \int_0^t [v_n(s) - u(s)]\, dH(t - s) \right| < M\delta \tag{5.11}$$

That is,

$$\left| -y_n(t) - \int_0^t u(s)\, dH(t - s) \right| < M\delta \tag{5.12}$$

and so as $n \to \infty$, $v_n(t) \to u(t)$, $0 \leqslant t \leqslant T$ and

$$y_n(t) \to -\int_0^t u(s)\, dH(t-s), \quad 0 \leqslant t \leqslant T \tag{5.13}$$

$$= \int_0^t H(t-s)\, du(s) \tag{5.14}$$

on integrating by parts and using $H(0) = u(0) = 0$.

On the other hand, the response y_δ of the system to the input $v_n - u$ satisfies, by linearity and assumption (ii),

$$|y_\delta(t)| = |y_n(t) - y(t)| < \frac{M_2\delta}{M_1} \tag{5.15}$$

where y is the response to u. So as $n \to \infty$, $y_n(t) \to y(t)$, $0 \leqslant t \leqslant T$. Using this with (5.14) it follows that the response to $u(t)$ is

$$y(t) = \int_0^t H(t-s)\, du(s) \tag{5.16}$$

Finally, it has to be established that for any $u \in D_R^1$, with $u(t) = 0$ for $t \leqslant 0$, there always exists a sequence $v_1(t), v_2(t), \ldots$ having the properties assumed above. From this it will follow that (5.16) is true whenever the assumptions about the system are fulfilled and u is in D_R^1 and is zero for $t \leqslant 0$.

Since $u \in D_R^1$, its derivative is bounded where it exists: say $|Du| < M_3$. Let P be a partition of $[0, T]$ containing all those points where u or Du is discontinuous, and with $|P| < \delta/M_3$. Let

$$0 = x_0 < x_1 < \cdots < x_n = T \tag{5.17}$$

be the points of P, and define v_n by

$$v_n(t) = \sum_{i=1}^{n} [u(x_i) - u(x_{i-1})]U(t - x_{i-1}) \tag{5.18}$$

as illustrated in Fig. 2. It will be shown that $v_n(t)$ has the desired property (5.10).

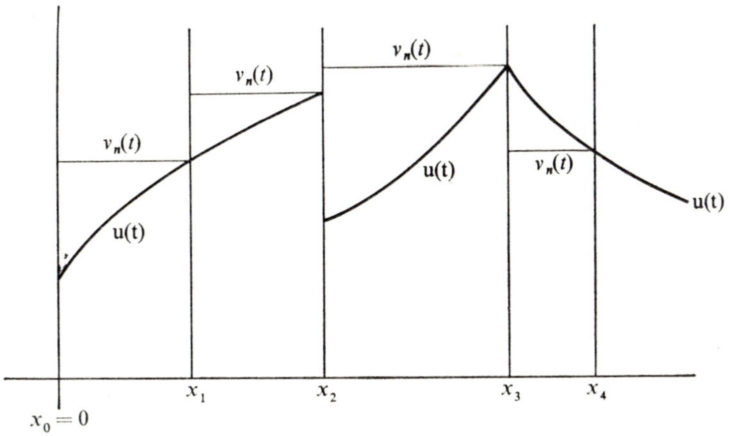

FIG. 2

First it follows from (5.18) that

$$v_n(x_i) - u(x_i) = 0, \quad i = 0, 1, \ldots, n \tag{5.19}$$

Also $v_n \in D_R^1$ and $u \in D_R^1$, whence $w_n = v_n - u \in D_R^1$, so $w_n(t)$ is everywhere continuous on the left and $w_n(x_i) = 0$. Let t_0 be any point in $[0, T]$. Then either t_0 is one of the x_i, when $w_n(t_0) = 0$, or t_0 is an interior point of an interval $[x_{k-1}, x_k]$. By the construction of P, w_n is then continuous on $[t_0, x_k]$, and Dw_n exists on (t_0, x_k) and satisfies $|Dw_n| < M_3$.

By the mean-value theorem

$$|w_n(t_0)| = |w_n(t_0) - w_n(x_k)| \leqslant M_3 \, |t_0 - x_k| < \delta \tag{5.20}$$

It follows that given $u(t)$ and δ, the required $v_n(t)$ satisfying (5.10) can be generated by (5.18) from any P having $|P| < \delta/M_3$. This concludes the proof.

Notice that the only point in the proof at which the time-invariance of the system was invoked is at equation (5.2). Consequently it is sufficient to assume that time-invariance holds when the input is a unit step. Time-invariance for arbitrary inputs then follows from equation (5.16). The result that has been proved can therefore be stated in the following way.

Theorem 5.1 Let a system be given with the following properties:

(a) It is linear.

(b) It is non-anticipative.

(c) If the input u satisfies $u(t) = 0$, $t \leqslant 0$, and $|u(t)| < M_1(T)$, $0 < t \leqslant T$, then the output y satisfies $|y(t)| < M_2(M_1, T)$, $0 < t \leqslant T$.

(d) If $H(t)$ is the response to $U(t)$, then $H(t - \tau)$ is the response to $U(t - \tau)$, $\tau > 0$.

(e) The response $H(t)$ to a unit step $U(t)$ satisfies $H \in D_R^1$.

Then the system is time-invariant and the response y to an input u satisfying $u(t) = 0$, $t \leqslant 0$; $u \in D_R^1$ is

$$y(t) = \int_0^t H(t - s) \, du(s) \tag{5.21}$$

Exercise 5.1 For the circuit shown in Fig. 3, the output voltage is given in terms of the input voltage by $y(t) = R_2 u(t)/(R_1 + R_2)$. Verify the conditions of Theorem 5.1 and write the equation for this circuit corresponding to equation (5.21). Verify that the equation gives the correct response when the input is $U(t - t_0)$, $t_0 > 0$. (Further examples will be found in Section 9.1 and in Chapter 6.)

Exercise 5.2 If condition (d) of Theorem 5.1 is not fulfilled, the response

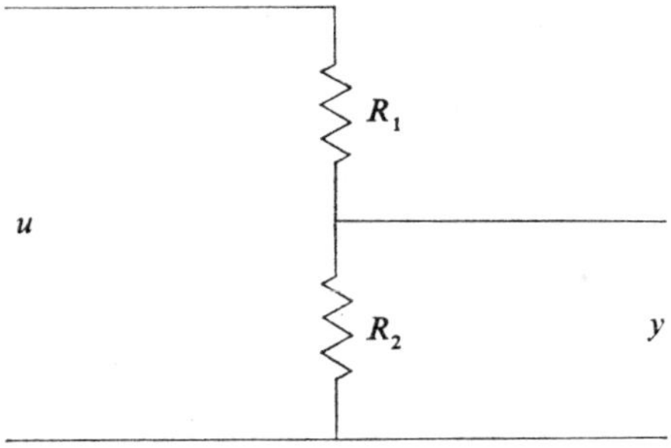

FIG. 3

$H(t, t_0)$ at t to a unit step $U(t - t_0)$ at t_0 will be a function of t and t_0 separately, and not simply of $t - t_0$. Let conditions (a), (b), and (c) be true, and let (e) be replaced by:

(e′) The response $H(t, t_0)$ at t to a unit step at t_0 is in D_L^1 as a function of t_0.

Show that equation (5.21) is then replaced by

$$y(t) = \int_0^t H(t, s)\, du(s) \tag{5.22}$$

[Note that $H(t)$ in (5.1) corresponds to $H(t, 0)$ in the notation of this exercise. The steps in the proof follow the same lines as before, with $H(t, s)$ in place of $H(t - s)$.]

EXERCISE 5.3 When a system has m inputs and l outputs, the place of the step response in Theorem 5.1 is taken by a matrix $H(t) = \big(H_{ij}(t)\big)$ in which $H_{ij}(t)$ is the step response at output i corresponding to a unit step at input j, all other inputs being zero. Find the equation corresponding to equation (5.21) which gives the output $y(t)$ (an m-vector) in terms of the input $u(t)$ (an l-vector). [Use the linearity of the system to superimpose the effects of the different inputs. The equation sought is formally identical with (5.21), but with H now representing a matrix, and u and y now representing vectors.]

6 Improper and infinite integrals

In Section 4, the integrals considered were on a finite interval $[a, b]$, and the conditions on f and α ensured that each of them was bounded on $[a, b]$. These conditions can sometimes be relaxed by allowing one or more of a, b, f to become infinite; the integral is then said to be *improper* and is defined by a limiting process. If it is a or b which becomes infinite, the improper integral is called an infinite integral. More specifically we have the following definitions.

Assume that $f \in R(\alpha)$ on an interval $[a, b]$, and write

$$I(b) = \int_a^b f\, d\alpha \tag{6.1}$$

If $I(b)$ tends to a finite limit I as $b \to \infty$ (with fixed a) then I is said to *converge* and is called an *infinite integral*: I is then written

$$I = \int_a^\infty f\, d\alpha \tag{6.2}$$

If $\int_a^\infty f\, d\alpha$ is said to *exist*, this implies that $\int_a^b f\, d\alpha$ exists for all $b > a$ and the integral converges. The infinite integral

$$I = \int_{-\infty}^b f\, d\alpha \tag{6.3}$$

is defined similarly.

Example 6.1 Define f and α by

$$f(x) = c^x, \qquad \alpha(x) = \sum_{r=1}^{\infty} (-)^r U(x - r) \tag{6.4}$$

Then f is continuous on any bounded open interval I and so $f \in D_L$ on $I \supset [0, b]$. Therefore by Theorem 4.1, $\int_0^b f\, d\alpha$ exists for all finite $b > 0$, and in fact its value obtained from the theorem is

$$I(b) = \sum_{r=0}^{[b]} (-)^r c^r \tag{6.5}$$

where $[b]$ is the greatest integer less than b. Hence

$$I(b) = \frac{1 - (-c)^{[b]+1}}{1 + c} \tag{6.6}$$

and if $|c| < 1$, $I(b) \to 1/(1 + c)$ as $b \to \infty$. So $\int_0^\infty f\, d\alpha$ exists when $|c| < 1$ and has the value $1/(1 + c)$. This example shows how an infinite series can be regarded as a particular case of an infinite integral.

Suppose that both the following infinite integrals exist:

$$\int_a^\infty f\, d\alpha, \qquad \int_{-\infty}^a f\, d\alpha \tag{6.7}$$

Then the infinite integral $\int_{-\infty}^\infty f\, d\alpha$ is defined by

$$\int_{-\infty}^\infty f\, d\alpha = \int_{-\infty}^a f\, d\alpha + \int_a^\infty f\, d\alpha \tag{6.8}$$

By equation (3.21) it follows that $\int_{-\infty}^\infty f\, d\alpha$ is independent of the choice of a.

Example 6.2 For any $f \in D_R$, the infinite integrals

$$\int_a^\infty f(x)\, dU(c - x), \qquad \int_{-\infty}^a f(x)\, dU(c - x) \tag{6.9}$$

both exist, and

$$\int_{-\infty}^\infty f(x)\, dU(c - x) = f(c) \tag{6.10}$$

One or other of the integrals (6.9) is zero according to the values of a and c.

If $\int_a^\infty f\, d\alpha$ and $\int_{-\infty}^a f\, d\alpha$ do not both exist it may still happen that

$$I(b) = \int_{-b}^b f\, d\alpha \tag{6.11}$$

tends to a finite limit I as $b \to \infty$. Then I is known as the *Cauchy principal value* of $\int_{-\infty}^\infty f\, d\alpha$.

Example 6.3 Define f and α by

$$f(x) = x, \quad \alpha(x) = \sum_{r=0}^{\infty} (-)^r U(x - r) - \sum_{r=0}^{-\infty} (-)^r U(r - x) \tag{6.12}$$

Then

$$\int_0^b f\, d\alpha = -1 + 2 - 3 + \cdots + (-)^{[b]}[b] \tag{6.13}$$

where $[b]$ is defined in Example 6.1. So $\int_0^\infty f\, d\alpha$ does not exist. But $\int_{-b}^b f\, d\alpha = 0$ for all b, so that the Cauchy principal value of $\int_{-\infty}^\infty f\, d\alpha$ exists and is zero.

Exercise 6.1 Show that the Cauchy principal value of $\int_{-\infty}^\infty x^3\, dx$ exists and is zero.

Suppose that $f(x) \to \infty$ as $x \to c$, and that $f \in R(\alpha)$ on $[a, b]$, $b < c$. If $I(b) = \int_a^b f\, d\alpha$ tends to a finite limit I as $b \to c$, then I is called an improper integral and is written

$$I = \int_a^c f\, d\alpha \tag{6.14}$$

Notice that the condition $b < c$ ensures that b tends to c from below. With the same f and α, the improper integral $\int_c^d f\, d\alpha$, $d > c$, is defined similarly.

If $f(x) \to \infty$ as $x \to c$, but $f \in R(\alpha)$ on $[a, b]$ and on $[d, e]$ for any $b < c < d$, and if both improper integrals $\int_a^c f\,d\alpha$ and $\int_c^e f\,d\alpha$ exist, then the improper integral $\int_a^e f\,d\alpha$ is defined by

$$\int_a^e f\,d\alpha = \int_a^c f\,d\alpha + \int_c^e f\,d\alpha \tag{6.15}$$

In the conditions of the previous definition, if $\int_a^c f\,d\alpha$ or $\int_c^e f\,d\alpha$ does not exist, but

$$I(\epsilon) = \int_a^{c-\epsilon} f\,d\alpha + \int_{c+\epsilon}^e f\,d\alpha \tag{6.16}$$

tends to a limit I as ϵ tends to zero through positive values, then I is called the *Cauchy principal value* of $\int_a^e f\,d\alpha$.

EXERCISE 6.2 Show that the Cauchy principal value of $\int_{-1}^2 (1/x)\,dx$ is $\log 2$. [Show first that $(\int_{-1}^{-\epsilon} + \int_{+\epsilon}^1)(1/x)\,dx = 0$.]

The above definitions do not exhaust the possibilities: for example if $f(x) \to \infty$ as $x \to c$, the improper integrals $\int_c^a f\,d\alpha$ and $\int_a^\infty f\,d\alpha$ may exist with $a > c$. Then an improper integral $\int_c^\infty f\,d\alpha$ can be defined by

$$\int_c^\infty f\,d\alpha = \int_c^a f\,d\alpha + \int_a^\infty f\,d\alpha \tag{6.17}$$

An improper integral may also arise because f or α fails to exist at a point, or from some other anomaly. For example, if f fails to exist at a, but $f \in R(\alpha)$ on $[a + \epsilon, b]$ for all $0 < \epsilon < b - a$, then the integral

$$I(\epsilon) = \int_{a+\epsilon}^b f\,d\alpha \tag{6.18}$$

is well defined. If $I(\epsilon) \to I$ as $\epsilon \to 0$, then I is the improper integral $\int_a^b f\,d\alpha$. In such cases it may be desirable to emphasize the improper nature of the integral by writing it $\int_{a+}^b f\,d\alpha$.

If an integral is not improper (and therefore also not infinite) it is termed *proper*.

7 Connection with the Riemann integral

The ordinary Riemann integral $\int_a^b f(x)\,dx$ corresponds to the special case in the definition of Section 2 where $\alpha(x) = x$. If a Riemann–Stieltjes integral $\int_a^b f(x)\,d\alpha(x)$ is given, it is sometimes possible to express it as a Riemann integral. The following theorem gives a result of this kind which will be useful later.

Theorem 7.1 Suppose $f \in D_L^{\frac{1}{2}}$ on an open interval $I \supset [a, b]$ and $\alpha \in D_R^{\frac{1}{2}}$ on I. As in equation (4.29) write

$$\alpha(x) = \gamma(x) + \beta(x) = \gamma(x) + \sum_{i=1}^{n} \alpha_i U(x - p_i) \tag{7.1}$$

where the $p_i \in [a, b]$ and the α_i are such that γ is continuous on $[a, b]$. Then

$$\int_a^b f\, d\alpha = \sum_{i=1}^{n, n-1} \alpha_i f(p_i) + \int_a^b f(x)\, D_{-}\alpha(x)\, dx \tag{7.2}$$

where i runs from 1 to n if $p_n < b$, and from 1 to $n - 1$ if $p_n = b$.

Proof As in Section 4,

$$\int_a^b f\, d\alpha = \sum_{i=1}^{n, n-1} \alpha_i f(p_i) + \int_a^b f(x)\, d\gamma(x) \tag{7.3}$$

Let T be a set of open disjoint intervals containing the discontinuities, finite in number, of $D\gamma$ in a bounded open interval $J \supset [a, b]$ and let W be the closure of $T \cap [a, b]$. Let S be the set of closed intervals such that

$$S \cup \{T \cap [a, b]\} = [a, b] \tag{7.4}$$

The scheme of proof will be to show first that the following integrals exist and are equal:

$$\int_S f\, d\gamma = \int_S f(x)\, D\gamma(x)\, dx = \int_S f(x)\, D_{-}\gamma(x)\, dx \tag{7.5}$$

Then it will be shown that the integrals $\int_W f\, d\gamma$ and $\int_W f(x)\, D_{-}\gamma(x)\, dx$ both exist. Thirdly, given any ϵ, it will be shown that there exists a δ such that if the total length of the intervals of W is less than δ,

$$\int_W f\, d\gamma < \frac{\epsilon}{2}, \qquad \int_W f(x)\, D_{-}\gamma(x)\, dx < \frac{\epsilon}{2} \tag{7.6}$$

Hence will follow

$$\begin{aligned} &\left| \int_a^b f\, d\gamma - \int_a^b f(x)\, D_{-}\gamma(x)\, dx \right| \\ &\quad = \left| \left\{ \int_S f\, d\gamma - \int_S f(x)\, D_{-}\gamma(x)\, dx \right\} + \int_W f\, d\gamma - \int_W f(x)\, D_{-}\gamma(x)\, dx \right| \\ &\quad < \epsilon \end{aligned} \tag{7.7}$$

which will complete the proof.

Consider one of the constituent intervals of S, say $[e, h]$. Because $f \in D_L^{\frac{1}{2}}$ and $\gamma \in D_R^{\frac{1}{2}}$ on $I \supset [e, h]$, $f \in R(\gamma)$ on $[e, h]$. By the construction of S, $D\gamma(x)$ exists and is continuous on an open interval containing $[e, h]$. Let P be a partition of $[e, h]$ and consider a Riemann–Stieltjes sum for $\int_e^h f\,d\gamma$,

$$S(P, f, \gamma) = \sum_{k=1}^{n} f(t_k)[\gamma(x_k) - \gamma(x_{k-1})] \tag{7.8}$$

and a Riemann sum for $\int_e^h f(x)\,D\gamma(x)\,dx$ using the same P and t_k,

$$S(P, f\,D\gamma, x) = \sum_{k=1}^{n} f(t_k)\,D\gamma(t_k)[x_k - x_{k-1}] \tag{7.9}$$

Apply the mean-value theorem to the function $g(x) = \gamma(x) - D\gamma(t_k)x$ on the interval $[x_{k-1}, x_k]$, so obtaining

$$|g(x_k) - g(x_{k-1})| \leqslant (x_k - x_{k-1}) \sup_{x_{k-1} < s < x_k} |Dg(s)| \tag{7.10}$$

$$= (x_k - x_{k-1}) \max_{x_{k-1} \leqslant s \leqslant x_k} |D\gamma(s) - D\gamma(t_k)| \tag{7.11}$$

where because Dg is continuous on $[x_{k-1}, x_k]$, it has a maximum on this interval, which is equal to the supremum on (x_{k-1}, x_k). Let the maximizing s be $s_k \in [x_{k-1}, x_k]$. Then (7.11) gives

$$|f(t_k)[g(x_k) - g(x_{k-1})]|$$
$$= |f(t_k)[\gamma(x_k) - \gamma(x_{k-1})] - f(t_k)\,D\gamma(t_k)[x_k - x_{k-1}]|$$
$$\leqslant |f(t_k)|\,|D\gamma(s_k) - D\gamma(t_k)|(x_k - x_{k-1}) \tag{7.12}$$
$$\leqslant M_1\,|D\gamma(s_k) - D\gamma(t_k)|(x_k - x_{k-1}) \tag{7.13}$$

where M_1 is a bound for $|f|$ on $[a, b]$, which exists because $f \in D_L^1$ on $I \supset [a, b]$. Hence

$$|S(P, f, \gamma) - S(P, f\,D\gamma, x)| < M_1 \sum_{k=1}^{n} |D\gamma(s_k) - D\gamma(t_k)|(x_k - x_{k-1}) \tag{7.14}$$

Now $D\gamma(x)$ is continuous on the closed interval $[e, h]$, and so is uniformly continuous on $[e, h]$. Choose δ so that $|D\gamma(s_k) - D\gamma(t_k)| < \epsilon/2M_1(h - e)$ whenever $|s_k - t_k| < \delta$. Then if P'_ϵ is a partition with $|P'_\epsilon| < \delta$, and P' is any refinement of P'_ϵ, it follows from (7.14) that

$$|S(P', f, \gamma) - S(P', f\,D\gamma, x)| < \epsilon/2 \tag{7.15}$$

But because $f \in R(\gamma)$ on $[e, h]$, there is a partition P''_ϵ such that for all refinements P'' of P''_ϵ

$$\left| S(P'', f, \gamma) - \int_e^h f\,d\gamma \right| < \epsilon/2 \tag{7.16}$$

From (7.15) and (7.16), if P_ϵ consists of all the points of P'_ϵ and P''_ϵ, and P is any refinement of P_ϵ (so that P is a refinement both of P'_ϵ and of P''_ϵ)

$$|S(P, f\,D\gamma, x) - \int_e^h f\,d\gamma| < \epsilon \tag{7.17}$$

from which it follows that $\int_e^h f(x)\, D\gamma(x)\, dx$ exists and is equal to $\int_e^h f(x)\, d\gamma(x)$. On summing for all the intervals of S, and noting $D_-\gamma = D\gamma$ on S, equation (7.5) follows.

Now because $f \in D_L^1$, $\gamma \in D_R^1$ on $I \supset [a, b]$, it follows that $\int_W f\, d\gamma$ exists. Also $|f| < M_1$ on $[a, b]$, and by Theorem 9.1 of Chapter 4 the variation of the continuous function γ on W does not exceed $M_2\eta$, where M_2 is a bound for $|D\gamma|$ where it exists on $[a, b]$, and η is the total length of the intervals of W. So by Exercise 4.1

$$\int_W f\, d\gamma < M_1 M_2 \eta \tag{7.18}$$

Also $f \in D_L^1$, $D_-\gamma \in D_R$ on $I \supset [a, b]$. Write

$$f(x) = \phi(x) - \sum_{i=1}^{m} f_i U(q_i - x) = \phi(x) + \zeta(x) \tag{7.19}$$

$$D_-\gamma(x) = \psi(x) + \sum_{i=1}^{l} \xi_i U(x - r_i) = \psi(x) + \theta(x) \tag{7.20}$$

in such a way that ϕ and ψ are continuous on an open interval $J \supset [a, b]$. Then

$$\int_W f(x)\, D_-\gamma(x)\, dx = \int_W [\phi\psi + \phi\theta + \psi\zeta + \zeta\theta]\, dx \tag{7.21}$$

and $\phi\psi$ is continuous, $\phi\theta \in D_R$, $\psi\zeta \in D_L$, and $\zeta\theta$ has bounded variation, on $J \supset W$. So by Exercises 4.2 and 4.3 each of the following integrals exists:

$$\int_W \phi\psi\, dx, \qquad \int_W \phi\theta\, dx, \qquad \int_W \psi\zeta\, dx, \qquad \int_W \zeta\theta\, dx \tag{7.22}$$

Then by Exercise 3.2

$$\int_W f(x)\, D_-\gamma(x)\, dx \tag{7.23}$$

exists. With $|f| < M_1$, $|D_-\gamma| < M_2$ on $[a, b]$, the integral in (7.23) is less than $M_1 M_2 \eta$.

Choose $\delta = \epsilon/2M_1M_2$. Then the inequalities (7.6) hold for $\eta < \delta$ and the proof is complete.

The formula for integration by parts of Riemann–Stieltjes integrals is

$$\int_a^b f\, d\alpha + \int_a^b \alpha\, df = f(b)\alpha(b) - f(a)\alpha(a) \tag{7.24}$$

If $f \in D_L^1$ and $\alpha \in D_R^1$ on $I \supset [a, b]$, each of the integrals in (7.24) can be transformed by equation (7.2) and the analogous equation which arises when the roles of $f \in D_L^1$ and $\alpha \in D_R^1$ are interchanged. For this purpose write $\alpha(x)$ as in equation (7.1) and $f(x)$ as in equation (7.19), when

$$\int_a^b f(x)\, D_-\alpha(x)\, dx + \int_a^b \alpha(x)\, D_+f(x)\, dx$$

$$= f(b)\alpha(b) - f(a)\alpha(a) - \sum_{i=1}^{n,n-1} \alpha_i f(p_i) - \sum_{i=1,2}^{m} f_i\alpha(q_i) \qquad (7.25)$$

In particular, if f and α both have continuous first derivatives on $J \supset [a, b]$, and are therefore continuous on $[a, b]$,

$$\int_a^b f(x)\, D\alpha(x)\, dx + \int_a^b \alpha(x)\, Df(x)\, dx = f(b)\alpha(b) - f(a)\alpha(a) \qquad (7.26)$$

Equation (7.26) is the usual formula for integration by parts of Riemann integrals.

EXERCISE 7.1 If f and α are both continuous on $[a, b]$ and Df, $D\alpha$ exist and are continuous on (a, b) show that if one of the two following improper integrals exists then the other exists and

$$\int_{a+}^{b-} f(x)\, D\alpha(x)\, dx + \int_{a+}^{b-} \alpha(x)\, Df(x)\, dx = f(b)\alpha(b) - f(a)\alpha(a) \qquad (7.27)$$

Apply the result to $\alpha(x) = \sqrt{x}, f(x) = 1, a = 0, b = 2$. [Notice that $|Df|$, $|D\alpha|$ need not be bounded on $[a, b]$ but only on $[p, q] \subset (a, b)$, which is ensured by the continuity of Df, $D\alpha$ on (a, b).]

8 The 'delta function'

In equation (7.2), if $f \in D_L^1$ and α is a unit step $U(x - c)$, the result of Theorem 4.1 is recovered

$$\int_a^b f(x)\, dU(x - c) = f(c), \quad a \leqslant c < b \qquad (8.1)$$

Now introduce a symbol $\delta(x)$ called the *Dirac delta function*, or *unit impulse*, by the formula

$$\int_a^b f(x)\delta(x - c)\, dx = \int_a^b f(x)\, dU(x - c) \qquad (8.2)$$

Moreover, if $\alpha(x) \in D_R^1$ is written as in (7.1)

$$\alpha(x) = \gamma(x) + \sum_{i=1}^{n} \alpha_i U(x - p_i) \qquad (8.3)$$

define $D\alpha$ (which does not exist according to the definition in Chapter 4) to mean

$$D\alpha(x) = D_{-}\alpha(x) + \sum_{i=1}^{n} \alpha_i\, \delta(x - p_i) \tag{8.4}$$

With this convention, equation (7.2) can be rewritten in the more compact form

$$\int_a^b f(x)\, d\alpha(x) = \int_a^b f(x)\, D\alpha(x)\, dx \tag{8.5}$$

The delta function introduced in this way may be regarded simply as a device for achieving neatness in the formulae. When $\delta(x)$ occurs, it is to be manipulated as though it were a function. When it occurs in an integral, it is to be interpreted according to equation (8.2). Any such manipulations can be replaced at will by the equivalent manipulations which have been justified earlier: for example (8.5) may be replaced by (7.2). This will be the attitude adopted in this book to the delta function. Other approaches are current, and these are explained briefly in Sections 9.1 and 9.2.

Note that with the approach described above, $\delta(x - c)$ and $\delta(c - x)$ have different properties. The first (which may be thought of as an 'impulse at $c+$') corresponds to $U(x - c)$ and is used with $f \in D_L$. The second (an 'impulse at $c-$') corresponds to $U(c - x)$ and is used with $f \in D_R$. Other approaches use an 'impulse at c' for which the first integral in (8.2) may only be defined (depending on the treatment) where f is continuous.

Finally it is emphasized that $\delta(x)$ is not a function in the sense of Chapter 1. That is, there is no way of associating a real number with each point of the interval $[a, b]$ which will produce a function having the properties of $\delta(x)$.

9 Impulse response of a system

In Theorem 5.1 it was shown that, subject to certain conditions on a system, the output $y(t)$ corresponding to an input $u(t) \in D_R^1$ with $u(t) = 0$, $t \leqslant 0$ could be written

$$y(t) = \int_0^t H(t - s)\, du(s) \tag{9.1}$$

where $H(t) \in D_R^1$ is the response to a unit step. Because $u(0) = H(0) = 0$, integration by parts gives

$$y(t) = -\int_0^t u(s)\, dH(t - s) \tag{9.2}$$

Write

$$h(x) = DH(x) \tag{9.3}$$

with the convention of Section 8; that is, $h(x)$ will contain a delta function wherever $H(x)$ is discontinuous. Equation (8.5) then gives

$$y(t) = -\int_0^t u(s)\,dH(t-s) = \int_0^t u(s)h(t-s)\,ds \tag{9.4}$$

The change in sign in (9.4) occurs because $H(t - s)$ is a function of s in the integral: compare equation (3.19) or Section 11. The last member of (9.4) expresses $y(t)$ by a *convolution integral.* To interpret h, let $u(s)$ be $\delta(s)$, so that using (8.2) and (8.1) in a formal way†

$$y(t) = \int_0^t \delta(s)h(t-s)\,ds = \int_0^t h(t-s)\,dU(s) = h(t) \tag{9.5}$$

That is, $h(t)$ is the 'response of the system to a unit impulse': it is known as the *impulse response.* Following the approach of Section 8, equation (9.5) is to be regarded as a purely formal result. That is, $h(t)$ is not to be sought by subjecting the physical system to an input $\delta(t)$; indeed, this is impossible because $\delta(t)$ is not a function. Rather, when $h(t)$ occurs, it is to be interpreted by reference to (9.4), in which $H(t)$ is a quantity which can be found (subject to the process of approximation and idealization explained in Section 7.4 of Chapter 4) by experiment on the system.

The historical approach to delta functions in engineering was rather different from the one explained above, as will be described in Section 9.1.

9.1 Historical treatment of delta functions

Delta functions were introduced into engineering in a way which was motivated very directly by physical considerations. A class of functions $\delta_\epsilon(x)$ was first defined: these were required to satisfy certain conditions, the following being a typical set

(a) $\delta_\epsilon(x) = 0, \quad |x| > \epsilon > 0.$
(b) $\delta_\epsilon(x) \geqslant 0, \quad |x| \leqslant \epsilon.$
(c) $\delta_\epsilon(x)$ has bounded variation on $[-\epsilon, \epsilon]$.
(d) $\int_{-\epsilon}^{\epsilon} \delta_\epsilon(x)\,dx = 1.$

Two possible forms of $\delta_\epsilon(x)$ are shown in Fig. 4. If f is continuous and has bounded variation on $[a, b]$, and if $a < c < b$, it is not difficult to show from the theory of the Riemann integral that subject to the above conditions on $\delta_\epsilon(x)$

$$\lim_{\epsilon\to 0} \int_a^b f(x)\,\delta_\epsilon(x-c)\,dx = f(c) \tag{9.6}$$

†Note that $\delta(t) \notin D_R$ as would be needed to apply these results directly. The interpretation of $h(t)$ is not entirely free from difficulties, as is shown by Problem 4. The difficulty in that problem is inherent, and persists if the theory of distributions is used.

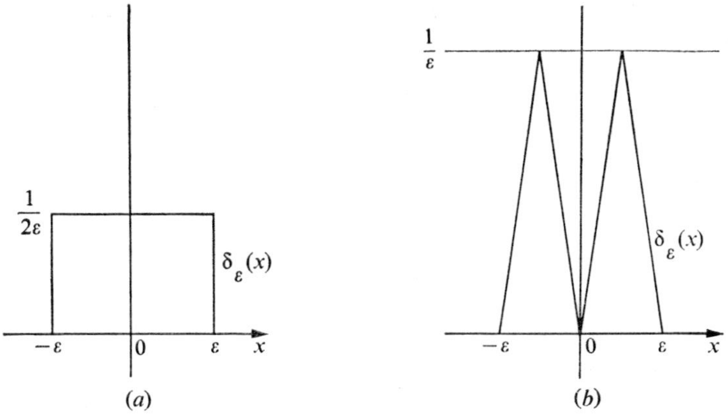

FIG. 4

this being true regardless of the particular form which may be chosen for $\delta_\epsilon(x)$.

With this approach any statement about a delta function $\delta(x)$ is to be interpreted as a statement about $\delta_\epsilon(x)$, and whenever $\delta_\epsilon(x)$ occurs in an integral the limit as $\epsilon \to 0$ is to be taken. So the equation

$$\int_a^b f(x)\,\delta(x - c)\,dx = f(c) \tag{9.7}$$

is simply a convenient way of writing (9.6).

This way of proceeding can be made rigorous (compare Lighthill, 1960). Elementary accounts, however, without any attempt at rigour, often proceed by regarding $\delta(x)$ as the 'limit of $\delta_\epsilon(x)$ as $\epsilon \to 0$'. Then if the order of the integration and the limiting process can be reversed in (9.6), the result is (9.7). The difficulties of this procedure can be appreciated from the fact that for the $\delta_\epsilon(x)$ shown in Fig. 4 (b),

$$\lim_{\epsilon \to 0} \delta_\epsilon(x) = 0, \quad \text{all } x \tag{9.8}$$

as is shown in Example 9.1 below. So for f continuous and of bounded variation on $[a, b]$, and for $a < c < b$,

$$\left.\begin{aligned} \lim_{\epsilon \to 0} \int_a^b f(x)\,\delta_\epsilon(x - c)\,dx &= f(c) \\ \int_a^b f(x) \lim_{\epsilon \to 0} \delta_\epsilon(x - c)\,dx &= 0 \end{aligned}\right\} \tag{9.9}$$

Example 9.1 For the function $\delta_\epsilon(x)$ shown in Fig. 4 (b),

$$\lim_{\epsilon \to 0} \delta_\epsilon(0) = \lim_{\epsilon \to 0} 0 = 0 \tag{9.10}$$

Let t be any non-zero value of x. Then for all $\epsilon < |t|$, $\delta_\epsilon(t) = 0$. So

$$\lim_{\epsilon \to 0} \delta_\epsilon(t) = 0 \tag{9.11}$$

and equation (9.8) is true for all x.

Example 9.2 The step response of a system is

$$H(t) = \begin{Bmatrix} 0, & t \leqslant 0 \\ e^{-t}, & t > 0 \end{Bmatrix} = U(t)e^{-t} \tag{9.12}$$

By equation (9.4) the impulse response is

$$h(t) = \begin{Bmatrix} 0 & t \leqslant 0 \\ \delta(t) - e^{-t}, & t > 0 \end{Bmatrix} = \delta(t) - U(t)e^{-t} \tag{9.13}$$

The response to an input

$$u(t) = \begin{cases} 0, & t \leqslant 0 \\ t, & t > 0 \end{cases} \tag{9.14}$$

according to the treatment of Section 5 is, by equation (5.16),

$$y(t) = \int_0^t U(t-s)e^{-(t-s)}\, ds \tag{9.15}$$

$$= \int_0^t e^{-(t-s)}\, ds$$

$$= e^{-t}[e^s]_0^t$$

$$= 1 - e^{-t}, \quad t \geqslant 0 \tag{9.16}$$

Equation (9.4) gives, alternatively,

$$y(t) = \int_0^t s[\delta(t-s) - U(t-s)e^{-(t-s)}]\, ds \tag{9.17}$$

and by definition, according to the treatment in Section 8,

$$\int_0^t s\, \delta(t-s)\, ds \triangleq \int_0^t s\, dU(t-s) = t \tag{9.18}$$

while

$$-\int_0^t sU(t-s)e^{-(t-s)}\, ds = -e^{-t}\int_0^t se^s\, ds \tag{9.19}$$

$$= -e^{-t}\left\{[se^s]_0^t - \int_0^t e^s\, ds\right\}$$

$$= -e^{-t}[te^t - e^t + 1]$$

$$= -t + 1 - e^{-t} \tag{9.20}$$

so that

$$y(t) = 1 - e^{-t} \tag{9.21}$$

as before.

To interpret equation (9.17) in accordance with equation (9.6), choose the $\delta_\epsilon(t)$ shown in Fig. 4 (a). Then using the interpretation of (9.7),

$$\int_0^t s\delta(t - s)\, ds \triangleq \lim_{\epsilon\to 0} \int_{t-\epsilon}^{t+\epsilon} \frac{s\, ds}{2\epsilon} \tag{9.22}$$

$$= \lim_{\epsilon\to 0} \frac{1}{2\epsilon}\left[\frac{s^2}{2}\right]_{t-\epsilon}^{t+\epsilon}$$

$$= \lim_{\epsilon\to 0} \frac{(t + \epsilon)^2 - (t - \epsilon)^2}{4\epsilon}$$

$$= \lim_{\epsilon\to 0} \frac{4t}{4} = t \tag{9.23}$$

This verifies (9.6) for a particular f and δ_ϵ, and agrees with (9.18). The remaining part of (9.17) is evaluated as before.

Notice that although the integral on the left-hand side of (9.22) runs from 0 to t, the integral on the right has to be extended to $t + \epsilon$. This is because the delta function must be included in the range of integration, and $\delta_\epsilon(x)$ in Fig. 4 (a) is symmetrical about the nominal point of occurrence $x = 0$ of the impulse $\delta(x)$.

EXERCISE 9.1 Prove the result given in equation (9.6).

9.2 Theory of distributions

One of several alternative ways of giving a rigorous foundation for formulae involving delta functions is provided by the theory of distributions. First, all integrals are defined in the Lebesgue sense. Then a set of *testing functions* ϕ is introduced, with the properties:

(a) ϕ has continuous derivatives of all orders at every point
(b) $\phi(x) = 0$, x not in some specified interval $[a, b]$.

Distributions are then defined by giving a value to an integral in which testing functions occur: for example, the distribution $\delta(x)$ is defined by the property

$$\int_{-\infty}^{\infty} \delta(x)\phi(x)\, dx = \phi(0), \quad \text{all } \phi \tag{9.24}$$

Two distributions are equal if they give the same result in their defining integral.

In this theory the unit step $U_D(x)$ is a distribution, and is defined by

$$\int_{-\infty}^{\infty} U_D(x)\phi(x)\, dx = \int_0^{\infty} \phi(x)\, dx, \quad \text{all } \phi \tag{9.25}$$

Evidently the unit step $U(x)$ as previously defined fulfils this definition, but by Section 1 [compare (1.4)] so also does the function $V(x)$ defined by

$$V(x) = \begin{cases} 0, & x \leqslant 0 \quad \text{or } x \text{ rational} \\ 1, & x > 0 \quad \text{and irrational} \end{cases} \tag{9.26}$$

This illustrates the fact that even when a distribution can be represented by a function, the function is not in general uniquely defined. In particular, the response of a system to a unit step (in the distribution sense) is again a distribution and so, if it can be represented by a function, this function is not in general uniquely defined. For this reason it is not possible without some further investigation to discuss, for example, the response of a 'sample and hold' circuit to the output of such a system.

9.3 Comparison of treatments

Three different treatments of the delta function have been sketched above. The physically motivated approach of Section 9.1 is undoubtedly attractive for informal work. It has the minor disadvantage that the limit in equation (9.6) does not exist where f is discontinuous; also the symmetrical nature of $\delta_\epsilon(t)$ in the usual treatment leads to some awkwardness over initial conditions and the non-anticipative nature of a system.

The use of the Riemann–Stieltjes integral, as in Section 5, allows results to be obtained in terms of the step response, for which a clear physical picture is available. A rigorous justification of these results can be obtained in elementary terms, and there is no difficulty over initial conditions. If they are desired, formulae involving delta functions can be obtained, as in Section 9, but as presented here these are only formal and must be interpreted in terms of step functions.

Both of these approaches to the delta function can be made mathematically rigorous in the sense that statements involving delta functions can be given a precise meaning and can be justified by the normal rules of mathematical proof. When this is done, the differences between these two approaches and the theory of distributions become rather minor; the theory of distributions stands in the relation of a general theory to which these two are particular ways of approach.

The theory of distributions has as its main advantage the generality and consistency which it brings to formulae. For example, distributions can always be differentiated, convergent series can always be differentiated term-by-term, and so on. Correspondingly, however, statements involving distributions rather than functions have a more slender content—the distribution zero cannot be said to be zero at any particular point, etc.

This last fact leads to some minor difficulties. There can be some awkwardness over initial conditions and the non-anticipative nature of a system. Also a distribution cannot meaningfully be sampled at a point—it is necessary to average first over some small interval. The major difficulties, however, arise in the physical interpretation of formulae involving delta functions.

In the application of mathematics to physical systems it is necessary to be sure that the mathematical treatment is soundly based. It is also necessary to be sure that the mathematics gives a valid description of the physical situation. This second point, which in practice is often the most troublesome, is particularly difficult to deal with when the theory of distributions is used. For example, no physical system is linear for arbitrarily large values of the variables. If, therefore, a physical system is characterized by its impulse response there is an immediate problem in interpreting this characterization in a meaningful way. The mathematical limiting process which must in some way be invoked cannot be matched by a corresponding physical picture.

Difficulties of this nature are always present when mathematical limiting processes are involved, but they are less troublesome with step functions than with impulses. This may seem to imply some contradiction because, for example, a step function of voltage will generally require an impulsive flow of current into a capacitor (possibly a stray). A moment's thought, however, will show that beyond a certain point the approximation to the step function will be improved by reducing the capacitance involved, rather than increasing the peak current. A similar point is made by Newcomb (1966), p. 37, who in a thoughtful account shows that network properties cannot be defined directly in terms of distributions.

It is chiefly for these reasons that impulses and the impulse response have been avoided in the main development of this book. As they are in widespread use, however, formulae involving the impulse response will be given where appropriate. In each case the formulae will appear merely as an alternative way of stating results obtained by other means.

10 Derivative of an integral

If the integral

$$F(x) = \int_a^x f(t)\, d\alpha(t) \tag{10.1}$$

exists, its behaviour may be studied as a function of x. The following theorem gives a special result of this type. The theorem is stated for scalar functions but is easily extended to vectors or matrices as in Section 2.

Theorem 10.1 Let $f(x)$ be continuous on $[a, b]$. Then $F(x)$ defined by

$$F(x) = \int_a^x f(t)\, dt \tag{10.2}$$

has the continuous derivative $DF(x) = f(x)$ for $x \in (a, b)$.

Proof Because f is continuous on $[a, b]$ it is uniformly continuous on $[a, b]$. So if c, t, x are points in (a, b), there exists a δ such that

$$|t - c| < \delta \Rightarrow |f(t) - f(c)| < \epsilon \tag{10.3}$$

whence

$$|x - c| < \delta \Rightarrow \left|\int_c^x [f(t) - f(c)]\, dt\right| < \epsilon\, |x - c| \tag{10.4}$$

By (10.2)

$$\int_c^x f(t)\, dt = F(x) - F(c) \tag{10.5}$$

and by the definition of the Riemann integral

$$f(c)\int_c^x dt = f(c)(x - c) \tag{10.6}$$

On using (10.5) and (10.6), (10.4) becomes

$$|x - c| < \delta \Rightarrow |F(x) - F(c) - f(c)(x - c)| < \epsilon\, |x - c| \tag{10.7}$$

That is, $DF(c)$ exists and is equal to $f(c)$, which is continuous on (a, b).

EXERCISE 10.1 Show that $D_+F(x)$ exists and is continuous on $[a, b)$ and has the value $f(x)$; and $D_-F(x)$ exists and is continuous on $(a, b]$ and has the value $f(x)$.

EXERCISE 10.2 Prove that $\int_0^t e^{A\tau}\, d\tau = A^{-1}[e^{At} - I]$. [First establish the matrix form of (10.2). Put $W(t) = \int_0^t e^{A\tau}\, d\tau$, and use Theorem 13.3 of Chapter 4 to prove that $D[AW(t) - e^{At}] = 0$.]

In (10.1) the integral depends on a parameter x, which is one of the 'limits of integration'. It is also possible for an integral to depend on a parameter through its occurrence in f:

$$F(x) = \int_a^b f(x, t)\, d\alpha(t) \tag{10.8}$$

The parallel case in which α and not f depends upon x is reduced to (10.8) on integrating by parts. When both f and α depend upon x, the chain rule can be used. Two theorems covering particular cases are now given.

Theorem 10.2 Let $f(x, t)$ have its first partial derivative with respect to x, $D_1f(x, t)$, continuous on the region in E_2 defined by $t \in [a, b]$, $x \in [e, h]$. Let α have bounded variation on $[a, b]$ and let the integral defined by (10.8) exist for each fixed $x \in (e, h)$. Then for all $x \in (e, h)$, $DF(x)$ exists and is equal to $\phi(x)$ defined by

$$\phi(x) = \int_a^b D_1f(x, t)\, d\alpha(t) \tag{10.9}$$

Proof Define a function $g(x, t)$ by

$$g(x, t) = f(x, t) - D_1 f(c, t)x \tag{10.10}$$

for $x \in (e, h)$ and some $c \in (e, h)$, $c \neq x$, and consider

$$\begin{aligned} I &= \int_a^b [g(x, t) - g(c, t)]\, d\alpha(t) \\ &= \int_a^b [f(x, t) - f(c, t) - D_1 f(c, t)(x - c)]\, d\alpha(t) \\ &= F(x) - F(c) - (x - c)\phi(c) \end{aligned} \tag{10.11}$$

where $\phi(c)$ is defined by (10.9) and exists because $D_1 f(c, t)$ is a continuous function of t and $\alpha(t)$ has bounded variation. By the mean-value theorem

$$|g(x, t) - g(c, t)| \leqslant |x - c| \max_{s \in [x, c]} |D_1 g(s, t)| \tag{10.12}$$

$$= |x - c| \max_{s \in [x, c]} |D_1 f(s, t) - D_1 f(c, t)| \tag{10.13}$$

where $[x, c]$ is to be replaced by $[c, x]$ if $c < x$.

Now $D_1 f(x, t)$ is continuous as a function of its two variables on the closed bounded region $t \in [a, b]$, $x \in [e, h]$, and so is uniformly continuous on that region. That is, given any ϵ there is a δ depending only on ϵ such that for (x_0, t_0) and (x_1, t_1) both in this region,

$$(x_1 - x_0)^2 + (t_1 - t_0)^2 < \delta^2 \Rightarrow |D_1 f(x_1, t_1) - D_1 f(x_0, t_0)| < \epsilon \tag{10.14}$$

Consequently, if $|x - c| < \delta$, (10.13) gives

$$|g(x, t) - g(c, t)| < |x - c|\, \epsilon \tag{10.15}$$

Also there is a bound M for the variation of α on $[a, b]$. By Exercise 4.1,

$$|x - c| < \delta \Rightarrow |I| < M\, |x - c|\, \epsilon \tag{10.16}$$

and so (10.11) gives

$$|x - c| < \delta \Rightarrow |F(x) - F(c) - (x - c)\phi(c)| < M\epsilon\, |x - c| \tag{10.17}$$

In other words, $DF(c)$ exists and is equal to $\phi(c)$.

EXERCISE 10.3 Extend Theorem 10.2 to a real-valued function f depending on a vector x and on t, $f(x, t)$. [In (10.9), $D_1 f(x, t)$ is replaced by a vector of partial derivatives of f with respect to the elements of x.]

EXERCISE 10.4 In the statement of Theorem 10.2, it is required that $D_1 f$ should be continuous at $x = e$ and at $x = h$, which implies that $D_1 f$ must exist for $x < e$ and $x > h$. Show that this requirement can be relaxed by replacing $D_1 f$ at $x = e$ by the right-hand partial derivative $D_{1+} f$ and at $x = h$ by $D_{1-} f$, and requiring the partial derivative thus modified to be continuous on $t \in [a, b]$, $x \in [e, h]$. Alternatively $D_1 f$ may be assumed to exist and be uniformly continuous on $t \in [a, b]$, $x \in (e, h)$. [Note that because x and c in the proof are in (e, h), the properties of $D_1 f$ at e and at h are needed only in the argument leading to (10.14).]

EXERCISE 10.5 Let $F(x) = \int_y^z f(x, t)\, dt$ where y, z are functions of x with bounded first derivative on $[a, b]$. Let f and its first partial derivative $D_1 f$ be continuous on the region R,

$$R\colon x \in [a, b],\ t \in [c, d]; \qquad c < \min_{x \in [a, b]} \{y(x), z(x)\}$$
$$d > \max_{x \in [a, b]} \{y(x), z(x)\}$$

Use the chain rule to obtain, for $x \in (a, b)$,

$$DF(x) = f\big(x, z(x)\big)\, Dz(x) - f\big(x, y(x)\big)\, Dy(x) + \int_{y(x)}^{z(x)} D_1 f(x, t)\, dt$$

Theorem 10.3 Let $H(t)$ and $u(t)$ both be zero for $t \leqslant 0$, and both belong to D_R^1 on an open interval containing $[0, T]$. Then y defined by

$$y(t) = \int_0^t H(t - s)\, du(s) \tag{10.18}$$

belongs to D_R^1 on an open interval containing $[0, T]$.

Proof Write

$$u(t) = \sum_{i=1}^{n} u_i U(t - p_i) + v(t) \tag{10.19}$$

where the u_i and $p_i \in [0, T]$ are such that v is continuous on a bounded open interval $J \supset [0, T]$. Similarly write

$$H(x) = \sum_{i=1}^{m} H_i U(x - q_i) + \eta(x) \tag{10.20}$$

where the H_i and $q_i \in [0, T]$ are such that η is continuous on J. Then because $H(x) = 0$ for $x \leqslant 0$, (10.18) gives for $t \in [-\epsilon, T + \epsilon] \subset J$, $\epsilon > 0$,

$$y(t) = \int_{-\epsilon}^{T+\epsilon} H(t - s) \left\{ \sum_{i=1}^{n} u_i\, dU(s - p_i) + dv(s) \right\} \tag{10.21}$$

$$= \sum_{i=1}^{n} u_i H(t - p_i) + \int_{-\epsilon}^{T+\epsilon} H(t - s)\, dv(s) \tag{10.22}$$

$$= \sum_{i=1}^{n} u_i H(t - p_i) - \int_{-\epsilon}^{T+\epsilon} v(s) \left\{ \sum_{i=1}^{m} H_i\, dU(t - s - q_i) + d\eta(t - s) \right\} \tag{10.23}$$

$$= \sum_{i=1}^{n} u_i H(t - p_i) + \sum_{i=1}^{m} H_i v(t - q_i) + \int_{-\epsilon}^{T+\epsilon} \eta(t - s)\, dv(s) \tag{10.24}$$

where (10.23) and (10.24) are each obtained by integrating by parts and using

$$v(s) = 0,\ s = -\epsilon; \quad U(t - s - q_i) = 0 = \eta(t - s),\ s = T + \epsilon,\ t \leqslant T + \epsilon \tag{10.25}$$

The two sums in (10.24) represent functions belonging to D_R^1 on J, so that only the integral need be considered: in this integral $\eta \in D_R^1$ and $v \in D_R^1$ on $J \supset [-\epsilon, T + \epsilon]$, and η and v are continuous on $[-\epsilon, T + \epsilon]$. Call the integral $I(t)$.

Because $\eta \in D_R^1$ it is bounded by M on $J \supset [-\epsilon, T + \epsilon]$, and similarly $v \in D_R^1$ has variation on $[-\epsilon, T + \epsilon]$ bounded by $M_1\,(T + 2\epsilon)$ where M_1 is a bound for $|Dv|$ where it exists on J. By Exercise 4.1, $I(t)$ is therefore bounded by $MM_1(T + 2\epsilon)$ on $[-\epsilon, T + \epsilon]$, and because $\epsilon > 0$ was arbitrary, I is bounded on J. Next we proceed to show that I has a derivative on J.

Let c, t be two points in $(-\epsilon, T + \epsilon)$, with $c \neq t$, and define

$$g(t, s) = \eta(t - s) - D_-\eta(c - s)t \tag{10.26}$$

Consider

$$\int_{-\epsilon}^{T+\epsilon} \{g(c, s) - g(t, s)\}\, dv(s)$$

$$= \int_{-\epsilon}^{T+\epsilon} \{\eta(c - s) - \eta(t - s) - D_-\eta(c - s)(c - t)\}\, dv(s) \tag{10.27}$$

$$= I(c) - I(t) - (c - t) \int_{-\epsilon}^{T+\epsilon} D_-\eta(c - s)\, dv(s) \tag{10.28}$$

Write

$$\eta(c - s) = \sum_{i=1}^{l} (c - s - r_i)\eta_i U(c - s - r_i) + \phi(c - s) \tag{10.29}$$

where the η_i and r_i are chosen so that $\phi(t)$ has continuous first derivative on J. Then

$$D_-\eta(c - s) = \sum_{i=1}^{l} \eta_i U(c - s - r_i) + D\phi(c - s) \tag{10.30}$$

and the function defined by the sum in (10.30) has bounded variation for $s \in [-\epsilon, T + \epsilon]$ while $D\phi(c - s)$ is continuous for $s \in [0, T + \epsilon]$. Since $v(s)$ is both continuous, and of bounded variation, on $[-\epsilon, T + \epsilon]$, and is zero for $s \leqslant 0$, Theorem 4.2 shows that the integral in (10.28) exists, and so therefore does the integral in (10.27). Our aim is to show that for any $\epsilon_1 > 0$ there is a δ such that the expression in (10.28) is less than $|c - t|\epsilon_1$ whenever $|c - t| < \delta$. This will establish the existence of $DI(c)$. To this end (10.28) is written as the sum of two integrals $I_1(c, t)$ and $I_2(c, t)$ which are defined below.

Define $I_1(c, t)$ by

$$I_1(c, t) = \sum_{i=1}^{l} \eta_i \int_{-\epsilon}^{T+\epsilon} \{[(c - s - r_i) - (c - t)]U(c - s - r_i) \\ -(t - s - r_i)U(t - s - r_i)\}\, dv(s) \quad (10.31)$$

$$= \sum_{i=1}^{l} \eta_i \int_{c-r_i}^{t-r_i} (t - s - r_i)\, dv(s) \quad (10.32)$$

The integrand in (10.32) is bounded by $|c - t|$, and the variation of the continuous function $v \in D_R^1$ is bounded by $M_1\,|c - t|$. Hence

$$|I_1(c, t)| < \sum_{i=1}^{l} |\eta_i|\, M_1(c - t)^2 \quad (10.33)$$

Similarly define $I_2(c, t)$ by

$$I_2(c, t) = \int_{-\epsilon}^{T+\epsilon} \{\phi(c - s) - \phi(t - s) - D\phi(c - s)(c - t)\}\, dv(s) \quad (10.34)$$

and apply the mean-value theorem to obtain

$$|\phi(c - s) - D\phi(c - s)c - [\phi(t - s) - D\phi(c - s)t]| \\ \leqslant |c - t| \max_{x\in[c,t]} |D\phi(x - s) - D\phi(c - s)| \quad (10.35)$$

where $[c, t]$ is to be replaced by $[t, c]$ if $t < c$. On using (10.35) in (10.34),

$$|I_2(c, t)| < M_1(T + 2\epsilon)\,|c - t| \max_{x\in[c,t]} |D\phi(x - s) - D\phi(c - s)| \quad (10.36)$$

On comparing (10.28), (10.33), and (10.36)

$$|I(c) - I(t) - (c - t) \int_{-\epsilon}^{T+\epsilon} D_{-}\eta(c - s)\, dv(s)| \\ = |I_1(c, t) + I_2(c, t)| \\ < M_1\,|c - t| \left\{ \sum_{i=1}^{l} |\eta_i(c - t)| + (T + 2\epsilon) \max_{x\in[c,t]} |D\phi(x - s) - D\phi(c - s)| \right\} \quad (10.37)$$

and because $D\phi$ is continuous it is possible, given any ϵ_1, to choose δ such that

$$|c - t| < \delta \Rightarrow \left\{ \sum_{i=1}^{l} |\eta_i(c - t)| + (T + 2\epsilon) \max_{x\in[c,t]} |D\phi(x - s) - D\phi(c - s)| \right\} < \frac{\epsilon_1}{M_1} \quad (10.38)$$

That is to say, by (10.37), the derivative DI exists on $(-\epsilon, T + \epsilon)$ and so on J and is given by

$$DI(c = \int_{-\epsilon}^{T+\epsilon} D_{-}\eta(c - s)\, dv(s) \quad (10.39)$$

The remaining properties which I must have if it is to belong to D_R^1 are now easily established. The existence of DI implies the continuity of I on $(-\epsilon, T + \epsilon)$ and so on J. Because $D_-\eta \in D_R$ on J it is bounded, and the variation of $v \in D_R^1$ is also bounded on J. Therefore (10.39) shows that DI is bounded on $(-\epsilon, T + \epsilon)$, and so on J.

Finally, on using (10.29) in (10.39),

$$DI(c) = \int_{-\epsilon}^{T+\epsilon} \left\{ \sum_{i=1}^{l} \eta_i U(c - s - r_i) + D\phi(c - s) \right\} dv(s) \tag{10.40}$$

and the first part of the integral in (10.40) is

$$\sum_{i=1}^{l} \eta_i \int_0^{c-r_i} dv(s) = \sum_{i=1}^{l} \eta_i [v(c - r_i) - v(0)] \tag{10.41}$$

$$= \sum_{i=1}^{l} \eta_i v(c - r_i) \tag{10.42}$$

Hence

$$DI(c) - DI(t) = \sum_{i=1}^{l} \eta_i [v(c - r_i) - v(t - r_i)] + \int_{-\epsilon}^{T+\epsilon} [D\phi(c - s) - D\phi(t - s)]\, dv(s) \tag{10.43}$$

and because v and $D\phi$ are continuous, and v has bounded variation, it follows that the right-hand side of (10.43) tends to zero as $t \to c$. That is, DI is continuous on $(-\epsilon, T + \epsilon)$ and so on J.

Collecting results, I and DI are continuous and bounded on J; that is, $I \in D_R^1$ on J. This completes the proof.

From (10.24) and (10.39) it follows that when Dy exists it is given by

$$Dy(t) = \sum_{i=1}^{n} u_i\, DH(t - p_i) + \sum_{i=1}^{m} H_i\, Dv(t - q_i) + \int_0^t D_-H(t - s)\, dv(s) \tag{10.44}$$

because $D_-H = D_-\eta$, and $D_-H(t - s) = 0,\ s \geqslant t$.

If H and u both have continuous derivatives on $J \supset [0, T]$, they are continuous on J and $D_-H = DH$. Equation (10.44) then becomes

$$Dy(t) = \int_0^t DH(t - s)\, du(s) \tag{10.45}$$

$$= \int_0^t h(t - s)\, Du(s)\, ds \tag{10.46}$$

where h is the impulse response. If h has continuous derivative on J, (10.46) can be integrated by parts using (7.27) to give

$$Dy(t) = \int_0^t Dh(t - s) u(s)\, ds \tag{10.47}$$

where a change of sign occurs because h is a function of s in the integral, whereas its derivative is with respect to $t - s$. The fact that $u(0) = 0 = DH(0)$ by continuity has also been used.

11 Change of variable

In Exercise 3.7 the following formula was given

$$\int_a^b f(s)\, d\beta(b - s) = -\int_a^b f(a + b - s)\, d\beta(s - a) \tag{11.1}$$

This is a direct consequence of the definition because the Riemann–Stieltjes sums for the two integrals, with a common partition for both, contain the same terms merely rearranged and changed in sign. Equation (11.1) is an example of a change of variable in the Riemann–Stieltjes integral.

Theorem 11.1 Let ϕ be a continuous one-to-one mapping of the closed bounded interval $[c, d]$ on to the closed bounded interval $[a, b]$, with $\phi(c) = a$. Let $\int_a^b f\, d\alpha$ exist and write for $y \in [c, d]$, $x = \phi(y)$ and $f(x) = f[\phi(y)] = g(y)$, $\alpha(x) = \alpha[\phi(y)] = \beta(y)$. Then

$$\int_c^d g(y)\, d\beta(y) = \int_a^b f(x)\, d\alpha(x) \tag{11.2}$$

Proof By Exercise 7.9, Chapter 4, ϕ has a continuous inverse ϕ^{-1} and both ϕ and ϕ^{-1} are strictly increasing. There is therefore a one-to-one correspondence between each partition P of $[a, b]$, $a = x_1 < x_2 < \cdots < x_n = b$, and a partition P' of $[c, d]$, $c = y_1 = \phi^{-1}(x_1) < y_2 = \phi^{-1}(x_2) < \cdots < y_n = \phi^{-1}(x_n) = b$. There is similarly a one-to-one correspondence between each Riemann–Stieltjes sum

$$S(P, f, \alpha) = \sum_{i=1}^{n} f(t_i)[\alpha(x_i) - \alpha(x_{i-1})] \tag{11.3}$$

and a sum

$$S(P', g, \beta) = \sum_{i=1}^{n} g(\theta_i)[\beta(y_i) - \beta(y_{i-1})] \tag{11.4}$$

where

$$y_i = \phi^{-1}(x_i) \tag{11.5}$$

$$\theta_i = \phi^{-1}(t_i) \tag{11.6}$$

and moreover the sums $S(P, f, \alpha)$ and $S(P', g, \beta)$ are equal. The existence of $\int_a^b f\, d\alpha$ means that given any ϵ there is a P_ϵ such that

$$\left| S(P, f, \alpha) - \int_a^b f\, d\alpha \right| < \epsilon \tag{11.7}$$

whenever P is a refinement of P_ϵ. It follows that there is a P'_ϵ (corresponding to P_ϵ) such that

$$\left| S(P', g, \beta) - \int_a^b f\, d\alpha \right| < \epsilon \tag{11.8}$$

whenever P' is a refinement of P'_ϵ [because corresponding to P' is P, a refinement of P_ϵ, and corresponding to $S(P', g, \beta)$ is $S(P, f, \alpha) = S(P', g, \beta)$].

Change of variable in the Riemann integral is used frequently, and a convenient formula is available.

Theorem 11.2 Let $[a, b]$ be a closed bounded interval. Let ϕ be a function having continuous first derivative $D\phi$ on $[c, d]$, and with $\phi(c) = a$, $\phi(d) = b$. Define p, q by $p < p_1 = \min_{y \in [c, d]} \phi(y)$, $q > q_1 = \max_{y \in [c, d]} \phi(y)$, and let f be continuous on $[p, q]$. Then if $f(x) = f[\phi(y)] = g(y)$,

$$\int_a^b f(x)\, dx = \int_c^d g(y)\, D\phi(y)\, dy \tag{11.9}$$

Proof Notice that $p_1 < a$ or $q_1 > b$ implies that $\phi(y) \notin [a, b]$ for some $y \in (c, d)$. The integral $\int_a^b f(x)\, dx$ exists by Theorem 4.2 because f is continuous on $[a, b]$ and $\alpha(x) = x$ has bounded variation on $[a, b]$. For $t \in [c, d]$ define

$$\psi(t) = \int_a^{\phi(t)} f(x)\, dx - \int_c^t g(y)\, D\phi(y)\, dy \tag{11.10}$$

Because ϕ has continuous first derivative on $[c, d]$, it is continuous on $[c, d]$ and therefore bounded on $[c, d]$, and so x has bounded variation on $[a, \phi(t)]$. Also, because $[a, \phi(t)] \in [p, q]$, f is continuous on $[a, \phi(t)]$. Therefore the first integral in (11.10) exists. Similarly, the composite function $g = f\phi$ of the two continuous functions f and ϕ is continuous on $[c, t]$, and so is the product of g with the continuous function $D\phi$. So the second integral also exists.

The derivative $D\psi$ is found from Theorem 10.1 and the chain rule to be

$$D\psi(t) = f[\phi(t)]D\phi(t) - g(t)\, D\phi(t) = 0, \quad t \in (c, d) \tag{11.11}$$

and $\psi(c) = 0$ because $\phi(c) = a$. By the mean-value theorem $\psi(d) = 0$, which is equation (11.9).

EXERCISE 11.1 The conditions of the theorem require that $D\phi$ should be continuous on $[c, d]$, and this implies that ϕ exists on an open interval $J \supset [c, d]$. Use the left- and right-hand derivatives $D_-\phi$ and $D_+\phi$ at d and c respectively to relax this condition.

EXERCISE 11.2 The conditions of the theorem demand $p < \min_{y\in[c,d]} \phi(y)$, $q > \max_{y\in[c,d]} \phi(y)$. If ϕ maps $[c, d]$ onto the bounded closed interval $[p_1, q_1]$, and if the values p_1, q_1 are achieved only a finite number of times by ϕ for $t \in [c, d]$, show that p, q can be replaced in the statement of the theorem by p_1, q_1. [When p_1 and q_1 are used in the statement of the theorem, the chain rule cannot be applied to find the derivative of $\int_a^{\phi(t)} f(x)\,dx$ at a point t_0 where $\phi(t_0) = p_1$ or $\phi(t_0) = q_1$, because $\int_a^{\xi} f(x)\,dx$ is then not defined on an open neighbourhood of $\xi = p_1$ or $\xi = q_1$. Enclose t_0 in an interval $t_0 - \epsilon, t_0 + \epsilon$ and study $|\psi(t_0 + \epsilon) - \psi(t_0 - \epsilon)|$ directly, noting that $D\phi(t_0) = 0$.]

Further reading

The treatment of the Riemann–Stieltjes integral given here follows closely that of Apostol (1957) which may be consulted for a more complete account of the theory. A more general approach to integration, including alternative treatments of the Riemann–Stieltjes integral and an account of Lebesgue integration, is given by Hildebrandt (1963).

Lighthill (1960) gives a particularly clear and simple account of the delta function on the lines of Section 9.1. Schwartz (1957, 1959) shows how the treatment in the introduction of Section 9 is related to the theory of distributions which he develops. An elementary account of this theory will be found in Zadeh and Desoer (1963), Appendix A. Another approach to the subject based on work by Mikusinski is given by Erdélyi (1962) and by Berg (1967).

References

APOSTOL, T. M., 1957, *Mathematical analysis*, Addison-Wesley, Reading, Mass.

BERG, L., 1967, *Introduction to the operational calculus*, North-Holland, Amsterdam.

ERDÉLYI, A., 1962, *Operational calculus and generalized functions*, Holt, Rinehart, and Winston, New York.

HILDEBRANDT, T. H., 1963, *Introduction to the theory of integration*, Academic Press, New York.

LIGHTHILL, M. J., 1960, *Introduction to Fourier analysis and generalized functions*, Cambridge University Press, London.

NEWCOMB, R. W., 1966, *Linear multiport synthesis*, McGraw-Hill, New York.

SCHWARTZ, L., 1957, *Théorie des distributions* (Tome I), Hermann, Paris.

SCHWARTZ, L., 1959, *Théorie des distributions* (Tome II), Hermann, Paris.

ZADEH, L. A. and C. A. DESOER, 1963, *Linear system theory*, McGraw-Hill, New York.

Problems

1 Show from the definition of the Riemann–Stieltjes integral that if m and n are positive integers, $x^m \in R(x^n)$ on $[a, b]$, and

$$\int_a^b x^m\, d(x^n) = \frac{n}{n+m}[b^{n+m} - a^{n+m}]$$

2 A system (consisting of two delay lines with their outputs summed) has step response

$$H(t) = U(t-1) + U(t-3)$$

Show from equation (5.16) that the response to an input

$$u(t) = tU(t-2)$$

is

$$y(t) = (t-1)U(t-3) - (t-3)U(t-5)$$

[Use $tU(t-2) = 2U(t-2) + (t-2)U(t-2)$; or integrate (5.16) by parts.]

3 A 'sample and hold' circuit takes an input $u(t)$, $t \geqslant 0$, and gives an output $y(t) = u(r)$, $r < t \leqslant r+1$, where r is a non-negative integer. Show that the step response for $0 \leqslant t, t_0 \leqslant n$, n an integer, satisfies the conditions of Exercise 5.2 and is

$$H(t, t_0) = \sum_{r=1}^{n-1} U(r - t_0)[U(t-r) - U(t-r-1)]$$

and so the response to $u \in D_R^1$ is

$$y(t) = \int_0^t \sum_{r=1}^{n-1} U(r-s)[U(t-r) - U(t-r-1)]\, du(s)$$

Check this result by integrating by parts.

4 Let the impulse response $h(t, t_0)$ of a time-varying linear system be defined by

$$h(t, t_0) = -\frac{d}{dt_0} H(t, t_0)$$

where d/dt_0 is interpreted in the distribution sense. Show that, subject to the conditions of Exercise 5.2, the response $y(t)$ can be written

$$y(t) = -\int_0^t u(s)\, dH(t, s) = \int_0^t u(s)h(t, s)\, ds$$

with the interpretation of Section 8. Apply this result to the system of Problem 3, and verify the resulting formula for $u(t) = tU(t)$. Show by considering $u(t) = \delta_\epsilon(t - t_0)$ that $h(t, t_0)$ cannot in this case be interpreted as the response to an impulse at $t = t_0$.

5 If $\phi \in D_L$ on $J \supset [a, b]$ and $\alpha \in D_R^1$ is continuous on $J \supset [a, b]$ show that $\phi \in R(\alpha)$ on $[a, b]$. Hence show that if $f \in D_L^1$, $D_- f \in R(\alpha)$. [Compare Section 10.]

6 Suppose $f \in D_L^1$ on $J \supset [a, b]$ and $\alpha \in D_R^1$ is continuous on $[a, b]$. Let T be a set of disjoint open intervals enclosing the discontinuities of f and of $D_- f$, and let W be the closure of $T \cap [a, b]$. Let S be a set of closed intervals such that $S \cup \{T \cap [a,b]\} = [a, b]$, and write $\int_S$ for the sum of integrals on the constituent intervals of S. Show that

$$\int_a^b D_- f\, d\alpha = \lim_{\epsilon \to 0} \int_S D_- f\, d\alpha = \lim_{\epsilon \to 0} \int_S Df\, d\alpha$$

where ϵ is the total length of the intervals of W.

7 Adopting the outlook of Section 8, let h in (10.47) arise from H defined in (10.20), and let u be defined by (10.19). Interpret the 'derivatives of delta functions' which occur in Dh by integrating by parts to obtain (10.39). Thence obtain from (7.2), (8.2), and further integration by parts,

$$Dy(t) = \sum_{i=1}^{n} u_i h(t - p_i) + \sum_{i=1}^{m} H_i\, Dv(t - q_i) + \int_0^t D_- H(t-s)\, dv(s)$$

Compare this result with (10.44).

8 Adopting the outlook of Section 8 show that if $\phi(t)$ has continuous derivatives of all orders and is zero outside $[a, b]$

$$\int_{-\infty}^{\infty} D\delta(t - c)\phi(t)\, dt = D\phi(c)$$

[This equation serves in the theory of distributions as the definition of $D\delta(t - c)$.]

9 Let $D\delta_\epsilon$ be defined by

$$D\delta_\epsilon(t) = \begin{cases} -\dfrac{1}{\epsilon^2}, & -\epsilon \leqslant t < 0 \\ \dfrac{1}{\epsilon^2}, & 0 \leqslant t \leqslant \epsilon \\ 0, & |t| > \epsilon \end{cases}$$

With ϕ as in Problem 8 show by considering

$$\frac{1}{\epsilon^2}\int_0^{\epsilon} [\phi(t) - \phi(0) - 2t\, D\phi(0)]\, dt$$

that

$$\lim_{\epsilon \to 0} \int_{-\infty}^{\infty} D\delta_\epsilon(t)\phi(t)\, dt = D\phi(0)$$

10 An analogue integrator has the step response $H(t) = tU(t)$, and the input is $u(t) \in D_R^1$, with $u(t) = 0$, $t \leqslant 0$. Show from equation (10.44) that where the derivative of the output exists it is

$$Dy(t) = \sum_{i=1}^{n} u_i U(t - p_i) + v(t) = u(t)$$

Deduce from Theorem 10.3 that $D_- y(t) = u(t)$ for all t. If u is continuous, derive Theorem 10.1.

6 Differential equations

1 Introduction

A differential equation is a relation between the values of the derivative Dx of a function and the values of the function x itself, for example

$$\dot{x}(t) = Dx(t) = f[x(t), t] \tag{1.1}$$

Here x may be a vector function of the scalar t, and (1.1) then stands for the set of simultaneous equations

$$\left.\begin{aligned} \dot{x}_1(t) &= f_1[x_1(t), x_2(t), \ \ldots, \ x_n(t), t] \\ \dot{x}_2(t) &= f_2[x_1(t), x_2(t), \ \ldots, \ x_n(t), t] \\ &\vdots \\ \dot{x}_n(t) &= f_n[x_1(t), x_2(t), \ \ldots, \ x_n(t), t] \end{aligned}\right\} \tag{1.2}$$

Usually (1.1) is supposed to hold everywhere in some open region $R \subseteq E_{n+1}$, for example

$$R = \{x, t \mid \|x\| < \alpha, |t| < \beta\} \tag{1.3}$$

Equation (1.1) is often written in the shorter form

$$\dot{x} = f(x, t) \tag{1.4}$$

or, if f does not depend explicitly on t,

$$\dot{x} = f(x) \tag{1.5}$$

but it should be noticed that this does not agree with the convention established in Chapter 1. According to that convention, $f(x)$ would represent a functional depending on the function x, whereas (1.5) or (1.4) represents a relation between values of the functions evaluated at a given t as in (1.1).

It is well known (and it will be proved below) that subject to certain conditions on f, there is a unique solution of (1.1) passing through a given point $x = c$, $t = t_0$ in R. This solution is written $\phi(t; c, t_0)$ to emphasize its dependence on t and on the condition $x(t_0) = c$. More briefly the solution is written $\phi(t)$. Often the solution is written $x(t)$, though this obscures the distinction between the differential equation (1.1), valid everywhere in R, and the solution determined by (1.1) taken with the condition $x(t_0) = c$.

An important mathematical problem is to give conditions on f which will ensure that (1.1) and the condition $x(t_0) = c$ will determine a solution and that this solution will be unique. Also, if R is

$$R = \{x, t \mid x \in E_n, \beta < t\} \tag{1.6}$$

it is important to know whether a solution passing through $(c, t_0) \in R$ can be continued for all t.

Example 1.1 By differentiating the function ϕ defined by

$$\phi(t) = \log(1 - t) \tag{1.7}$$

it will be found that for $t < 1$ it satisfies

$$\dot{\phi}(t) = \frac{1}{t - 1}; \qquad \phi(0) = 0 \tag{1.8}$$

The solution ϕ, however, cannot be carried forward continuously from $t = 0$ beyond the point $t = 1$. The solution is said to have a *finite escape time.*

It is usual to express the behaviour of physical systems by differential equations such as (1.1). When the approach of Chapter 4 is adopted, however, and variables are allowed to be discontinuous, the derivatives may not exist at certain points. The physical situation is then better described by equations such as

$$D_{-}x(t) = f[x(t), t] \tag{1.9}$$

together with the condition that x is continuous.

Example 1.2 Let a current $i(t)$ flow into a capacitance C, and let the charge $q(t)$ in the capacitance be zero at $t = 0$. The charge at time t is then

$$q(t) = \int_0^t i(\tau)\, d\tau \tag{1.10}$$

and the voltage across the capacitance is $v(t) = q(t)/C$. Suppose $i \in D_R^1$ and write

$$i(t) = \sum_k i_k U(t - q_i) + w(t) \tag{1.11}$$

where $w(t)$ is continuous. Then (1.10) gives

$$q(t) = \sum_k (t - q_i) i_k U(t - q_i) + \int_0^t w(\tau)\, d\tau \tag{1.12}$$

Theorem 10.1 in Chapter 5 shows that

$$D_{-} \int_0^t w(\tau)\, d\tau = D \int_0^t w(\tau)\, d\tau = w(t) \tag{1.13}$$

and so

$$D_{-}q(t) = \sum_k i_k U(t - q_i) + w(t) = i(t) \tag{1.14}$$

and $v(t)$ obeys the equation

$$D_{-}v(t) = i(t)/C \tag{1.15}$$

Also it is clear from (1.12) and (1.13) that q, and therefore v, is continuous. Where i is continuous, (1.12) and (1.13) show that v obeys the differential equation

$$Dv(t) = i(t)/C \tag{1.16}$$

but (1.12) shows that $Dq(t)$, and therefore $Dv(t)$, does not exist where i is discontinuous. Physically, the continuity of v follows from the fact that the current is finite, while the occurrence of $D_{-}v$ in (1.15) reflects the non-anticipative nature of the system.

Some existence theorems will now be given. Their form is appropriate to the applications which will be made.

2 Linear time-invariant differential equations

The following is an existence theorem for linear time-invariant differential equations. This is given separately, first because its form is particularly adapted to the requirements of many dynamical systems, and secondly because it can be proved more simply than the more general theorem of Section 3.

Theorem 2.1 Let there be given a constant $n \times n$ matrix A; a constant $n \times l$ matrix B; a vector function $u \in D_R$ with domain E_1 and range $R_u \subseteq E_l$; and a constant n-vector c. Then there is a unique continuous function ϕ with domain E_1 and range $R_\phi \subseteq E_n$ such that $\phi(t_0) = c$ and for all t

$$D_{-}\phi(t) = A\phi(t) + Bu(t) \tag{2.1}$$

If u is continuous at $t = t_1$ then $D\phi(t_1)$ exists. The function ϕ is given explicitly by

$$\phi(t) = e^{A(t-t_0)}c + \int_{t_0}^{t} e^{A(t-\tau)}Bu(\tau)\,d\tau \tag{2.2}$$

and $\phi \in D_R^1$.

Proof By Theorem 13.3 of Chapter 4,

$$D[e^{At}c] = [De^{At}]c = Ae^{At}c \tag{2.3}$$

Write

$$u(\tau) = \sum_i u_i U(\tau - p_i) + v(\tau) \tag{2.4}$$

where the $p_i \in [t_0, t]$ are such that $v(\tau)$ is continuous on an open interval J containing $[t_0, t]$. Then

$$\int_{t_0}^{t} e^{A(t-\tau)} B u_i U(\tau - p_i)\, d\tau = \left\{\int_{p_i}^{t} e^{A(t-\tau)} d\tau\right\} B u_i \tag{2.5}$$

$$= A^{-1}\{e^{A(t-p_i)} - I_n\}\, B u_i U(t - p_i) \tag{2.6}$$

by Exercise 10.2, Chapter 5. Hence

$$D_- \int_{t_0}^{t} e^{A(t-\tau)} B u_i U(\tau - p_i) d\tau = e^{A(t-p_i)} B u_i U(t - p_i) \tag{2.7}$$

Also by Theorem 10.1, Chapter 5,

$$D \int_{t_0}^{t} e^{A(t-\tau)} B v(\tau)\, d\tau$$

$$= [De^{At}] \left\{\int_{t_0}^{t} e^{-A\tau} B v(\tau)\, d\tau\right\} + e^{At}.e^{-At} B v(t) \tag{2.8}$$

$$= A \int_{t_0}^{t} e^{A(t-\tau)} B v(\tau)\, d\tau + B v(t) \tag{2.9}$$

Equations (2.4), (2.7), and (2.9) give

$$D_- \int_{t_0}^{t} e^{A(t-\tau)} B u(\tau)\, d\tau$$

$$= \sum_i e^{A(t-p_i)} B u_i U(t - p_i) + A \int_{t_0}^{t} e^{A(t-\tau)} B v(\tau)\, d\tau + B v(t) \tag{2.10}$$

$$= A \int_{t_0}^{t} e^{A(t-\tau)} B\left[\sum_i u_i U(\tau - p_i) + v(\tau)\right] d\tau + B\left[\sum_i u_i U(t - p_i) + v(t)\right] \tag{2.11}$$

$$= A \int_{t_0}^{t} e^{A(t-\tau)} B u(\tau)\, d\tau + B u(t) \tag{2.12}$$

where (2.11) uses equation (2.6). Then on using (2.3) and (2.12) in (2.2),

$$D_-\phi(t) = A\phi(t) + B u(t) \tag{2.13}$$

so that ϕ defined by (2.2) satisfies (2.1) on J.

Equations (2.3) and (2.9) show that the derivatives of $e^{At}c$ and of

$$\int_{t_0}^{t} e^{A(t-\tau)} B v(\tau)\, d\tau$$

exist, and so these functions are continuous. Also because

$$e^{A(t-p_i)} = I_n, \quad t = p_i \tag{2.14}$$

it is easy to verify from (2.6) that

$$\int_{t_0}^{t} e^{A(t-\tau)} B \sum_i u_i U(\tau - p_i)\, d\tau$$

is continuous. It follows that the function ϕ defined by (2.2) is continuous on J. Because $J \supset [t_0, t_1]$ is arbitrary, ϕ is continuous on a closed bounded interval $K \supset J$, and so is bounded on J.

When $t = t_0$, $e^{A(t-t_0)} = I_n$, and the integral in (2.2) is zero. Hence

$$\phi(t_0) = c \tag{2.15}$$

Also in (2.7), D_- can be replaced by D except at the points $t = p_i$. Except at these points, (2.13) then becomes

$$D\phi(t) = A\phi(t) + Bu(t) \tag{2.16}$$

The function ϕ, by (2.16) and (2.13), has continuous first derivative on J except at the points $t = p_i$ where $D\phi$ is discontinuous but $D_-\phi$ exists. It follows that $D_-\phi$ is bounded on J, and hence $\phi \in D_R^1$.

Finally it will be shown that if ψ is any continuous function satisfying

$$D_-\psi(t) = A\psi(t) + Bu(t); \qquad \psi(t_0) = c \tag{2.17}$$

then for all t,

$$g(t) \triangleq e^{-At}[\psi(t) - \phi(t)] = 0 \tag{2.18}$$

By (2.1) and (2.17),

$$D_-g(t) = -Ae^{-At}[\psi(t) - \phi(t)] + e^{-At}A[\psi(t) - \phi(t)] = 0 \tag{2.19}$$

and also $g(t)$ is continuous. Suppose, contrary to what is to be proved, that $g(t_2) = 3\epsilon > 0$ for some $t_2 > t_0$. The case $t_2 < t_0$ is treated similarly. Then the function h defined by

$$h(t) = g(t) - \epsilon\left[1 + \frac{t - t_0}{t_2 - t_0}\right] \tag{2.20}$$

is continuous and satisfies

$$h(t_0) = -\epsilon, \qquad h(t_2) = \epsilon \tag{2.21}$$

Consequently there is at least one $t \in (t_0, t_2)$ for which

$$h(t) = 0 \tag{2.22}$$

Let t_3 be the infimum of those t in the interval $[t_0, t_2]$ satisfying (2.22). Because h is continuous,

$$h(t_3) = 0 \tag{2.23}$$

and (2.21) shows that $t_3 > t_0$ and

$$h(t) < 0, \qquad t_0 \leqslant t < t_3 \tag{2.24}$$

Then

$$\lim_{\eta \to 0+} \frac{h(t_3) - h(t_3 - \eta)}{\eta} = D_{-}h(t_3) \geqslant 0 \tag{2.25}$$

and on using (2.19) and (2.20), (2.25) gives

$$0 \leqslant D_{-}g(t_3) - \frac{\epsilon}{t_2 - t_0} = -\frac{\epsilon}{t_2 - t_0} < 0 \tag{2.26}$$

which is a contradiction. Therefore

$$g(t) = e^{-At}[\psi(t) - \phi(t)] = 0, \quad \text{all } t \tag{2.27}$$

and because e^{At} is non-singular for all A and t (Chapter 4, Theorem 13.4) this shows that

$$\psi(t) = \phi(t), \quad \text{all } t \tag{2.28}$$

This completes the proof.

If u is continuous, the condition that ϕ should be continuous can be dropped if (2.1) is replaced by (2.16). This gives the following more usual form of the existence theorem.

Theorem 2.2 Let A, B, and c be given as in Theorem 2.1. Let u be a continuous vector function with domain E_1 and range $R_u \subset E_l$. Then there is a unique function ϕ on E_1 such that $\phi(t_0) = c$ and for all t

$$D\phi(t) = A\phi(t) + Bu(t) \tag{2.29}$$

The function ϕ has continuous first derivative and is given explicitly by

$$\phi(t) = e^{A(t-t_0)}c + \int_{t_0}^{t} e^{A(t-\tau)}Bu(\tau)\,d\tau \tag{2.30}$$

Proof Let ϕ be any function satisfying (2.29) for all t. Then $D\phi(t)$ exists for all t and so ϕ is continuous everywhere. Also because $D\phi$ exists, $D_{-}\phi$ exists and is equal to $D\phi$. So ϕ satisfies the conditions of Theorem 2.1. Conversely, if ϕ satisfies the conditions of Theorem 2.1 and u is continuous, ϕ satisfies (2.16) for all t, and (2.16) is (2.29). Therefore when u is continuous the conditions in the statements of the two theorems are equivalent. The existence and uniqueness of ϕ, and the validity of (2.30), follow from Theorem 2.1. The fact that ϕ has continuous first derivative follows from (2.29), the continuity of ϕ, and the continuity of u.

EXERCISE 2.1 In Theorem 2.2, replace the condition that u is continuous everywhere by the condition that u is continuous on the open interval (a, b) where $a < t_0 < b$. Show that all the conclusions hold if t is restricted to the interval (a, b).

Example 2.1 Let A be the companion matrix

$$A = \begin{pmatrix} 0 & 0 & \dots & 0 & -a_0 \\ 1 & 0 & \dots & 0 & -a_1 \\ \vdots & \vdots & & \vdots & \vdots \\ 0 & 0 & \dots & 1 & -a_{n-1} \end{pmatrix} \tag{2.31}$$

and let B be

$$B = \begin{pmatrix} b_0 \\ b_1 \\ \vdots \\ b_{n-1} \end{pmatrix} \tag{2.32}$$

Then the differential equations corresponding to (2.29) can be written [omitting the argument t of $x(t)$ and $u(t)$ as in (1.5)]

$$\left.\begin{aligned} Dx_1 &= \quad -a_0x_n + b_0u \\ Dx_2 &= x_1 \quad -a_1x_n + b_1u \\ &\vdots \\ Dx_{n-1} &= x_{n-2} - a_{n-2}x_n + b_{n-2}u \\ Dx_n &= x_{n-1} - a_{n-1}x_n + b_{n-1}u \end{aligned}\right\} \tag{2.33}$$

where u is a continuous real-valued function of t. The last of these equations gives

$$(Dx_n + a_{n-1}x_n - b_{n-1}u) = x_{n-1} \tag{2.34}$$

and the last equation but one in (2.33) shows that x_{n-1} has a derivative. Hence the expression on the left-hand side of (2.34) has a derivative, and

$$D(Dx_n + a_{n-1}x_n - b_{n-1}u) + a_{n-2}x_n - b_{n-2}u = x_{n-2} \tag{2.35}$$

Proceeding in this way we obtain

$$0 = a_0x_n - b_0u + D\{a_1x_n - b_1u + D[a_2x_n - b_2u + D(\dots)]\} \tag{2.36}$$

Equation (2.36) is usually written

$$a_0x_n + a_1\,Dx_n + \cdots + a_{n-1}\,D^{n-1}x_n + D^nx_n = b_0u + b_1\,Du + \cdots + b_{n-1}\,D^{n-1}u \tag{2.37}$$

but notice that u was assumed only to be continuous, and so $Du, D^2u, \ldots, D^{n-1}u$ may not exist. Equations (2.33) and (2.36) therefore have a more general application than (2.37). [In the theory of distributions, a meaning can be given to Du, D^2u, etc. at points where these derivatives do not exist in the ordinary sense, so that (2.37) becomes legitimate. In many applications, however, the equations of a system† arise in the form (2.33), or its equivalent, and we may avoid the difficulties inherent in (2.37). Compare Hamilton's equations in dynamics, or the equations for heat and material balances in chemical engineering, etc.]

EXERCISE 2.2 In Example 2.1, let $n = 2$, $a_0 = a_1 = b_1 = 1$, $b_0 = 0$, $u(t) = tU(t)$, so that equations (2.33) become

$$\begin{aligned} Dx_1 &= \quad -x_2 \\ Dx_2 &= x_1 - x_2 + tU(t) \end{aligned} \tag{2.38}$$

and let the initial condition be $x(0) = 0$. By direct verification show that equations (2.38) are satisfied by

$$x_1(t) = U(t)\left\{1 - t - e^{-t/2}\left[\cos\frac{t\sqrt{3}}{2} - \frac{1}{\sqrt{3}}\sin\frac{t\sqrt{3}}{2}\right]\right\} \tag{2.39}$$

$$x_2(t) = U(t)\left\{1 - e^{-t/2}\left[\cos\frac{t\sqrt{3}}{2} + \frac{1}{\sqrt{3}}\sin\frac{t\sqrt{3}}{2}\right]\right\} \tag{2.40}$$

The function $tU(t)$ has no derivative at $t = 0$, so that (2.37) has no meaning in the ordinary sense. On the other hand, (2.36) gives

$$D(Dx_2 + x_2 - u) + x_2 = 0 \tag{2.41}$$

Show that the bracketed quantity in (2.41) has a derivative everywhere. [Expand $Dx_2 + x_2 - u$ in the form $U(t)(\alpha + \beta t + \gamma t^2 + \cdots)$ and show $\alpha = \beta = 0$.]

3 A more general existence theorem

The existence theorems in Section 2 relate to the special case of linear differential equations with constant coefficients. If the differential equation is

$$Dx(t) = f[x(t), t]; \qquad x(t_0) = c \tag{3.1}$$

where x is an n-vector, then the existence and uniqueness of a solution can be proved under much less restrictive conditions on f than those used in Section 2.

Reference to Theorem 2.1 will show that even in the special case considered there, the function f in equation (3.1) had to be continuous (through u) as a function of t if the derivative $Dx(t)$ was to exist. So we may expect to need the condition in (3.1) that f is continuous. If $f(x, t)$ is discontinuous as a function of

†If u may be discontinuous, equation (2.1) is to be substituted for (2.33); or the theory of distributions must again be invoked.

t alone at a finite number of points in any finite interval, (3.1) must be replaced at these points by a condition that the solution is continuous (or by whatever other condition is imposed by the physics of the problem). This possibility will not be considered here, and it will be assumed that f is continuous as a function of x and t on a domain D of E_{n+1},

$$D = \{x, t \mid \|x - c\| \leqslant \alpha, \quad \beta_1 \leqslant t - t_0 \leqslant \beta_2\} \tag{3.2}$$

where $\beta_1 \leqslant 0 \leqslant \beta_2$.

It can be shown (Coddington and Levinson, 1955) that continuity of f is sufficient to ensure the existence of a solution of (3.1) in D. It is not enough to ensure uniqueness, as is shown by the following example.

Example 3.1 (Coddington and Levinson, 1955) In equation (3.1), put $f(x) = \frac{3}{2}x^{1/3}$, which is continuous, and let $x(0) = 0$. Then if η is any positive number the function ϕ defined by

$$\left.\begin{aligned} \phi(t) &= 0, \quad t \leqslant \eta \\ \phi(t) &= (t - \eta)^{3/2}, \quad t > \eta \end{aligned}\right\} \tag{3.3}$$

is a solution of (3.1). For

$$D\phi(t) = 0 = \tfrac{3}{2}[\phi(t)]^{1/3}, \quad t < \eta \tag{3.4}$$

$$D\phi(t) = \tfrac{3}{2}(t - \eta)^{1/2} = \tfrac{3}{2}[\phi(t)]^{1/3}, \quad t > \eta \tag{3.5}$$

while

$$D_{-}\phi(\eta) = \lim_{h\to 0+} \frac{\phi(\eta) - \phi(\eta - h)}{h} = 0 \tag{3.6}$$

$$D_{+}\phi(\eta) = \lim_{h\to 0+} \frac{\phi(\eta) - \phi(\eta + h)}{h} \tag{3.7}$$

$$= \lim_{h\to 0+} \frac{1}{h}\{-h^{3/2}\} \tag{3.8}$$

$$= \lim_{h\to 0+} \{-h^{1/2}\} \tag{3.9}$$

$$= 0 \tag{3.10}$$

and so $D\phi(\eta)$ exists and is zero. Since η can be chosen arbitrarily greater than zero, the solution is not unique.

A convenient condition on f, in addition to continuity, which ensures uniqueness of the solution of (3.1) is the *Lipschitz condition*, which is defined in the following way.

Let the function f have domain $D \subset E_{n+1}$ as defined by (3.2), and range

$R \subset E_n$. Let there exist a constant $k > 0$ such that for every $(x, t) \in D$ and $(y, t) \in D$

$$\|f(y, t) - f(x, t)\| \leqslant k \|y - x\| \tag{3.11}$$

Then f is said to satisfy a Lipschitz condition with respect to x on D, written $f \in \text{Lip}$ on D.

If the function f is continuous on D and satisfies a Lipschitz condition with respect to x on D, the notation $f \in (C, \text{Lip})$ will be used.

EXERCISE 3.1 Show that if f is a continuous function of t on D and satisfies a Lipschitz condition with respect to x on D, then f is continuous on D and so $f \in (C, \text{Lip})$. [Write

$$\|f(x_2, t_2) - f(x_1, t_1)\| \leqslant \|f(x_2, t_2) - f(x_2, t_1)\| + \|f(x_2, t_1) - f(x_1, t_1)\|]$$

EXERCISE 3.2 Show that the real function f defined by

$$f(x, t) = x^2 t^{1/3} \tag{3.12}$$

satisfies $f \in (C, \text{Lip})$ on $D = \{x, t \mid |x| \leqslant 1, \quad |t| \leqslant 1\}$. [Notice that f does not satisfy a Lipschitz condition with respect to t on D: this is not required by the definition of (C, Lip).]

EXERCISE 3.3 Let the function f have continuous and bounded first partial derivatives on an open region including the domain D defined by (3.2). Then by Theorem 8.7 of Chapter 4, f has a derivative everywhere on D and so is continuous on D. Use Corollary 2 of Theorem 8.3 in Chapter 4 to show that $f \in (C, \text{Lip})$ on D. [This is a convenient sufficient condition for $f \in (C, \text{Lip})$. Notice that the proof uses the convexity properties of D.]

Because f is continuous on D, it follows that $\|f\|$ is continuous on D. Then because D is closed and bounded, $\|f\|$ has a maximum value M on D. Define $\gamma_1 \leqslant t_0$ and $\gamma_2 \geqslant t_0$ by

$$\gamma_1 = \max\left\{t_0 - \frac{\alpha}{M}, t_0 + \beta_1\right\}, \qquad \gamma_2 = \min\left\{t_0 + \frac{\alpha}{M}, t_0 + \beta_2\right\} \tag{3.13}$$

where α, β_1, β_2 are as in (3.2). Then it will be shown below that there exists a unique solution $\phi(t)$ of equation (3.1) for $t \in [\gamma_1, \gamma_2]$, and this solution lies in D. The solution will in fact be obtained as the limit of a sequence of functions ϕ_r defined by

$$\left.\begin{aligned} \phi_0(t) &= c \\ \phi_{r+1}(t) &= c + \int_{t_0}^{t} f[\phi_r(\tau), \tau]\, d\tau, \quad r = 0, 1, 2, \ldots; \quad t \in [\gamma_1, \gamma_2] \end{aligned}\right\} \tag{3.14}$$

First the existence of a solution will be proved.

Theorem 3.1 If $f \in (C, \text{Lip})$ on D, then the functions ϕ_r, $r = 0, 1, 2, \ldots$, exist on $[\gamma_1, \gamma_2]$ and are continuous. As $r \to \infty$, the functions ϕ_r converge uniformly on $[\gamma_1, \gamma_2]$ to a function ϕ which satisfies (3.1) on (γ_1, γ_2). This solution ϕ of (3.1) is continuous on $[\gamma_1, \gamma_2]$ and has continuous first derivative on (γ_1, γ_2).

Proof The function ϕ_0 has the following properties

(a) ϕ_0 exists on $[\gamma_1, \gamma_2]$
(b) ϕ_0 is continuous on $[\gamma_1, \gamma_2]$
(c) $\|\phi_0(t) - c\| \leqslant M\,|t - t_0|$ for $t \in [\gamma_1, \gamma_2]$.

Assume, as the basis for an induction, that ϕ_r has these three properties. Then $f[\phi_r(t), t]$ is defined and continuous on $[\gamma_1, \gamma_2]$, and so ϕ_{r+1} defined by (3.14) exists on this interval by Exercise 4.2, Chapter 5. Also ϕ_{r+1} is continuous on $[\gamma_1, \gamma_2]$ by the result of Exercise 4.6, Chapter 5. Further,

$$\|\phi_{r+1}(t) - c\| = \left\| \int_{t_0}^{t} f[\phi_r(\tau), \tau]\, d\tau \right\| \tag{3.15}$$

$$\leqslant \left| \int_{t_0}^{t} \|f[\phi_r(\tau), \tau]\|\, d\tau \right| \tag{3.16}$$

$$\leqslant M\,|t - t_0| \tag{3.17}$$

so that ϕ_{r+1} also has the three properties listed. By induction these properties therefore are possessed by each ϕ_r, $r = 0, 1, 2, \ldots.$

Let $m > 0$ and $n > m$ be integers. Then for $t \in [\gamma_1, \gamma_2]$,

$$\|\phi_n(t) - \phi_m(t)\| \leqslant \sum_{j=1}^{n-m} \|\phi_{m+j}(t) - \phi_{m+j-1}(t)\| \tag{3.18}$$

But because $f \in (C, \text{Lip})$,

$$\|\phi_{r+1}(t) - \phi_r(t)\| \leqslant \left| \int_{t_0}^{t} \|f[\phi_r(\tau), \tau] - f[\phi_{r-1}(\tau), \tau]\|\, d\tau \right| \tag{3.19}$$

$$\leqslant k \left| \int_{t_0}^{t} \|\phi_r(\tau) - \phi_{r-1}(\tau)\|\, d\tau \right| \tag{3.20}$$

and by an induction on (3.20), starting with

$$\|\phi_1(t) - c\| \leqslant M\,|t - t_0| \tag{3.21}$$

where $c = \phi_0(t)$, it follows that

$$\|\phi_r(t) - \phi_{r-1}(t)\| \leqslant \frac{M}{k} \cdot \frac{(k\,|t - t_0|)^r}{r!} \tag{3.22}$$

Using (3.22) in (3.18),

$$\|\phi_n(t) - \phi_m(t)\| \leqslant \sum_{j=1}^{n-m} \frac{M}{k} \cdot \frac{(k\,|t - t_0|)^{m+j}}{(m+j)!} \tag{3.23}$$

$$\leqslant \frac{M}{k} \sum_{j=1}^{n-m} \frac{[k(\gamma_2 - \gamma_1)]^{m+j}}{(m+j)!} \tag{3.24}$$

for $t \in [\gamma_1, \gamma_2]$. Now the series

$$\sum_{r=0}^{\infty} \frac{[k(\gamma_2 - \gamma_1)]^r}{r!} = e^{k(\gamma_2 - \gamma_1)} \tag{3.25}$$

has been shown in Example 12.2 of Chapter 4 to be convergent for all $k(\gamma_2 - \gamma_1)$. By Cauchy's condition this implies that given any $\epsilon > 0$ there exists an m_0 such that for all $m > m_0$ and all $n > m$

$$\sum_{j=1}^{n-m} \frac{[k(\gamma_2 - \gamma_1)]^{m+j}}{(m+j)!} < \frac{k\epsilon}{M} \tag{3.26}$$

This in turn, by (3.24), implies the uniform convergence of $\phi_m(t)$ as $m \to \infty$ to a limit $\phi(t)$ for each t in $[\gamma_1, \gamma_2]$. In particular, $\phi(t_0) = c$. The function $\phi(t)$ is continuous on $[\gamma_1, \gamma_2]$ by Theorem 11.1 of Chapter 4.

Because every ϕ_m satisfies $\|\phi_m(t) - c\| \leqslant M\,|t - t_0|$ for $t \in [\gamma_1, \gamma_2]$, it follows that ϕ also satisfies this condition, and so $[\phi(t), t] \in D$ for $t \in [\gamma_1, \gamma_2]$, and $f[\phi(t), t]$ is defined and continuous for $t \in [\gamma_1, \gamma_2]$, and so is Riemann integrable on this interval. Then for $t \in [\gamma_1, \gamma_2]$,

$$\left\| \int_{t_0}^{t} \{f[\phi(\tau), \tau] - f[\phi_m(\tau), \tau]\}\, d\tau \right\|$$

$$\leqslant \left| \int_{t_0}^{t} \|f[\phi(\tau), \tau] - f[\phi_m(\tau), \tau]\|\, d\tau \right| \tag{3.27}$$

$$\leqslant k \left| \int_{t_0}^{t} \|\phi(\tau) - \phi_m(\tau)\|\, d\tau \right| \tag{3.28}$$

As $m \to \infty$, the right-hand side of (3.28) tends to zero, and so as $m \to \infty$

$$\int_{t_0}^{t} f[\phi_m(\tau), \tau]\, d\tau \to \int_{t_0}^{t} f[\phi(\tau), \tau]\, d\tau \tag{3.29}$$

Letting r tend to infinity in (3.14), and using (3.29),

$$\phi(t) = c + \int_{t_0}^{t} f[\phi(\tau), \tau]\, d\tau \tag{3.30}$$

The right-hand side of (3.30) has a derivative for $t \in (\gamma_1, \gamma_2)$, which is continuous and satisfies

$$D\phi(t) = f[\phi(t), t] \tag{3.31}$$

This completes the proof of existence.

It will be noticed that the function ϕ defined by this theorem is a 'solution' of the differential equation in a slightly more extended sense than was originally envisaged. If $\beta_1 = 0$ or $\beta_2 = 0$ in (3.2), $(\phi(t_0), t_0)$ is in the boundary of D. Similarly, the solution may be in the boundary of D at $t = t_0 - \alpha/M$ or at $t = t_0 + \alpha/M$. As we have only assumed f to exist and have the required properties on D, it is clearly impossible to ask that ϕ should satisfy the differential equation at the boundary of D. However, ϕ is continuous at such points, and it is easy to verify from (3.30) that $D_-\phi$ or $D_+\phi$ (one of which will always exist) is equal to $f(\phi, t)$ at such a point. (See Problem 3.)

Example 3.2 Let ϕ satisfy

$$D\phi(t) = f[\phi(t)]; \qquad \phi(0) = 0 \tag{3.32}$$

where

$$f(x) = \begin{cases} 1, & |x| \leqslant 1 \\ 0, & |x| > 1 \end{cases} \tag{3.33}$$

Then for

$$D = \{x, t \mid |x| \leqslant 1, \quad 0 \leqslant t \leqslant 2\} \tag{3.34}$$

the conditions of the theorem are satisfied, with $M = 1$, $\alpha = 1$. The functions ϕ_r are

$$\begin{aligned} \phi_0(t) &= 0 \\ \phi_1(t) &= \phi_2(t) = \cdots = t \end{aligned} \tag{3.35}$$

Consequently $\phi(t) = t$, which is continuous on $[0, \alpha/M] = [0, 1]$ and satisfies (3.31) on (0, 1). At $t = 1$, $D_-\phi$ exists and has the value 1, but the solution ϕ cannot be extended in such a way that it satisfies (3.31) on an interval $(1 - \epsilon, 1 + \epsilon)$.

EXERCISE 3.4 Prove the last statement in Example 3.2. [If $D\phi(1)$ exists, it has the value 1. So $\phi(1 + \eta) > 1$ for some $\eta > 0$. Use the mean-value theorem to establish a contradiction.]

The theorem shows that a solution ϕ certainly exists on $[\gamma_1, \gamma_2]$. If at γ_1 or at γ_2, ϕ lies inside D, the solution can be extended over a larger interval. Suppose that ϕ remains always in D_1,

$$D_1 = \{x, t \mid \|x - c\| \leqslant \alpha_1 < \alpha, \quad \beta_1 \leqslant t \leqslant \beta_2\} \tag{3.36}$$

Then at any t_1, the solution can certainly be extended forward to

$$t_1 + \min\left\{\frac{\alpha - \alpha_1}{M}, \quad \beta_2 - t_1\right\}$$

Consequently, the solution ϕ can be carried forward by successive stages to $t = \beta_2$. Similarly it can be carried back to $t = \beta_1$.

3.1 Uniqueness

In proving uniqueness, it is necessary to take account of the generalization of the idea of a 'solution' made above.

Theorem 3.2 Let ψ be a function which satisfies the differential equation (3.1) at every interior point of D where it exists, which is continuous at each boundary point of D where it exists, and which satisfies $\psi(t_0) = c$. Then on $[\gamma_1, \gamma_2]$, $\psi(t) = \phi(t)$, where $\phi(t)$ is defined in Theorem 3.1.

Proof Assume first that (c, t_0) is an interior point of D. At each interior point of D where it exists, ψ has a derivative and is therefore continuous. But ψ passes through the point $(c, t_0) \in D$ and therefore lies in D on some interval $[\omega_1, \omega_2]$ including t_0. By the mean value theorem, if $(\psi(t), t) \in D$ for $t \in [\omega_1, \omega_2]$,

$$\|\psi(t) - c\| \leqslant M\,|t - t_0|, \quad t \in (\omega_1, \omega_2) \tag{3.37}$$

It follows that $[\omega_1, \omega_2]$ may be identified with $[\gamma_1, \gamma_2]$.

Define a function g on $[\gamma_1, \gamma_2]$ by

$$g(t) = \phi(t) - \psi(t) \tag{3.38}$$

Then because ϕ and ψ each have a derivative on (γ_1, γ_2), they are continuous on this interval. They are also continuous at γ_1 and γ_2, because these are either interior points of D or boundary points of D. Hence g is continuous on $[\gamma_1, \gamma_2]$. Because f is continuous on D,

$$Dg(t) = f[\phi(t), t] - f[\psi(t), t] \tag{3.39}$$

is continuous on (γ_1, γ_2), and it follows that

$$\int_{t_0}^{t} Dg(\tau)\, d\tau$$

exists for $t \in (\gamma_1, \gamma_2)$. An easy application of equation (7.26) of Chapter 5 gives

$$\int_{t_0}^{t} Dg(\tau)\, d\tau = g(t), \quad t \in (\gamma_1, \gamma_2) \tag{3.40}$$

because $g(t_0) = 0$. Then

$$\|g(t)\| \leqslant \left| \int_{t_0}^{t} \|Dg(\tau)\|\, d\tau \right| \tag{3.41}$$

$$= \left| \int_{t_0}^{t} \|f[\phi(\tau), \tau] - f[\psi(\tau), \tau]\|\, d\tau \right| \tag{3.42}$$

$$\leqslant k \left| \int_{t_0}^{t} \|\phi(\tau) - \psi(\tau)\|\, d\tau \right| \tag{3.43}$$

$$= k \left| \int_{t_0}^{t} \|g(\tau)\|\, d\tau \right| \tag{3.44}$$

The function $\|g\|$, continuous on the closed interval $[\gamma_1, \gamma_2]$, has a maximum value m on this interval. So (3.44) gives

$$\|g(t)\| \leqslant mk\,|t - t_0| \tag{3.45}$$

and using this again in (3.44)

$$\|g(t)\| \leqslant \tfrac{1}{2}mk^2(t - t_0)^2 \tag{3.46}$$

whence (using induction), for $t \in (\gamma_1, \gamma_2)$,

$$\|g(t)\| \leqslant \frac{mk^n\,|t - t_0|^n}{n!} \leqslant \frac{mk^n\,|\gamma_2 - \gamma_1|^n}{n!} \tag{3.47}$$

By the continuity of $\|g\|$, (3.47) is true for $t \in [\gamma_1, \gamma_2]$. Given any $\epsilon > 0$, there is an n_0 such that the last member of (3.47) is less than ϵ for all $n > n_0$. From this it follows that $g(t) = 0$ for all $t \in [\gamma_1, \gamma_2]$. That is, $\psi(t) = \phi(t)$ for all $t \in [\gamma_1, \gamma_2]$.

If (c, t_0) is a boundary point of D, ψ is continuous at t_0 by hypothesis. Therefore ψ lies in D on some interval $[t_0, \omega_2]$ or $[\omega_1, t_0]$. Suppose the former case; the latter is treated similarly. As in (3.37), for $t \in (t_0, \omega_2)$, $t_1 \in (t_0, \omega_2)$,

$$\|\psi(t) - \psi(t_1)\| \leqslant M\,|t - t_1|$$

Let $t_1 \to t_0$; then by the continuity of ψ at t_0,

$$\|\psi(t) - c\| \leqslant M\,|t - t_0|, \quad t \in (t_0, \omega_2) \tag{3.48}$$

whence ω_2 may be identified with γ_2. Similarly

$$\int_{t_1}^{t} Dg(\tau) = g(t) - g(t_1); \quad t, t_1 \in (t_0, \gamma_2) \tag{3.49}$$

and by the continuity of g on $[t_0, \gamma_2]$,

$$\lim_{t_1 \to t_0} \int_{t_1}^{t} Dg(\tau)\,d\tau = g(t), \quad t \in (t_0, \gamma_2) \tag{3.50}$$

That is, (3.40) is true if the integral is treated as an improper one at the lower limit of integration. The remainder of the proof is easily completed.

3.2 Continuity in initial conditions

The solution ϕ of (3.1) takes a value $\phi(t)$ at some given t where ϕ exists. This value of ϕ is a function of the initial condition c, as is expressed more clearly by the notation $\phi(t; c, t_0)$. We may therefore ask what properties ϕ has as a function of c. To determine this question we need the following preliminary result.

Theorem 3.3 Let $\phi_1(t; c_1, t_0)$ and $\phi_2(t; c_2, t_0)$ be two solutions of the differential equation (3.1) with different initial conditions at t_0. Let each of these solutions remain in D on $\xi_1 \leqslant t - t_0 \leqslant \xi_2$ where $\xi_1 \leqslant t_0 \leqslant \xi_2$. Then for all $t - t_0 \in [\xi_1, \xi_2]$,

$$\|\phi_2(t) - \phi_1(t)\| \leqslant \|c_2 - c_1\| \, e^{k|t-t_0|} \tag{3.51}$$

Proof By equation (3.30)

$$\phi_1(t) = c_1 + \int_{t_0}^{t} f[\phi_1(\tau), \tau] \, d\tau \tag{3.52}$$

$$\phi_2(t) = c_2 + \int_{t_0}^{t} f[\phi_2(\tau), \tau] \, d\tau \tag{3.53}$$

so that

$$\|\phi_2(t) - \phi_1(t)\| = \left\| c_2 - c_1 + \int_{t_0}^{t} \{f[\phi_2(\tau); \tau] - f[\phi_1(\tau), \tau]\} \, d\tau \right\| \tag{3.54}$$

$$\leqslant \|c_2 - c_1\| + k \left| \int_{t_0}^{t} \|\phi_2(\tau) - \phi_1(\tau)\| \, d\tau \right| \tag{3.55}$$

Now $\|\phi_2(t) - \phi_1(t)\|$ is continuous on $\xi_1 \leqslant t - t_0 \leqslant \xi_2$ so that

$$w(t) \triangleq \int_{t_0}^{t} \|\phi_2(\tau) - \phi_1(\tau)\| \, d\tau \tag{3.56}$$

has a derivative on $\xi_1 < t - t_0 < \xi_2$ given by

$$Dw(t) = \|\phi_2(t) - \phi_1(t)\| \tag{3.57}$$

Assume first that $t \geqslant t_0$. Then equation (3.55) becomes

$$Dw(t) \leqslant \|c_2 - c_1\| + kw(t) \tag{3.58}$$

$$\frac{kDw(t)/\|c_2 - c_1\|}{1 + kw(t)/\|c_2 - c_1\|} \leqslant k \tag{3.59}$$

whence, using $w(t_0) = 0$,

$$\log\left\{1 + \frac{kw(t)}{\|c_2 - c_1\|}\right\} \leqslant k(t - t_0) \tag{3.60}$$

or

$$\|c_2 - c_1\| + kw(t) \leqslant \|c_2 - c_1\| \, e^{k(t-t_0)} \tag{3.61}$$

which with (3.55) gives, for $\xi_1 < t - t_0 < \xi_2$

$$\|\phi_2(t) - \phi_1(t)\| \leqslant \|c_2 - c_1\| \, e^{k(t-t_0)} \tag{3.62}$$

A similar result is obtained when $t \leqslant t_0$, but with $e^{-k(t-t_0)}$ in place of $e^{k(t-t_0)}$. Combining the results and using the continuity of ϕ_1 and ϕ_2 on $\xi_1 \leqslant t - t_0 \leqslant \xi_2$ gives the theorem.

With this result available, the following theorem can be proved.

Theorem 3.4 Let $\phi(t; c, t_0)$ be the solution of (3.1), and let ϕ satisfy $\|\phi(t) - c\| \leqslant a_1 < \alpha$ for $\beta_1 \leqslant \zeta_1 \leqslant t - t_0 \leqslant \zeta_2 \leqslant \beta_2$, where $\zeta_1 \leqslant t_0 \leqslant \zeta_2$ and α, β_1, β_2 are the real numbers defining D as in (3.2). Then given any $\epsilon > 0$ there is a $\delta > 0$ such that $\|c_1 - c\| < \delta$ implies that $\phi_1(t; c_1, t_0)$ exists on $\zeta_1 \leqslant t - t_0 \leqslant \zeta_2$ and $\|\phi_1(t; c_1, t_0) - \phi(t; c, t_0)\| < \epsilon$ for all $\zeta_1 \leqslant t - t_0 \leqslant \zeta_2$.

Proof Choose $\epsilon_1 \leqslant \min\{\epsilon, \alpha - \alpha_1\}$. Then $\phi(t; c, t_0) + x \in D$ whenever $\zeta_1 \leqslant t - t_0 \leqslant \zeta_2$ and $\|x\| < \epsilon_1$. Then choose

$$\delta = \epsilon_1 e^{-k(\zeta_2 - \zeta_1)} \tag{3.63}$$

For all c_1 satisfying $\|c_1 - c\| < \delta$, equation (3.51) shows that $\|\phi_1(t; c_1, t_0) - \phi(t; c, t_0)\| < \epsilon_1 \leqslant \epsilon$ for all $\zeta_1 \leqslant t - t_0 \leqslant \zeta_2$, provided that ϕ_1 exists and remains in D on this interval. This implies that ϕ_1 exists and $\|\phi_1(t; c_1, t_0) - \phi(t; c, t_0)\| < \epsilon$ for all $\zeta_1 \leqslant t - t_0 \leqslant \zeta_2$; because the opposite assumption, with the continuity of ϕ_1 where it exists in D, leads to a contradiction.

In words, this theorem shows that the solutions of (3.1) starting 'near' c at $t = t_0$ remain 'near' $\phi(t; c, t_0)$ for $\zeta_1 \leqslant t - t_0 \leqslant \zeta_2$. Moreover, because the solution through any point in D is unique, these adjacent solutions cannot meet each other. The differential equation, together with the conditions on f, therefore places a severe restriction on the behaviour of its solutions.

4 Linear time-dependent systems

Linear time-invariant differential equations could be treated separately in Section 2 because a convenient formula, equation (2.2), was available for the solution. A generalization of this formula is available for the linear time-dependent differential equation

$$Dx(t) = A(t)x(t) + B(t)u(t); \quad x(t_0) = c \tag{4.1}$$

where x is an n-vector and u an l-vector. The generalized formula, however, assumes the existence of solutions of (4.1), so this is first established.

Theorem 4.1 Let $A(t)$, $B(t)$ and $u(t)$ be continuous functions of t on $\beta_1 \leqslant t - t_0 \leqslant \beta_2$. Then there exists a function satisfying $\phi(0) = c$ and such that for $t - t_0 \in (\beta_1, \beta_2)$

$$D\phi(t) = A(t)\phi(t) + B(t)u(t) \tag{4.2}$$

This solution ϕ of (4.1) is unique, is continuous on $\beta_1 \leqslant t - t_0 \leqslant \beta_2$, and has continuous first derivative on $\beta_1 < t - t_0 < \beta_2$.

Proof It is only necessary to show that the conditions of Theorem 3.1 are satisfied. Given any α, it is clear that

$$f(x, t) \triangleq A(t)x(t) + B(t)u(t) \tag{4.3}$$

is continuous on D defined by

$$D = \{x, t \mid \|x - c\| \leqslant \alpha, \quad \beta_1 \leqslant t - t_0 \leqslant \beta_2\} \tag{4.4}$$

Also if $(x, t) \in D$ and $(y, t) \in D$,

$$\|f(y, t) - f(x, t)\| = \|A(t)y(t) - A(t)x(t)\| \tag{4.5}$$

$$\leqslant \|A(t)\|_c \, \|y(t) - x(t)\| \tag{4.6}$$

because the cubic norm $\|A\|_c$ has been shown in Section 10 of Chapter 3 to be consistent with the euclidean norm $\|x\|$. Now $\|A(t)\|_c$ is clearly a continuous function of its elements, and therefore of t, on $\beta_1 \leqslant t - t_0 \leqslant \beta_2$, and so has a maximum value k on this interval. This k qualifies as a Lipschitz constant, and so $f \in (C, \text{Lip})$ on D.

The continuous function $\|B(t)u(t)\|$ has a maximum value m on $\beta_1 \leqslant t - t_0 \leqslant \beta_2$. Choose

$$\alpha = \|c\| + m/k \tag{4.7}$$

Then the maximum value M of $\|f(x, t)\|$ in D is

$$\max_{(x,t) \in D} \|A(t)x(t) + B(t)u(t)\| \leqslant \max_{t \in D} \|A(t)\|_c \max_{x \in D} \|x\| + \max_{t \in D} \|B(t)u(t)\| \tag{4.8}$$

$$= k(\|c\| + \alpha) + m \tag{4.9}$$

Theorem 3.1 shows that a solution of (4.1) exists in D and may be carried forward to $t - t_0 = \beta_2$ or to

$$t - t_0 = \frac{\alpha}{M} = \frac{1}{2k} \tag{4.10}$$

whichever is the smaller. If $(1/2k) < \beta_2$, the point $\phi(t_0 + 1/2k)$ on the solution ϕ can be treated as a new initial point in place of c, and a new α may be defined: in this way the solution may be carried forward successively to $t_0 + 1/2k, t_0 + 2/2k, t_0 + 3/2k, \ldots$ until finally it reaches $t = t_0 + \beta_2$. Similarly the solution can be carried back to $t = t_0 + \beta_1$. This ends the proof.

Knowing that equation (4.1) has a solution ϕ on $\beta_1 \leqslant t - t_0 \leqslant \beta_2$, it is not difficult to give an explicit formula for ϕ in terms of a fundamental set of solutions. Let ψ_1 be the solution of (4.1) when $u = 0$ and $c = e_1$, let ψ_2 be the solution when $u = 0$ and $c = e_2$, etc.; here the column vector e_i has zero elements except in the ith row, where it has 1. Let Ψ be the matrix constructed from the ψ_i,

$$\Psi(t) = \{\psi_1(t), \psi_2(t), \ldots, \psi_n(t)\} \tag{4.11}$$

This matrix is called the *transition matrix* and it is the solution of the matrix differential equation

$$D\Psi(t) = A(t)\Psi(t); \qquad \Psi(t_0) = I \tag{4.12}$$

With this notation we have the following theorem.

Theorem 4.2 The solution ϕ of (4.1) on $\beta_1 \leqslant t - t_0 \leqslant \beta_2$ can be written

$$\phi(t) = \Psi(t)c + \Psi(t)\int_{t_0}^{t} \Psi^{-1}(\tau)B(\tau)u(\tau)\, d\tau \tag{4.13}$$

Proof First, Ψ exists and is continuous on $\beta_1 \leqslant t - t_0 \leqslant \beta_2$ by Theorem 4.1. Secondly, the determinant $|\Psi|$ of Ψ obeys the differential equation (see Exercise 4.1 below)

$$D\,|\Psi(t)| = [\text{trace } A(t)]\,|\Psi(t)| \tag{4.14}$$

whence

$$|\Psi(t)| = \exp\left\{\int_{t_0}^{t} \text{trace } A(\tau)\, d\tau\right\} \tag{4.15}$$

so that $|\Psi(t)|$ is non-zero for $t - t_0 \in [\beta_1, \beta_2]$. Hence $\Psi^{-1}(t)$ exists, and by Cramer's formula and the continuity of $\Psi(t)$, $\Psi^{-1}(t)$ is continuous on $\beta_1 \leqslant t - t_0 \leqslant \beta_2$. This shows that the integral in (4.13) exists and can be differentiated with respect to t on $\beta_1 < t - t_0 < \beta_2$.

Now differentiating the expression given in (4.13) and using (4.12)

$$D\phi(t) = A(t)\Psi(t)c + A(t)\Psi(t)\int_{t_0}^{t} \Psi^{-1}(\tau)B(\tau)u(\tau)\, d\tau + \Psi(t)\Psi^{-1}(t)B(t)u(t) \tag{4.16}$$

$$= A(t)\phi(t) + B(t)u(t) \tag{4.17}$$

so that ϕ satisfies the differential equation on $\beta_1 < t - t_0 < \beta_2$. From (4.13) and the continuity of the ψ_i, $i = 1, 2, \ldots, n$, it follows that ϕ is continuous on $\beta_1 \leqslant t - t_0 \leqslant \beta_2$. Also, putting $t = t_0$ in (4.13) and using (4.12)

$$\phi(t_0) = c \tag{4.18}$$

so that ϕ defined by (4.13) is the solution of (4.1).

EXERCISE 4.1 Prove equation (4.14). [If $A = (a_{ij})$ and $\Psi = (\psi_{ij})$, the derivative of Ψ is the sum of n determinants of the form

$$\begin{vmatrix} D\psi_{11} & D\psi_{12} & \cdots & D\psi_{1n} \\ \psi_{21} & \psi_{22} & \cdots & \psi_{2n} \\ \vdots & \vdots & & \vdots \\ \psi_{n1} & \psi_{n2} & \cdots & \psi_{nn} \end{vmatrix}, \quad \begin{vmatrix} \psi_{11} & \psi_{12} & \cdots & \psi_{1n} \\ D\psi_{21} & D\psi_{22} & \cdots & D\psi_{2n} \\ \vdots & \vdots & & \vdots \\ \psi_{n1} & \psi_{n2} & \cdots & \psi_{nn} \end{vmatrix}, \text{ etc.}$$

Use (4.12) to evaluate $D\psi_{11}$, etc., and simplify by subtracting multiples of the other rows.]

EXERCISE 4.2 When A and B are constant, show that (4.13) reduces to equation (2.2).

5 Stability

An important question in the study of dynamical systems concerns the behaviour of solutions of a differential equation as $t \to \infty$. It is assumed that solutions exist for initial conditions c_1 such that $\|c_1 - c\| \leqslant r$, and that these solutions can be carried forward to any finite time t. At any given finite t, Theorem 3.4 shows that $c_1 \to c$ implies $\phi_1(t; c_1, t_0) \to \phi(t; c, t_0)$. However, if c_1 and c are fixed and $c_1 \neq c$, it may happen as $t \to \infty$ that $\|\phi_1(t; c_1, t_0) - \phi(t; c, t_0)\| \to \infty$ however small $\|c_1 - c\|$ may be. This type of behaviour is often undesirable and the property of *stability* defined in the following way is therefore often a requirement placed on the solution ϕ.

A solution $\phi(t; c, t_0)$ of equation (3.1) is stable if, given any $\epsilon > 0$, there exists a $\delta > 0$ such that $\|c_1 - c\| < \delta$ implies $\|\phi_1(t; c_1, t_0) - \phi(t; c, t_0)\| < \epsilon$ for all $t \geqslant t_0$.

In addition, it may be required that solutions starting near ϕ at $t = t_0$ tend to ϕ as $t \to \infty$. The solution ϕ is then said to be *asymptotically stable*, the definition being as follows.

A solution $\phi(t; c, t_0)$ of equation (3.1) is asymptotically stable if it is stable and if in addition there is an $r > 0$ such that $\|c_1 - c\| < r$ implies $\|\phi_1(t; c_1, t_0) - \phi(t; c, t_0)\| \to 0$ as $t \to \infty$.

There are many refinements of these ideas, but they will not be pursued here. An extensive account is given by Willems (1970).

EXERCISE 5.1 Show that the solution $\phi(t; 0, 0) = 0$ of

$$Dx(t) = -x(t); \qquad x(0) = 0 \tag{5.1}$$

is asymptotically stable.

EXERCISE 5.2 Show that the solution $\phi(t; 0, 0) = 0$ of

$$Dx(t) = Ax(t); \qquad x(0) = 0 \tag{5.2}$$

where A is an $n \times n$ matrix having distinct eigenvalues, is asymptotically stable if and only if every eigenvalue of A has negative real part. [Write $Hx = y$ where H is a non-singular constant matrix such that $H^{-1}AH$ is diagonal.]

EXERCISE 5.3 Show that the conclusion of Exercise 5.2 remains true when A has repeated eigenvalues. [Bring A to Jordan form.]

Further reading

More extended accounts of the theory of differential equations will be found in Coddington and Levinson (1955), and in Nemitskii and Stepanov (1960). The stability properties of differential equations are treated from the point of view of dynamical systems theory by Willems (1970) and from a more purely mathematical viewpoint by Bellman (1953), by LaSalle and Lefschetz (1961), and by Cesari (1963).

References

BELLMAN, R., 1953, *Stability theory of differential equations*, McGraw-Hill, New York.
CESARI, L., 1963, *Asymptotic behaviour and stability problems in ordinary differential equations*, Springer Verlag, Heidelberg.
CODDINGTON, E. A. and N. LEVINSON, 1955, *Theory of ordinary differential equations*, McGraw-Hill, New York.
LASALLE, J. P. and S. LEFSCHETZ, 1961, *Stability by Liapunov's direct method*, Academic Press, New York.
NEMITSKII, V. V. and V. V. STEPANOV, 1960, *Qualitative theory of differential equations*, Princeton University Press, Princeton, N.J.
WILLEMS, J. L., 1970, *Stability theory of dynamical systems*, Nelson, London.

Problems

1 Obtain equations (2.39) and (2.40) by evaluating equation (2.2).

2 By producing an example in which the equality holds show that the inequality (3.37) is the best obtainable.

3 Let the solution ϕ of (3.1) lie in the boundary of D at $t = \gamma_1$ and at $t = \gamma_2$, where γ_1 and γ_2 are defined by (3.13). Show that $D_+\phi(\gamma_1)$ and $D_-\phi(\gamma_2)$ exist and satisfy

$$D_+\phi(\gamma_1) = f[\phi(\gamma_1), \gamma_1]$$
$$D_-\phi(\gamma_2) = f[\phi(\gamma_2), \gamma_2]$$

[Use the result of Exercise 10.1, Chapter 5.]

4 Let the solution ϕ of (3.1) lie in the boundary of D at $t = t_0 + \beta_1$, where β_1 is as in (3.2). Show that the functions ϕ_k of Section 3 are unchanged if the values of $f(x, t_0)$ are changed to any other (finite) values so that f becomes discontinuous as a function of t at t_0. Hence show that a continuous solution ϕ can be defined on $[t_0, \gamma_2]$, but that $D_+\phi(t_0)$ no longer equals $f[\phi(t_0), t_0]$. [Use the result of Exercise 4.4, Chapter 5, and compare Problem 3.]

5 In Section 4, let $A(t)$, $B(t)$, and $u(t)$ belong to D_R on $\beta_1 \leqslant t - t_0 \leqslant \beta_2$. Divide this interval into a finite number of sub-intervals $(\gamma_i, \gamma_{i+1}]$ on which A, B, and u are continuous. Use the result of Problem 4 to show that (4.1) has a solution $\phi \in D_R^1$ on $\beta_1 \leqslant t - t_0 \leqslant \beta_2$.

6 In (3.1), let $f(x, t) = 2tx$, $t_0 = 0$, $c = 1$. Show that the functions ϕ_k of (3.14) generate the partial sums of the series for e^{t^2}.

7 Two substances take part in a chemical reaction, their concentrations at time t being $x_1(t)$ and $x_2(t)$. For physical reasons $0 \leqslant x_1 \leqslant 1$, $0 \leqslant x_2 \leqslant 1$, and $x_1 + x_2 \leqslant 1$. Let D be

$$D = \{x, t \mid 0 \leqslant x_1, \quad 0 \leqslant x_2, \quad x_1 + x_2 \leqslant 1; \quad 0 \leqslant t\}$$

At interior points of D the x_i satisfy

$$Dx_1 = -k_1 x_1$$
$$Dx_2 = k_1 x_1 - k_2 x_2$$

where $0 < k_1, k_2$, and $x_1(t)$, $x_2(t)$ are continuous on the boundary of D. Show, without explicitly solving the equations, that the solution starting from $x_1(0) = x_2(0) = 0$ is $x_1(t) = x_2(t) = 0$, and that solutions starting from any other point on the boundary of D enter the interior of D; hence no solution can leave D, in agreement with the physical requirement.

8 By finding the explicit solution of the differential equations in Problem 7, show that all solutions starting in D tend asymptotically to $x_1 = x_2 = 0$ as $t \to \infty$.

9 By considering $D(x_1^2 + x_2^2)$ show that all solutions of

$$Dx_1 = -x_1 + x_2$$
$$Dx_2 = -x_1 - x_2$$

tend to the origin as $t \to \infty$. Deduce that the eigenvalues of

$$\begin{pmatrix} -1 & 1 \\ -1 & -1 \end{pmatrix}$$

have negative real parts, and verify this.

10 By considering $D(x_1 + x_2) = -kx_2$, deduce the conclusion of Problem 8 without using the explicit solution. [By the result of Problem 7, x_2 cannot remain equal to zero if $x_1 \neq 0$. Therefore $x_1 + x_2$ is decreasing, and is bounded below by 0. Hence $x_1 + x_2$ tends to a limit: show that it cannot be other than 0.]

11 A linear system is defined by the equations

$$\dot{x} = Ax + Bu; \qquad x(0) = 0$$
$$y = Cx$$

where A, B, C are constant matrices, respectively $n \times n$, $n \times l$, $m \times n$. Show that the system satisfies the conditions of Theorem 5.1, Chapter 5, and that the step response is

$$H(t) = CA^{-1}[e^{At} - I]B$$

Hence show that

$$y(t) = \int_0^t CA^{-1}[e^{A(t-\tau)} - I]B\, du(\tau)$$

Integrate by parts in this formula and reconcile the result with equation (2.2).

7 Complex analysis

1 Complex numbers

The idea of a complex number was introduced in Chapter 1 and some of its simple properties were discussed there. This chapter is concerned with complex analysis and begins with a further discussion of complex numbers.

1.1 Geometric interpretation

As an extension of the idea of representing a real number as a point on a line it is possible to represent a complex number as a point in a plane. Thus the number $z = x + iy$ can be represented as the point with coordinates (x, y) with respect to a set of rectangular cartesian coordinates as shown in Fig. 1. A diagram such as Fig. 1 is called an *Argand diagram* after its originator. The x axis is called the

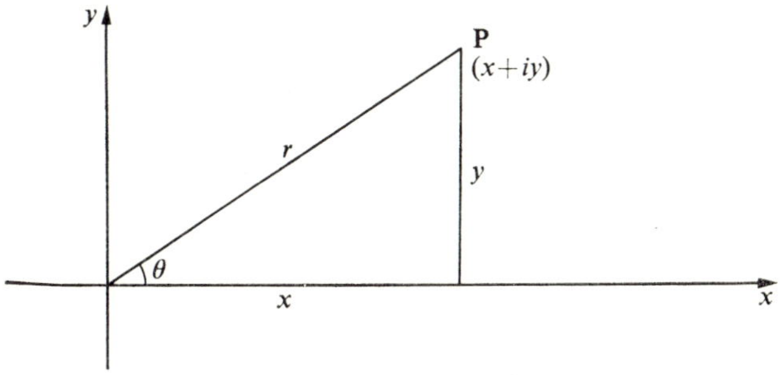

FIG. 1

real axis and the y axis the imaginary axis. The *modulus* (or *absolute value*) of z is $|x + iy| = \sqrt{(x^2 + y^2)} = r$ and is just the distance of the point P from the origin of coordinates O. The angle θ (measured in radians) that the vector OP makes with the positive direction of the x axis is called the *argument* (or *phase*) of z. The polar coordinates (r, θ) of the point P are therefore the modulus and argument of $x + iy$. From the diagram it is easily seen that

$$\begin{aligned} x &= r \cos \theta \\ y &= r \sin \theta \end{aligned} \qquad (1.1)$$

and it is obvious that the argument of any complex number is only determined up to an even multiple of π. An alternative representation of the complex number z is the *polar form*

$$z = r(\cos\theta + i\sin\theta) \tag{1.2}$$

with

$$r = |z| \tag{1.3}$$

and

$$\theta = \tan^{-1}\frac{y}{x} \tag{1.4}$$

In (1.4) it is usual (to avoid ambiguity) to take $-\pi < \theta \leqslant \pi$ in which case θ is called the *principal value* of the argument.

If $z_1 = x_1 + iy_1$ is represented by P_1 in Fig. 2 and $z_2 = x_2 + iy_2$ is represented by P_2, then $z_1 + z_2 = (x_1 + x_2) + i(y_1 + y_2)$ and the sum $z_1 + z_2$ is

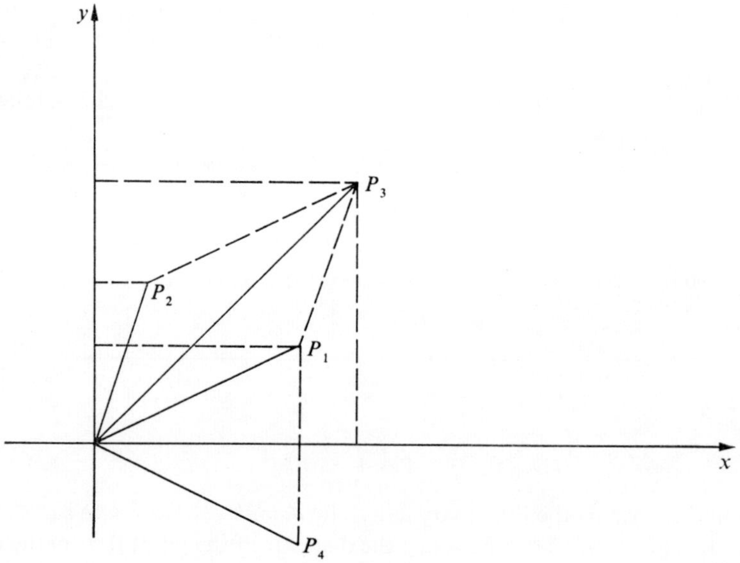

FIG. 2

therefore obtained in the Argand diagram by the usual parallelogram rule for addition of vectors and is represented by P_3 in Fig. 2. The complex conjugate $\bar{z}_1$ of z_1 is $x_1 - iy_1$ which is obtained by reflecting the point P_1 in the x axis to obtain P_4 in Fig. 2.

EXERCISE 1.1 Prove that the modulus of the product of two complex numbers is equal to the product of their moduli and that the argument of the product is equal to the sum of the arguments of the factors. Extend this result to products of more than two factors.

1.2 De Moivre's theorem

Theorem 1.1 If p is any integer, positive or negative, and $z = x + iy = r(\cos\theta + i\sin\theta)$ then

$$z^p = r^p(\cos p\theta + i\sin p\theta) \tag{1.5}$$

Proof When p is zero the result is trivial and for $p > 0$ it follows directly from Exercise 1.1. Now when $p = -1$ a direct evaluation gives

$$\begin{aligned} z^{-1} &= \frac{\bar{z}}{z\bar{z}} \\ &= \frac{1}{r}(\cos\theta - i\sin\theta) \\ &= \frac{1}{r}[\cos(-\theta) + i\sin(-\theta)] \end{aligned}$$

because $\cos\theta = \cos(-\theta)$ and $\sin\theta = -\sin(-\theta)$. If, finally, $p < -1$ set $p = -q$ where $q > 1$. Then

$$z^p = z^{-q} = (z^q)^{-1}$$

so that

$$\begin{aligned} z^p &= [r^q(\cos q\theta + i\sin q\theta)]^{-1} \\ &= r^{-q}[\cos(-q\theta) + i\sin(-q\theta)] \end{aligned}$$

and the result follows on replacing $-q$ by p.

Example 1.1 If $z = \cos\theta + i\sin\theta$, then by De Moivre's theorem

$$z^4 = \cos 4\theta + i\sin 4\theta \tag{1.6}$$

But also

$$\begin{aligned} z^4 &= (\cos\theta + i\sin\theta)^4 \\ &= \cos^4\theta + 4\cos^3\theta\, i\sin\theta + 6\cos^2\theta\, i^2\sin^2\theta + 4\cos\theta\, i^3\sin^3\theta + i^4\sin^4\theta \end{aligned}$$

which gives

$$z^4 = (\cos^4\theta - 6\cos^2\theta\sin^2\theta + \sin^4\theta) + i(4\cos^3\theta\sin\theta - 4\cos\theta\sin^3\theta) \tag{1.7}$$

Equating real parts in (1.6) and (1.7) and using $\sin^2\theta = 1 - \cos^2\theta$ gives $\cos 4\theta = 8\cos^4\theta - 8\cos^2\theta + 1$, which expresses $\cos 4\theta$ as a polynomial in $\cos\theta$.

EXERCISE 1.2 Give a geometric construction in the Argand diagram for obtaining the point $1/z$.

1.3 Roots of complex numbers

A number w satisfying

$$w^n = z$$

where n is a positive integer is said to be an *nth root* of z and is written $z^{1/n}$. Write

$$w = p(\cos\phi + i\sin\phi)$$
$$z = r(\cos\theta + i\sin\theta)$$

Then, by De Moivre's theorem,

$$p^n(\cos n\phi + i\sin n\phi) = r(\cos\theta + i\sin\theta)$$

so that

$$p = r^{1/n} \quad (\text{since } r > 0 \text{ and } p > 0) \tag{1.8}$$

and

$$\left.\begin{aligned}\cos(n\phi) &= \cos\theta\\ \sin(n\phi) &= \sin\theta\end{aligned}\right\} \tag{1.9}$$

It follows from equations (1.9) that

$$\phi = \frac{\theta + 2k\pi}{n} \tag{1.10}$$

where k is any positive or negative integer or zero. If, however, the values $k = 0, 1, \ldots, n-1$ are taken there can arise no other values of w, because any other value of k gives a ϕ differing from one of the previous values by some multiple of 2π. Also the values of w arising from any two of these values of k must be distinct because the difference in argument between them is non-zero and less than 2π. Thus the following theorem has been proved.

Theorem 1.2 Any non-zero complex number $z = r(\cos\theta + i\sin\theta)$ has n distinct nth roots (n a positive integer) given by

$$r^{1/n}\left\{\cos\left(\frac{\theta + 2\pi k}{n}\right) + i\sin\left(\frac{\theta + 2\pi k}{n}\right)\right\} \text{ for } k = 0, 1, \ldots, n-1$$

If θ is the principal argument of z then

$$r^{1/n}\left\{\cos\frac{\theta}{n} + i\sin\frac{\theta}{n}\right\}$$

is taken as the *principal value* of $z^{1/n}$.

Example 1.2

$$\begin{aligned}(-1)^{1/5} &= (-1 + 0i)^{1/5}\\ &= (\cos\pi + i\sin\pi)^{1/5}\end{aligned}$$

so that the 5 values of $(-1)^{1/5}$ are

$$\cos\frac{\pi}{5} + i\sin\frac{\pi}{5} = \frac{\sqrt{5}+1}{4} + i\sqrt{\left(\frac{5-\sqrt{5}}{8}\right)}$$

$$\cos\frac{3\pi}{5} + i\sin\frac{3\pi}{5} = -\frac{\sqrt{5}-1}{4} + i\sqrt{\left(\frac{5+\sqrt{5}}{8}\right)}$$

$$\cos\pi + i\sin\pi = -1 + i\,0$$

$$\cos\frac{7\pi}{5} + i\sin\frac{7\pi}{5} = -\frac{\sqrt{5}-1}{4} - i\sqrt{\left(\frac{5+\sqrt{5}}{8}\right)}$$

$$\cos\frac{9\pi}{5} + i\sin\frac{9\pi}{5} = \frac{\sqrt{5}+1}{4} - i\sqrt{\left(\frac{5-\sqrt{5}}{8}\right)}$$

EXERCISE 1.3 Prove that the nth roots of $r(\cos\theta + i\sin\theta)$ can be represented in the Argand diagram by the vertices of a regular n-sided polygon inscribed in a circle of radius $r^{1/n}$ with centre at the origin.

EXERCISE 1.4 Prove that the nth roots of unity can be written $1, w, w^2, \ldots, w^{n-1}$ where $w = \cos(2\pi/n) + i\sin(2\pi/n)$.

A direct extension of De Moivre's theorem is now possible.

Theorem 1.3 Let the p/q power of a non-zero complex number z, where p and q are integers and q is positive, be defined as the pth power of any one of the qth roots of z. Then there are exactly q such p/q powers given by

$$z^{p/q} = r^{p/q}\left\{\cos p\left(\frac{\theta + 2\pi k}{q}\right) + i\sin p\left(\frac{\theta + 2\pi k}{q}\right)\right\} \tag{1.10}$$

for $k = 0, 1, \ldots, q - 1$.

Proof This is trivial on applying De Moivre's theorem to the result for $z^{1/q}$ given in Theorem 1.2.

Example 1.3

$$(-1) = (-1 + 0i)$$
$$= (\cos \pi + i \sin \pi)$$

Thus the 4 values of $(-1)^{3/4}$ are

$$\cos \frac{3}{4}(\pi + 2k\pi) + i \sin \frac{3}{4}(\pi + 2k\pi), \quad k = 0, 1, 2, 3$$

This gives

$$\cos \frac{3}{4}\pi + i \sin \frac{3}{4}\pi = -\frac{1}{\sqrt{2}} + \frac{1}{\sqrt{2}} i$$

$$\cos \frac{1}{4}\pi + i \sin \frac{1}{4}\pi = \frac{1}{\sqrt{2}} + \frac{1}{\sqrt{2}} i$$

$$\cos \frac{7}{4}\pi + i \sin \frac{7}{4}\pi = \frac{1}{\sqrt{2}} - \frac{1}{\sqrt{2}} i$$

$$\cos \frac{5}{4}\pi + i \sin \frac{5}{4}\pi = -\frac{1}{\sqrt{2}} - \frac{1}{\sqrt{2}} i$$

Example 1.4 Express $x^{2n} + 1$ as a product of n real quadratic factors. Write

$$x = (-1)^{1/2n}$$
$$= \cos\left(\frac{\pi + 2k\pi}{2n}\right) + i \sin\left(\frac{\pi + 2k\pi}{2n}\right), \quad k = 0, 1, \ldots, 2n - 1$$

When

$$k = 2n - r - 1$$

it follows that

$$x = \cos\left(\frac{4n - 2r - 1}{2n}\right)\pi + i \sin\left(\frac{4n - 2r - 1}{2n}\right)\pi$$
$$= \cos\left(\frac{\pi + 2r\pi}{2n}\right) - i \sin\left(\frac{\pi + 2r\pi}{2n}\right)$$

so that

$$x = \cos \frac{2k + 1}{2n}\pi \pm i \sin \frac{2k + 1}{2n}\pi, \quad k = 0, 1, \ldots, n - 1$$

and the factors of $x^{2n} + 1$ are therefore

$$x - \cos \frac{2k + 1}{2n}\pi \pm i \sin \frac{2k + 1}{2n}\pi, \quad k = 0, 1, \ldots, n - 1$$

Now take the products of the complex conjugate factors to obtain

$$x^{2n} + 1 = \prod_{k=0}^{n-1} \left(x^2 - 2x \cos \frac{2k+1}{2n} \pi + 1 \right)$$

For example

$$x^4 + 1 = (x^2 - \sqrt{2}\, x + 1)(x^2 + \sqrt{2}\, x + 1)$$

2 Complex functions

There is nothing new in the concept of a complex function of a complex variable: it is simply a mapping from C_1 to C_1, where the elements of C_1 are complex numbers and the scalars of C_1 are also complex numbers. The ideas of continuity and limit and the theorems concerned with these ideas in Chapter 4 carry forward without modification to complex functions. However, it is equally possible to regard such functions as mappings from E_2 to E_2, where the elements of E_2 are the real and imaginary parts of the complex numbers, and the scalars of E_2 are real numbers. Much of the importance of complex function theory arises from the relations between these two ways of regarding the functions.

EXERCISE 2.1 Prove that a complex function $f(z) = u + iv$ is continuous if and only if $u(x, y)$ and $v(x, y)$ are continuous.

Example 2.1 Let $f(z) = |z|^2$ (notice that this is, in fact, a real-valued function of a complex variable, that is, a mapping of C_1 into E_1). For any point z_0 in the complex plane

$$|f(z) - f(z_0)| = \big| |z|^2 - |z_0|^2 \big| < \epsilon$$

provided that $|z - z_0| < \delta$, where $\delta = (|z_0|^2 + \epsilon)^{1/2} - |z_0|$.

Example 2.2 Let $f(z) = z^2$ (this is now a complex function of a complex variable). In this case it is evident that for any z_0

$$\lim_{z \to z_0} f(z) = \lim_{z \to z_0} (z^2) = {z_0}^2$$

so that z^2 is continuous everywhere in the complex plane.

2.1 Analytic functions

Since the complex numbers form a field in which division is defined (Section 4, Chapter 1), it is possible to introduce the derivative of a function of a complex variable in the following way (compare Chapter 4, Section 8). Let $f(z)$ be defined on $N(z_0)$ and let the following limit exist:

$$f'(z_0) = \lim_{z \to z_0} \frac{f(z) - f(z_0)}{z - z_0} \tag{2.1}$$

Then f is said to be *differentiable* at z_0 and to have *derivative* $f'(z_0)$. A function $f(z)$ with domain an open set $S \subseteq C_1$ is said to be *analytic* on S (sometimes *regular* or *holomorphic*) if it possesses a finite derivative at each point of S. A function $f(z)$ is *analytic at a point* $z_0 \in S$ if it is analytic on some neighbourhood $N(z_0) \subset S$. A function which is analytic on every finite region of the complex plane is called an *entire function* (or *integral function*).

In equation (2.1) it is important to notice that f is regarded as a mapping from C_1 to C_1. Provided that this interpretation is preserved, most† of the results for derivatives of mappings from E_1 to E_1 carry over, and they generalize in a similar way to mappings from C_n to C_m. For example the derivative Df of a function f from C_n to C_m is an $m \times n$ matrix, and the existence of $Df(z_0)$ implies the continuity of f at z_0. However, if f: $C_1 \to C_1$ is regarded as a mapping from E_2 to E_2, its derivative is a 2×2 matrix: that is, if

$$z = x + iy \tag{2.2}$$

$$f(z) = u(x, y) + iv(x, y) \tag{2.3}$$

and

$$w = \begin{pmatrix} x \\ y \end{pmatrix}, \qquad g(w) = \begin{pmatrix} u(w) \\ v(w) \end{pmatrix} \tag{2.4}$$

then when Dg exists it is given by

$$Dg = \begin{pmatrix} \dfrac{\partial u}{\partial x} & \dfrac{\partial u}{\partial y} \\ \dfrac{\partial v}{\partial x} & \dfrac{\partial v}{\partial y} \end{pmatrix} \tag{2.5}$$

In this Chapter derivatives in the sense of equation (2.1) will be denoted by the dash, f', and not by Df. This distinction is not a standard one.

The necessary conditions for a function $f(z) = u + iv$ to be analytic will now be shown to impose quite restrictive conditions on the real and imaginary parts u and v.

Theorem 2.1 Let the function $f(z)$ have domain an open set $S \subseteq C_1$ and assume that $f'(z_0)$ exists and is finite at some point $z_0 \in S$. Then, if $f(z) = u(x, y) + iv(x, y)$, u and v are continuous at $z_0 = x_0 + iy_0$ and the four partial derivatives $\partial u/\partial x$, $\partial u/\partial y$, $\partial v/\partial x$, $\partial v/\partial y$, exist at z_0 and satisfy the equations (known as the *Cauchy–Riemann equations*)

$$\left.\begin{aligned} \frac{\partial u}{\partial x} &= \frac{\partial v}{\partial y} \\ \frac{\partial v}{\partial x} &= -\frac{\partial u}{\partial y} \end{aligned}\right\} \tag{2.6}$$

†But not those which depend on the ordering of the real numbers: for example, those relating to D_-f or D_+f.

Proof The expression for the derivative in (2.1) can be written

$$f'(z_0) = \lim_{h \to 0} \frac{f(z_0 + h) - f(z_0)}{h} \tag{2.7}$$

The existence of this finite limit implies that $f(z_0 + h) \to f(z_0)$ as $h \to 0$; whence $u(x, y)$ and $v(x, y)$ are continuous at $z_0 = x_0 + iy_0$. If real values are chosen for h in (2.7) the derivative becomes the partial derivative with respect to x and

$$f'(z) = \frac{\partial f}{\partial x} = \frac{\partial u}{\partial x} + i\frac{\partial v}{\partial x} \tag{2.8}$$

On the other hand, if h is taken purely imaginary the following result is obtained

$$f'(z) = -i\frac{\partial f}{\partial y} = -i\frac{\partial u}{\partial y} + \frac{\partial v}{\partial y} \tag{2.9}$$

The limit in (2.7) must be the same, whatever the manner in which $h \to 0$, and the result follows directly from equations (2.8) and (2.9).

It is interesting to notice that, by equations (2.6)

$$|f'(z)|^2 = \frac{\partial u}{\partial x}\cdot\frac{\partial v}{\partial y} - \frac{\partial u}{\partial y}\cdot\frac{\partial v}{\partial x} \tag{2.10}$$

The expression on the right-hand side of equation (2.10) is the determinant of the matrix in equation (2.5).

Exercise 2.2 Prove that if any of the real part, imaginary part, modulus, or argument of an analytic function is a constant, then the function is itself constant.

Exercise 2.3 Show by considering the function $f(z) = \sqrt{|xy|}$ that although the Cauchy–Riemann equations are necessary for differentiability they are not sufficient.

Equally the existence of the derivative Dg, where g is defined in (2.4), does not ensure that f' exists, because the partial derivatives may not satisfy the Cauchy–Riemann equations. However, if the derivative Dg exists and the partial derivatives satisfy the Cauchy–Riemann equations, f' exists.

Theorem 2.2 Let $f(z) = u(x, y) + iv(x, y)$, where $z = x + iy$, and let $w = \begin{pmatrix} x \\ y \end{pmatrix}$, $g(w) = \begin{pmatrix} u(w) \\ v(w) \end{pmatrix}$. If g has a derivative Dg at $w_0 = \begin{pmatrix} x_0 \\ y_0 \end{pmatrix}$, and if the partial derivatives $\partial u/\partial x$, $\partial u/\partial y$, $\partial v/\partial x$, $\partial v/\partial y$ obey the Cauchy–Riemann equations at w_0, then f has a derivative $f'(z_0)$ at $z_0 = x_0 + iy_0$.

Proof Because $Dg(w_0)$ exists,

$$\lim_{w \to w_0} \|g(w) - g(w_0) - Dg(w_0).(w - w_0)\| \div \|w - w_0\| = 0 \tag{2.11}$$

which because Dg is given by (2.5) gives

$$\lim_{w \to w_0} \sqrt{\left(\left\{\left[u(w) - u(w_0) - \frac{\partial u}{\partial x}(x - x_0) - \frac{\partial u}{\partial y}(y - y_0)\right]^2 + \left[v(w) - v(w_0) - \frac{\partial v}{\partial x}(x - x_0) - \frac{\partial v}{\partial y}(y - y_0)\right]^2\right\} \div \{(x - x_0)^2 + (y - y_0)^2\}\right)} = 0 \tag{2.12}$$

But on using the Cauchy–Riemann equations and (2.8)

$$|f(z) - f(z_0) - f'(z_0)(z - z_0)| \div |z - z_0|$$
$$= \sqrt{\left(\left\{\left[u(x, y) - u(x_0, y_0) - \frac{\partial u}{\partial x}(x - x_0) - \frac{\partial u}{\partial y}(y - y_0)\right]^2 + \left[v(x, y) - v(x_0, y_0) - \frac{\partial v}{\partial x}(x - x_0) - \frac{\partial v}{\partial y}(y - y_0)\right]^2\right\} \div \{(x - x_0)^2 + (y - y_0)^2\}\right)} \tag{2.13}$$

from which it follows that $f'(z_0)$ exists.

EXERCISE 2.4 Prove the converse of Theorem 2.2.

A sufficient condition for a function to be analytic in an open set is contained in the following theorem.

Theorem 2.3 Let $f(z) = u(x, y) + i\,v(x, y)$ have for its domain an open set $S \subseteq C_1$ and assume that $\partial u/\partial x, \partial u/\partial y, \partial v/\partial x, \partial v/\partial y$ exist and are continuous on S. Then $f(z)$ is analytic at any point $z_0 \in S$ where the Cauchy–Riemann equations are satisfied.

Proof The proof follows directly from Theorem 2.2 above and Theorem 8.7 of Chapter 4, and its details are left as an exercise for the reader.

Example 2.3 (Exponential function e^z) The exponential function e^z has already been defined as a power series and is continuous in the whole complex plane (Chapter 4, Example 12.2). Indeed, by Theorem 12.2 of Chapter 4, it has derivatives of all orders everywhere on C_1 and it is equal to its own derivative (Chapter 4, Example 12.3).

Another way of proceeding with the exponential function is as follows. Consider exp (iy) where y is real,

$$\exp(iy) = 1 + iy - \frac{y^2}{2!} - i\frac{y^3}{3!} + \frac{y^4}{4!} + \cdots \tag{2.14}$$

Then on collecting real and imaginary parts

$$\exp(iy) = \cos y + i \sin y \tag{2.15}$$

Hence by Exercise 12.1 of Chapter 4, it follows that

$$\exp(z) = \exp(x + iy) = e^x(\cos y + i \sin y) \tag{2.16}$$

Now consider the conditions required in Theorem 2.3. The four partial derivatives are given by

$$\frac{\partial u}{\partial x} = e^x \cos y, \qquad \frac{\partial u}{\partial y} = -e^x \sin y \tag{2.17}$$

$$\frac{\partial v}{\partial x} = e^x \sin y, \qquad \frac{\partial v}{\partial y} = e^x \cos y \tag{2.18}$$

It is easily verified that these are continuous and satisfy the Cauchy–Riemann equations. Consequently exp (z) is analytic everywhere in the complex plane: that is, exp (z) is an entire function.

EXERCISE 2.5 Let $w = f(z)$ be a one-to-one function with domain S and range R both in C_1 and suppose that $f^{-1}(w) = g(w)$ is continuous on R. Prove that if $f(z)$ is differentiable at $z_0 \in S$ then $g(w)$ is differentiable at $w_0 = f(z_0) \in R$ and $g'(w_0) = 1/[f'(z_0)]$, provided $f'(z_0) \neq 0$. [Use the continuity of $f^{-1}(w)$ and the fact that

$$\frac{f^{-1}(w_2) - f^{-1}(w_1)}{w_2 - w_1} = \frac{z_2 - z_1}{w_2 - w_1} = \left(\frac{w_2 - w_1}{z_2 - z_1}\right)^{-1}\Bigg]$$

2.2 Higher derivatives of analytic functions

The derivative of a complex function is itself a function that may or may not be differentiable. Hence the possibility arises of forming second and third and higher order derivatives of an analytic function. In Chapter 4, Section 12 it was shown that a power series has derivatives of all orders inside its circle of convergence. It follows that a complex function which can be represented by a power series is an analytic function with derivatives of all orders.

It will be shown later that any function analytic on an open set S possesses derivatives of all orders on S and can be expressed as a power series. First, however, it is necessary to consider complex integration.

3 Complex integration

In Chapter 5 the Riemann–Stieltjes integral $\int_a^b f\,d\alpha$ was defined for real-valued functions f and α on an interval $[a, b]$. There is no difficulty in allowing f and α to be complex-valued functions from E_1 to C_1 defined on an interval $[a, b]$: the Riemann–Stieltjes sums remain well defined and the definition of the integral needs no change. It is easy to verify that if $f = u + iv$, and $\alpha = \beta + i\gamma$, the complex integral can be expressed in the following way in terms of real integrals

$$\int_a^b f\,d\alpha = \int_a^b u\,d\beta - \int_a^b v\,d\gamma + i\int_a^b u\,d\gamma + i\int_a^b v\,d\beta \tag{3.1}$$

which is valid when the four integrals on the right exist. Equation (3.1) allows the properties of the complex integral to be obtained directly from those of the real integrals.

EXERCISE 3.1 Evaluate $\int_0^{\pi/2} (1 + it)\,d(e^{it})$.

3.1 Contour integrals

If f is defined as a function of the complex variable z, the definition of an integral given above cannot be used. However, if z itself is made a function of the real parameter t, then f becomes a function of t, and the definition can be applied. Let $z = x + iy$, where x and y are continuous real-valued functions of the real variable t on the interval $[t_a, t_b]$ and possess continuous derivatives on (t_a, t_b). Then as t varies from t_a to t_b the complex variable

$$z(t) = x(t) + iy(t) \tag{3.2}$$

will describe the arc $C(z)$ of a curve from the point $a = z(t_a)$ to $b = z(t_b)$.

Let $f(z) = u(z) + iv(z)$ where z is given by (3.2). Then the *contour integral* of f on an arc C is defined to be the complex Riemann–Stieltjes integral

$$I = \int_{C(z)} f(z)\,dz = \int_{t_a}^{t_b} f[z(t)]\,dz(t) \tag{3.3}$$

$$= \int_{t_a}^{t_b} u[z(t)]\,dx(t) - \int_{t_a}^{t_b} v[z(t)]\,dy(t) + i\int_{t_a}^{t_b} u[z(t)]\,dy(t) + i\int_{t_a}^{t_b} v[z(t)]\,dx(t) \tag{3.4}$$

It now follows directly from Chapter 5, Theorem 11.1, that if w is another complex function with properties of continuity and differentiability similar to those of z, and describing the same arc C as the parameter θ goes from θ_a to θ_b, then

$$\int_{C(z)} f(z)\,dz = \pm\int_{C(w)} f(w)\,dw \tag{3.5}$$

In (3.5) the positive sign is taken if the two parametrizations of C describe C in the same direction, and the negative sign is taken if C is described in opposite directions.

Notice also that since x and y have continuous derivatives on (t_a, t_b) it follows from Chapter 5, Section 7, that each of the integrals in (3.4) can be expressed as a Riemann integral, whence

$$I = \int_{t_a}^{t_b} f[z(t)]z'(t)\, dt \tag{3.6}$$

For our purposes it is only necessary to consider arcs which are straight lines or arcs of circles. These will be called *elementary arcs*, although this is not a standard terminology. An *elementary contour* will be defined here as a continuous curve, made up of a finite set of elementary arcs, which does not cross itself but may be closed. A closed elementary contour clearly divides the plane into two disjoint open sets having the contour as common boundary: namely the open bounded interior set and the open unbounded exterior set. [For a more general discussion of contours see Markushevich (1965) or Apostol (1957).]

The following simple properties of contour integrals are easily verified by the relevant definitions and theorems of Chapter 5 and their proofs are left as exercises.

EXERCISE 3.2 The integral of a complex function on a contour C is the sum of the contour integrals (provided they exist) of the function on the constituent arcs of C. (Notice that this result applies equally to a contour C comprising two contours C_1 and C_2, provided the end point of one constituent contour coincides with the starting point of the other.)

EXERCISE 3.3 Show that if f is a continuous function of z on an elementary contour C, then $\int_C f(z)\, dz$ exists. In particular this contour integral exists if $f'(z)$ exists at each z on C. [Use Exercise 2.4 and Section 8.5 of Chapter 4.]

EXERCISE 3.4 If the contour C when described in the opposite direction is denoted by $-C$, then show that

$$\int_C f(z)\, dz = -\int_{-C} f(z)\, dz \tag{3.7}$$

EXERCISE 3.5 Define the *length* of the elementary arc C described by the complex variable $z = x(t) + iy(t)$ as t varies throughout the interval $[t_a, t_b]$, by $L = \int_{t_a}^{t_b} [\{Dx(t)\}^2 + \{Dy(t)\}^2]^{1/2}\, dt$, and let M be the supremum of $|f(z)|$ on C. Then

$$\left| \int_C f(z)\, dz \right| \leqslant ML \tag{3.8}$$

Example 3.1 Consider the integral $\int_C dz/z$, where C is a circle with radius r and centre the origin. In polar coordinates

$$z = r(\cos\theta + i\sin\theta) \tag{3.9}$$

$$\frac{\partial z}{\partial\theta} = r(-\sin\theta + i\cos\theta) \tag{3.10}$$

$$= rie^{i\theta} \tag{3.11}$$

On the contour C, z is a function of θ only, and the formula for change of variable given in Section 11 of Chapter 5 is easily applied to give

$$\int_C \frac{dz}{z} = \int_0^{2\pi} \frac{rie^{i\theta}\,d\theta}{re^{i\theta}} = 2\pi i \tag{3.12}$$

EXERCISE 3.6 Prove that if C is any closed elementary contour both $\int_C dz$ and $\int_C z\,dz$ vanish.

3.2 Cauchy's theorem

A very important theorem concerning the value of a contour integral on a closed contour will now be stated and proved. It is interesting that it is this theorem concerning an integral which will lead to the proof that an analytic function has derivatives of all orders.

Theorem 3.1 (Cauchy's theorem) If a function $f(z)$ is analytic on an open set $S \subseteq C_1$ and the interior and boundary of a closed elementary contour C lie within S, then $\int_C f(z)\,dz = 0$.

Proof Notice, first of all, that the integral exists by Exercise 3.3. Before proceeding to the proof of the theorem the following lemma is required.

Lemma It is possible to divide the interior of C into a finite number of complete or incomplete squares (by lines parallel to the x and y axes) such that within each square there is a point z_0 for which

$$f(z) - f(z_0) = f'(z_0)(z - z_0) + \epsilon_1 |z - z_0| \tag{3.13}$$

for all z in the square, where $|\epsilon_1| < \epsilon$ and ϵ is a given positive number.

Proof Suppose that the lemma is false. Enclose C in a large square Q, quadrisect this square and select a quarter, Q_1 say, in which the lemma fails. If this process is repeated, selecting at each stage a quarter square Q_n in which the lemma fails, it will give a point of accumulation a inside or on C as $n \to \infty$. Since $a \in S, f(z)$ is differentiable there, and hence [from the definition of $f'(a)$]

$$f(z) - f(a) = f'(a)(z - a) + \epsilon_1 |z - a|$$

and $|\epsilon_1| < \epsilon$ provided $|z - a|$ is taken less than some suitable $\delta > 0$.

By the method of construction of the point a, however, all of the squares Q_n will, for n sufficiently large, lie within a circle Γ centred on a and such that $|z - a| < \delta$ for all z in Γ. Therefore (3.13) is satisfied for the point $z_0 = a$ contradicting the assumption that the lemma is false.

Returning now to the proof of Theorem 3.1, the interior of C is assumed to be divided up into squares in accordance with the lemma. Let γ denote the boundary of one of the squares (or part squares) described in the same sense as C itself. Then

$$\int_C f(z)\, dz = \sum \int_\gamma f(z)\, dz \tag{3.14}$$

by Exercises 3.2 and 3.4, where the sum is over all the squares or part squares. Substitute (3.13) into the right-hand side of (3.14) and use Exercise 3.6 to obtain

$$\int_C f(z)\, dz = \sum \int_\gamma \epsilon_1 |z - z_0|\, dz \tag{3.15}$$

Hence

$$\left| \int_C f(z)\, dz \right| \leqslant \sum \left| \int_\gamma \epsilon_1 |z - z_0|\, |dz| \right| \tag{3.16}$$

Now consider the incomplete squares in the mesh into which C has been divided. Denote the residual parts of the sides of such a square by γ_1 and the part of the contour C, intercepted by the square, by γ_2. Then if L is the length of the contour C it follows from Exercise 3.5 that

$$\sum \int_{\gamma_2} \epsilon_1 |z - z_0|\, |dz| < \epsilon d L \tag{3.17}$$

where d is the diagonal of the square Q, the sum is taken over all the parts γ_2 of C, and ϵ_1 has its appropriate value in each square.

For a complete square, or for the part γ_1 of an incomplete square, Exercise 3.5 gives

$$\left| \int_\gamma |z - z_0|\, dz \right| < 4\sqrt{2}.l^2 \tag{3.18}$$

where l is the length of the side and z_0 is in the square. Hence

$$\sum \left| \int_\gamma \epsilon_1 |z - z_0|\, |dz| \right| < 4\sqrt{2}.\epsilon\, (\text{Area } Q) \tag{3.19}$$

where the sum is taken over all complete squares and the parts γ_1 of the incomplete squares.

From equations (3.16), (3.17), and (3.18) it follows that $\left|\int_C f(z)\,dz\right| < K\epsilon$ where K is a constant. This proves the theorem.

EXERCISE 3.7 Let $f(z)$ be analytic everywhere inside and on the boundary of the annular region between two closed elementary contours C_1 and C_2, where C_1 lies entirely within C_2. Prove that $\int_{C_1} f(z)\,dz = \int_{C_2} f(z)\,dz$ when the contours are described in the same direction. [Use Theorem 3.1 noting that by the definition of analyticity at a point, in Section 2.1, $f(z)$ is analytic on an open region containing the annulus.]

A direct consequence of Cauchy's theorem is that the integral $F(z) = \int_{z_0}^{z} f(w)\,dw$, where z and z_0 lie in the region S in which $f(z)$ is analytic, does not depend on the path joining z and z_0 provided it lies in S.

It will now be shown that $F(z)$ is itself analytic on S and that $f(z)$ is its derivative (compare Section 10 of Chapter 5).

Theorem 3.2 Let $f(z)$ be analytic on an open set S and let $F(z) = \int_{z_0}^{z} f(w)\,dw$ with the path of integration any contour $\Gamma \subset S$. Then $F(z)$ is also analytic on S and $F'(z) = f(z)$.

Proof Choose h so that the circle with centre z and radius h lies completely within S. Since the integral of an analytic function is independent of the path of integration, we may write

$$F(z+h) - F(z) = \int_{z}^{z+h} f(w)\,dw$$

and may integrate along the straight line from z to $z + h$. Then because

$$\int_{z}^{z+h} f(z)\,dw = hf(z)$$

it follows that

$$\frac{F(z+h) - F(z)}{h} - f(z) = \frac{1}{h}\int_{z}^{z+h} \{f(w) - f(z)\}\,dw \tag{3.20}$$

Because $f(z)$ is analytic on S it is continuous on S, so that given $\epsilon > 0$, there exists a δ such that

$$|f(w) - f(z)| < \epsilon \tag{3.21}$$

for $|w - z| < \delta$. If h is taken so that $0 < |h| < \delta$ it is seen from (3.20) and (3.21) that $|[F(z+h) - F(z)]/h - f(z)| < \epsilon$ which proves the theorem.

$F(z)$ is called the *indefinite integral* or *primitive* (sometimes *anti-derivative*) of $f(z)$ on S.

EXERCISE 3.8 Let $f(z)$ be analytic inside and on a closed elementary contour C. Prove that if $F(z)$ is any primitive of $f(z)$ then

$$\int_{z_0}^{z} f(w)\,dw = F(z) - F(z_0)$$

where z and z_0 are on or inside C. [Note that by the definition in Section 2.1 of analyticity at a point, $f(z)$ is analytic on an open region containing C.]

EXERCISE 3.9 Prove that

$$\int_C \sum_{n=0}^{\infty} f_n(z)\,dz = \sum_{n=0}^{\infty} \int_C f_n(z)\,dz$$

if $\sum f_n(z)$ is a uniformly convergent series of analytic functions and the contour C lies within the region of uniform convergence. [Use the result of Exercise 3.5, and Theorem 6.1 of Chapter 4.]

As stated previously, Cauchy's theorem leads to results concerning the higher derivatives of an analytic function. The starting point is the formula known as *Cauchy's integral formula*.

Theorem 3.3 If $f(z)$ is analytic inside and on a closed elementary contour C, and if z is any point inside C, then

$$f(z) = \frac{1}{2\pi i}\int_C \frac{f(w)}{w - z}\,dw \tag{3.22}$$

Proof The point z lies within C: it is possible, therefore, to take r sufficiently small to ensure that the circle $C_1 = \{w|\ |z - w| = r\}$ lies entirely within C. Then by the result of Exercise 3.7

$$\int_C \frac{f(w)\,dw}{w - z} = \int_{C_1} \frac{f(w)}{w - z}\,dw \tag{3.23}$$

and from Example 3.1 (with a simple transformation)

$$\int_{C_1} \frac{dw}{w - z} = 2\pi i \tag{3.24}$$

Because $f(w)$ is continuous it is possible to take r so small that for any given $\epsilon > 0$

$$|f(w) - f(z)| < \epsilon$$

whenever w lies on C_1. Then

$$\int_{C_1} \frac{f(w)}{w - z}\, dw - f(z) \int_{C_1} \frac{dw}{w - z} = \int_{C_1} \frac{f(w) - f(z)}{w - z}\, dw \tag{3.25}$$

and by Exercise 3.5 the integral on the right-hand side is less than $2\pi\epsilon$ in modulus. On using (3.23), (3.24), and (3.25),

$$\left| \int_C \frac{f(w)}{w - z}\, dz - 2\pi i f(z) \right| < 2\pi\epsilon$$

proving the theorem.

The value of $f(z)$ at a point within a closed contour has thus been expressed in terms of its values on the contour. The corresponding result for $f'(z)$ is given by the next theorem.

Theorem 3.4 If $f(z)$ is analytic inside and on a closed elementary contour C its derivative at a point z inside C is given by

$$f'(z) = \frac{1}{2\pi i} \int_C \frac{f(w)\, dw}{(w - z)^2} \tag{3.26}$$

Proof The theorem follows from the possibility of differentiating under the integral sign in equation (3.22) (Chapter 5, Section 10) but it is interesting to give a direct proof.

From equation (3.22) it follows that

$$\begin{aligned} \frac{f(z + h) - f(z)}{h} &= \frac{1}{2\pi i} \int_C \frac{1}{h} \left\{ \frac{1}{w - z - h} - \frac{1}{w - z} \right\} f(w)\, dw \\ &= \frac{1}{2\pi i} \int_C \left\{ \frac{f(w)}{(w - z)^2} + \frac{hf(w)}{(w - z)^2(w - z - h)} \right\} dw \end{aligned} \tag{3.27}$$

Consider the last integral on the right-hand side of equation (3.27). Because $f(w)$ is analytic in and on C it is continuous, and so $|f(w)| \leqslant M$ on C. Let l be the length of C and let m be the infimum of $|w - z|$ for all w on C; then if h is taken so that $|h| < m/2$

$$I = \left| \int_C \frac{f(w)\, dw}{(w - z)^2(w - z - h)} \right| < \frac{1}{2\pi} \frac{Ml}{m^2 . \frac{1}{2}m}$$

It follows that hI tends to zero as h tends to zero, thus proving the theorem.

This process can obviously be repeated, whence by induction the following general theorem is obtained.

Theorem 3.5 If $f(z)$ is analytic inside and on the closed elementary contour C it has derivatives of all orders at all points inside C and

$$f^{(n)}(z) = \frac{n!}{2\pi i} \int_C \frac{f(w)\, dw}{(w - z)^{n+1}} \tag{3.28}$$

EXERCISE 3.10 Prove that a function which is analytic at a point z_0 has derivatives of all orders at z_0. [Note the definition of analyticity at a point in Section 2.1.]

EXERCISE 3.11 Prove that if $f(z) = u(x, y) + iv(x, y)$ is analytic at $z_0 = x_0 + iy_0$ then u and v have partial derivatives of all orders in a neighbourhood of (x_0, y_0).

EXERCISE 3.12 If $f(z) = u(x, y) + iv(x, y)$ is analytic inside and on a closed elementary contour C, show that u and v satisfy Laplace's equation inside C, that is,

$$D_{11}u(x, y) + D_{22}u(x, y) = 0$$
$$D_{11}v(x, y) + D_{22}v(x, y) = 0$$

It is now possible to prove the converse of the result on the differentiability of power series. (Chapter 4, Theorem 12.2.)

Theorem 3.6 (Taylor's series) Let $f(z)$ be analytic inside and on a closed elementary contour C. Then if the point a lies inside C

$$f(z) = \sum_{n=0}^{\infty} \frac{f^{(n)}(a)}{n!} (z - a)^n \tag{3.29}$$

and the radius of convergence of the series is not less than the least distance from a to C.

Proof Denote the least distance from a to C by d. Then if z lies inside a circle C_1, centre a and radius $r < d$,

$$f(z) = \frac{1}{2\pi i} \int_{C_1} \frac{f(w)}{w - z}\, dw$$

It is possible to write $1/(w - z)$ in the following way as a geometric series, which converges uniformly for w inside C_1 (Chapter 4, Theorem 12.1),

$$\frac{1}{w - z} = \frac{1}{w - a} + \frac{z - a}{(w - a)^2} + \cdots + \frac{(z - a)^n}{(w - a)^{n+1}} + \cdots \tag{3.30}$$

By Exercise 3.9 it is possible to multiply equation (3.30) on both sides by $f(w)/2\pi i$ and then integrate around C_1 to get

$$f(z) = \frac{1}{2\pi i}\int_{C_1} \frac{f(w)}{w - a}\,dw + \frac{z - a}{2\pi i}\int_{C_1} \frac{f(w)}{(w - a)^2}\,dw + \cdots$$
$$+ \frac{(z - a)^n}{2\pi i}\int_{C_1} \frac{f(w)}{(w - a)^{n+1}}\,dw + \cdots$$

The result now follows directly from equation (3.28).

Example 3.2 The equation $e^x = y$, $x > 0$, in real variables has the unique solution $x = \log y$. In complex variables, however, $e^w = z$, z not zero, has infinitely many solutions w, each of which is defined to be a logarithm of z. This, of course, is at variance with our definition of a function in Chapter 1. 'Functions' exhibiting this kind of multiple-valued behaviour occur constantly throughout the theory of functions of a complex variable, and it is possible to resolve their anomalous behaviour by the use of *Riemann surfaces.* To deal with Riemann surfaces is beyond the scope of this book and the interested reader is referred to Ahlfors (1966) for details.

To return to the logarithm, if $w = x + iy$ then $e^x(\cos y + i \sin y) = z$. Thus y must be one of the values of arg z and $e^x = \log |z|$. If arg z is given its principal value then the *principal value of log z* is obtained. This principal value of the logarithm is written log z and hence

$$\log z = \log |z| + i\,(\text{principal value arg } z)$$

If the principal value of the logarithm is understood, and the complex plane, with the origin and the negative real axis excluded, is taken as domain, then log z is a function in accordance with the definition of Chapter 1. Since

$$\log z = \log \sqrt{(x^2 + y^2)} + i \arctan \frac{y}{x}$$

it is easy to verify that the conditions of Theorem 2.2 are satisfied and so log z is analytic. Furthermore, from the Cauchy–Riemann equations,

$$\frac{d}{dz}\log z = \frac{\partial u}{\partial x} - i\frac{\partial u}{\partial y}$$

where $u = \log \sqrt{(x^2 + y^2)}$ giving

$$\frac{d}{dz}\log z = \frac{x - iy}{x^2 + y^2} = \frac{1}{z} \tag{3.31}$$

A direct proof of the continuity of log z, and therefore of analyticity by Exercise 2.5, can be found in Ahlfors (1966), p. 71.

The function $\log(1 + z)$ is analytic in the circle $|z| < 1$ and from (3.31)

$$\frac{d^n}{dz^n} \log(1 + z) = (-1)^{n-1} \frac{(n-1)!}{(1 + z)^n}, \quad n \geqslant 2$$

From Theorem 3.6, with $a = 0$, it follows that

$$\log(1 + z) = z - \frac{z^2}{2} + \frac{z^3}{3} - \frac{z^4}{4} + \cdots + (-1)^{n-1}\frac{z^n}{n} + \cdots \tag{3.32}$$

since $\log 1 = 0$ in accordance with real-variable theory.

EXERCISE 3.13 Derive the Taylor series for $\log(1 + z)$, equation (3.32), by considering the expansion of

$$f'(z) = \frac{1}{z} = \frac{1}{(z-1) + 1} \tag{3.33}$$

and integrating term by term.

Example 3.3 (Compare Example 2.3) We have

$$\exp(z) = 1 + \frac{z}{1!} + \frac{z^2}{2!} + \cdots + \frac{z^n}{n!} + \cdots \tag{3.34}$$

If *sin* (z) and *cos* (z) are defined by the equations

$$\sin z = \frac{e^{iz} - e^{-iz}}{2i} \tag{3.35}$$

$$\cos z = \frac{e^{iz} + e^{-iz}}{2} \tag{3.36}$$

it follows that

$$\sin z = z - \frac{z^3}{3!} + \frac{z^5}{5!} - \cdots \tag{3.37}$$

$$\cos z = 1 - \frac{z^2}{2!} + \frac{z^4}{4!} - \cdots \tag{3.38}$$

and the series both have infinite radius of convergence. It also follows directly from the definition that

$$\frac{d}{dz} \sin z = \cos z \tag{3.39}$$

$$\frac{d}{dz} \cos z = -\sin z \tag{3.40}$$

Now consider the function $\sin^2 z$. This is analytic everywhere in the complex plane and therefore possesses a Taylor expansion in any circle about the origin. Since

$$\frac{d}{dz}\sin^2 z = 2\sin z\cos z \tag{3.41}$$

it is easy to compute the coefficients in this expansion to give

$$\sin^2 z = \frac{2}{2!}z^2 - \frac{8}{4!}z^4 + \frac{32}{6!}z^6 - \cdots + (-1)^{n-1}\frac{2^{2n-1}}{(2n)!}z^{2n} + \cdots \tag{3.42}$$

EXERCISE 3.14 Verify the results of Example 3.3 directly, using the formula for the product of two power series.

To complete this section a useful result known as Morera's theorem will be given.

Theorem 3.7 (Morera's theorem) Let $f(z)$ be continuous inside the circle C_1: $|z| < R$ and suppose that

$$\int_C f(z)\,dz = 0 \tag{3.43}$$

for all closed contours C inside C_1. Then $f(z)$ is analytic inside C_1.

Proof Let z and z_0 be points inside C_1; then from (3.43),

$$g(z) = \int_{z_0}^{z} f(w)\,dw$$

does not depend on the path of integration which may therefore be taken as the straight line joining the two points. By hypothesis $f(w)$ is continuous and it follows that

$$\frac{g(z+h) - g(z)}{h} - f(z) = \frac{1}{h}\int_{z}^{z+h} \{f(w) - f(z)\}\,dw$$

tends to zero as $h \to 0$, showing that $g'(z) = f(z)$. By Theorem 3.5, the derivative of an analytic function is also analytic and the theorem is proved.

EXERCISE 3.15 Prove that a function $f(z)$ which is analytic for all finite values of z and satisfies $|f(z)| < M$ for all z (that is, $f(z)$ is bounded) must be a constant. [Use Theorems 3.3 and 3.5. This result is known as *Liouville's theorem*.]

4 Laurent series

If a function $f(z)$ is not analytic at a point z_0 but is analytic in some deleted neighbourhood of z_0 then z_0 is called an *isolated singularity* of $f(z)$. It is obviously not possible to expand $f(z)$ in a Taylor series about z_0 but it is of interest to inquire whether a series expansion can be made in some annular region formed by concentric circles about z_0.

It will now be shown that if a function is analytic in the annular region $r \leqslant |z - z_0| \leqslant R$ it can be expressed as the sum of an infinite series of positive and negative powers of $(z - z_0)$. In fact the following theorem holds.

Theorem 4.1 If the function f is analytic inside and on the boundary of the annulus

$$r \leqslant |z - z_0| \leqslant R$$

then $f(z)$ is given by the *Laurent series*

$$f(z) = \sum_{n=0}^{\infty} a_n(z - z_0)^n + \sum_{n=1}^{\infty} b_n(z - z_0)^{-n} \tag{4.1}$$

where

$$a_n = \frac{1}{2\pi i}\int_{C_2} \frac{f(w)\,dw}{(w - z_0)^{n+1}} \tag{4.2}$$

$$b_n = \frac{1}{2\pi i}\int_{C_1} (w - z_0)^{n-1} f(w)\,dw \tag{4.3}$$

and z is any point inside the annulus. The circle $|z - z_0| = r$ has been denoted by C_1 and $|z - z_0| = R$ by C_2.

Proof Consider the closed contour indicated by the arrows in Fig. 3, starting at the point A, describing the circle C_2 in a positive direction, crossing along the

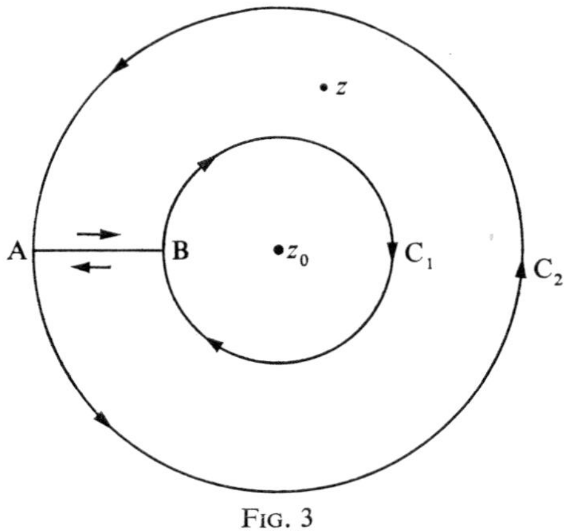

FIG. 3

diameter from A to B, describing circle C_1 in the negative direction and finally crossing back from B to A.

By Theorem 3.3, if AB is chosen so that it does not pass through z, the integral $\frac{1}{2\pi i}\int \frac{f(w)}{w - z}\,dz$ round the above contour gives

$$f(z) = \frac{1}{2\pi i}\int_{C_2} \frac{f(w)}{w - z}\,dw - \frac{1}{2\pi i}\int_{C_1} \frac{f(w)}{w - z}\,dw \tag{4.4}$$

because the integrals on the line AB cancel each other. The sign of the second integral has been reversed and the circle C_1 is described in a positive direction.

As in the proof of Theorem 3.6 it follows that

$$\frac{1}{2\pi i}\int_{C_2} \frac{f(w)}{w - z}\,dw = \sum_{n=0}^{\infty} a_n(z - z_0)^n \tag{4.5}$$

where the a_n are as in equation (4.2).

Consider the expansion

$$\frac{1}{z - w} = \frac{1}{z - z_0} + \frac{w - z_0}{(z - z_0)^2} + \cdots + \frac{(w - z_0)^{n-1}}{(z - z_0)^n} + \cdots \tag{4.6}$$

This is a geometric series with common ratio $(w - z_0)/(z - z_0)$ and therefore is uniformly convergent on C_1 (Chapter 4, Theorem 12.1). It is possible therefore to multiply both sides of equation (4.6) by $f(w)/2\pi i$ and integrate term by term around C_1 to give

$$\begin{aligned} -\frac{1}{2\pi i}\int_{C_1} \frac{f(w)}{w - z}\,dw &= \frac{1}{z - z_0}\cdot\frac{1}{2\pi i}\int_{C_1} f(w)\,dw + \cdots \\ &\quad \cdots + \frac{1}{(z - z_0)^n}\cdot\frac{1}{2\pi i}\int_{C_1} (w - z_0)^{n-1} f(w)\,dw + \cdots \\ &= \sum_{n=1}^{\infty} b_n(z - z_0)^{-n} \end{aligned} \tag{4.7}$$

where the b_n are as in equation (4.3). The theorem now follows from equations (4.4), (4.5), and (4.7).

Clearly, the series $\sum a_n(z - z_0)^n$ is convergent for any z inside the circle C_2 and the series $\sum b_n(z - z_0)^{-n}$ is convergent for any z outside the circle C_1. If $f(z)$ is analytic within C_1 then the b_n all vanish by Cauchy's theorem and the Laurent series becomes a Taylor series. Notice also that the integrals involved in the evaluation of a_n and b_n can be taken around any closed contour in the annulus between C_1 and C_2.

EXERCISE 4.1 Prove that the Laurent series for a function analytic in an annulus is unique. [Assume two expansions and compare the expressions for their coefficients.]

Example 4.1

$$\left(\frac{1}{z^2}+\frac{2}{z}\right)e^z = \left(\frac{1}{z^2}+\frac{2}{z}\right)\left(1+z+\frac{z^2}{2!}+\cdots+\frac{z^n}{n!}+\cdots\right)$$

$$= \frac{1}{z^2}+\frac{1}{z}+\frac{1}{2!}+\frac{z}{3!}+\cdots+\frac{z^n}{(n+2)!}+\cdots$$

$$+\frac{2}{z}+2+\frac{2z}{2!}+\cdots+\frac{2z^n}{(n+1)!}+\cdots$$

$$= \frac{1}{z^2}+\frac{3}{z}+\frac{5}{2}+\cdots+\frac{2n+5}{(n+2)!}z^n+\cdots$$

Example 4.2 If $z \neq 0$ then

$$\exp\frac{1}{z} = 1+\frac{1}{z}+\frac{1}{2!z^2}+\cdots+\frac{1}{n!z^n}+\cdots$$

Example 4.3 If $z \neq 0$ then

$$\frac{\sin z}{z} = \frac{1}{z}\left\{\sum_{n=0}^{\infty}(-1)^n\frac{z^{2n+1}}{(2n+1)!}\right\}$$

$$= 1-\frac{z^2}{3!}+\frac{z^4}{5!}-\cdots$$

4.1 Poles and zeros

If the function $f(z)$ has an isolated singularity at a point z_0 then it is possible to construct an annular region A, centred on z_0, in which $f(z)$ is analytic. It is then possible to form the Laurent expansion

$$f(z) = \sum_{n=1}^{\infty} b_n(z-z_0)^{-n} + \sum_{n=0}^{\infty} a_n(z-z_0)^n \tag{4.8}$$

for $f(z)$ in A. The first series on the right-hand side of equation (4.8) is known as the *principal part* of $f(z)$ at $z = z_0$.

If every term in the principal part vanishes, the singularity at z_0 is said to be a *removable singularity*, since by defining the value of $f(z)$ at $z = z_0$ to be a_0 the singularity is removed. This was the case in Example 4.3; $(\sin z)/z$ is discontinuous at $z = 0$ (being undefined there) but the singularity can be removed by defining $(\sin z)/z$ to be 1 at $z = 0$.

Now suppose that $b_n = 0$ for n greater than some integer m where $m \geqslant 1$. The principal part becomes $\sum_{k=1}^{m} b_k/(z-z_0)^k$ and the singularity is called *a pole of order m*. Notice that in this case $(z-z_0)^m f(z)$ has a removable singularity at z_0. The value of b_1 (which may be zero) is called the *residue* of $f(z)$ at the pole $z = z_0$.

When $m = 1$ it is easily seen that the residue is $b_1 = \lim_{z \to z_0} (z - z_0)f(z)$. In general for a pole of mth order

$$\lim_{z \to z_0} (z - z_0)^m f(z) = b_m \tag{4.9}$$

from which it follows that if $f(z)$ has a pole at z_0 then $|f(z)| \to \infty$ as $z \to z_0$. In Example 4.1 there is a pole of order 2 at the origin with residue 3. A function having no singularities other than poles in any finite region of the complex plane is said to be *meromorphic*.

Another possibility is that the principal part of $f(z)$ consists of an infinite series and in this case there is said to be an *essential singularity* at z_0. The value of b_1 is called the residue of the essential singularity. Thus (Example 4.2) $\exp(1/z)$ has an essential singularity at the origin with residue 1.

A function $f(z)$ which is analytic at a point z_0 has a Taylor expansion

$$f(z) = \sum_{n=0}^{\infty} a_n(z - z_0)^n \tag{4.10}$$

about z_0. If the first term which does not vanish in (4.10) is a_m then $f(z)$ is said to have a *zero of order m* at z_0.

EXERCISE 4.2 If $f(z) = (z - z_0)^m g(z)$ and $f(z)$ is analytic at z_0 and has a zero of order m at z_0, show that $g(z)$ is analytic at z_0. [Use the differentiability properties of power series, Chapter 4, Section 12.]

Theorem 4.2 If the function $f(z)$ is analytic at the point z_0 and has a zero of finite order at z_0, then there is some neighbourhood $N(z_0, h)$ which contains no other zero of $f(z)$ (that is, zeros are *isolated*).

Proof Assume the zero at z_0 is of order m and write $f(z) = (z - z_0)^m g(z)$ where, by Exercise 4.2, $g(z)$ is analytic in $|z - z_0| < H$, say. Since the zero is of order m $g(z_0) = a_m \neq 0$. But as $g(z)$ is continuous it is possible to select $h > 0$ so that $|z - z_0| < h$ implies $|g(z) - g(z_0)| < |a_m/2|$. It follows that, for $|z - z_0| < h$, $|g(z)| \geqslant |\{|g(z_0)| - |g(z) - g(z_0)|\}| > |a_m/2|$ and $g(z)$ is not zero in $N(z_0, h)$.

An important result connecting poles and zeros will now be proved.

Theorem 4.3 If the function $f(z)$ is analytic, and has a zero of order m, at the point z_0 then $1/f(z)$ has a pole of order m at z_0.

Proof Since $f(z)$ is analytic and has a zero at z_0 it is analytic in some neighbourhood $N(z_0, h)$ and $f(z) \neq 0$ in $N'(z_0, h)$. It follows that $1/f(z)$, which has derivative $-f'/f^2$ where $f \neq 0$, is also analytic in $N'(z_0, h)$ and therefore has an isolated singularity at z_0.

Now write $f(z) = (z - z_0)^m g(z)$. Then $g(z) = \sum_{n=0}^{\infty} a_{n+m}(z - z_0)^n$ and $g(z)$ is analytic and does not vanish in $N(z_0, h)$. The function $1/g(z)$ is analytic at z_0 and therefore has a Taylor expansion

$$\frac{1}{g(z)} = \sum_{n=0}^{\infty} A_n(z - z_0)^n$$

in $N(z_0, h)$, where $A_0 = \dfrac{1}{g(z_0)} \neq 0$. Then in $N'(z_0, h)$.

$$\frac{1}{f(z)} = \frac{A_0}{(z - z_0)^m} + \frac{A_1}{(z - z_0)^{m-1}} + \cdots + A_m + \sum_{n=1}^{\infty} A_{m+n}(z - z_0)^n$$

and so $1/f(z)$ has a pole of order m at z_0.

EXERCISE 4.3 Prove the converse of Theorem 4.3; that is, if $f(z)$ has a pole of order m at z_0, then $1/f(z)$ can be suitably defined at z_0 to make it analytic there and have a zero of order m.

EXERCISE 4.4 Prove that a limit point of a sequence of zeros of a non-constant function $f(z)$ must be an essential singularity of $f(z)$. [Show first that if $f(z)$ is analytic at a limit point of a sequence of its zeros it must vanish identically; now show that if $f(z)$ is not identically zero, such a limit point cannot be a pole.]

EXERCISE 4.5 Prove that the limit point of a sequence of isolated singularities of a function which is analytic except at these singularities is itself a singularity but cannot be a pole. Such a limit point is said to be a *non-isolated essential singularity*. [The limit point is not isolated.]

An essential singularity of a function $f(z)$ exhibits much more complex behaviour than a pole. It can in fact be shown that if z_0 is an essential singularity of $f(z)$, then $f(z)$ can be made to tend to any arbitrary limiting value by letting $z \to z_0$ in some suitable way. (See for example Titchmarsh, 1939, p. 93.)

If $f(z)$ is a rational function (that is, the ratio of two polynomials) then it is analytic for all values of z for which its denominator does not vanish. The behaviour of a complex function at infinity is investigated by means of the transformation $z = 1/w$ and the properties of $f(z)$ at infinity are the corresponding properties of $f(1/w)$ at $w = 0$. The *point at infinity* in the complex plane is defined to be the point which corresponds to the origin under the transformation $z = 1/w$. Notice that there is only one such point; compare Chapter 4, Section 2. It is now possible to prove the following theorem regarding rational functions.

Theorem 4.4 A function $f(z)$ which has no essential singularity for finite or infinite z is a rational function.

Proof The point at infinity is not an essential singularity by hypothesis so there is a neighbourhood $N(\infty)$ in which the point at infinity is a pole of $f(z)$ or in which $f(z)$ is analytic. Hence the principal part of $f(z)$ at infinity must be $\sum_{k=1}^{m} b_k z^k$ for some finite value of m.

It is also possible to take a region about the origin in such a way that the point at infinity is the only singularity lying outside the region. Inside this region there can be only a finite number of poles since otherwise there would be a limit point which would be a singularity other than a pole.

Denote this finite set of poles by $z_1, z_2, \ldots, z_n$ and let the principal parts of $f(z)$ at these poles be $\sum_{k=1}^{p_j} b_{kj}/(z - z_j)^k$ for $j = 1, 2, \ldots, n$ respectively, where the pole at z_j has order p_j. Now the function

$$g(z) = f(z) - \sum_{j=1}^{n} \sum_{k=1}^{pj} \frac{b_{kj}}{(z - z_j)^k} - \sum_{k=1}^{m} b_k z^k$$

is analytic in the whole complex plane (including ∞) and is therefore bounded there. It follows from Liouville's theorem, Exercise 3.15, that $g(z)$ is a constant, so proving that $f(z)$ is a rational function.

EXERCISE 4.6 Prove that a function with a pole of order m at infinity and no other singularities must be a polynomial of degree m.

Example 4.4 It has already been shown that $\exp(1/z)$ has an essential singularity at $z = 0$. It follows that e^z has an essential singularity at infinity. Notice that for any non-zero complex number $z_0 = r_0(\cos\theta_0 + i\sin\theta_0)$ the infinite sequence $w_n = \log r_0 + i(\theta_0 + 2\pi n)$ gives $\exp(w_n) = z_0$ for all values of n.

Example 4.5 Consider the rational function

$$g(z) = \frac{z(z - 1)}{(z + 1)(z - 2)(z - 3)}$$

Put $z = 1/w$; then

$$g(w) = \frac{w(1 - w)}{(1 + w)(1 - 2w)(1 - 3w)}$$

which is analytic at the origin since the denominator does not vanish there. Hence $g(z)$ is analytic at infinity. The function $1/[g(z)]$ has zeros of the first order at $z = -1$, $z = 2$ and $z = 3$ and therefore $g(z)$ has simple poles at these points, with residues equal to $\frac{1}{6}$, $-\frac{2}{3}$ and $\frac{3}{2}$ respectively, by equation (4.9). It is easy to confirm that $g(z)$ can be written in partial fractions,

$$g(z) = \frac{1}{6(z + 1)} - \frac{2}{3(z - 2)} + \frac{3}{2(z - 3)}$$

5 The calculus of residues

Cauchy's theorem can be used as a powerful tool for the evaluation of certain definite integrals. The following theorem forms the basis for this application.

Theorem 5.1 (Cauchy's residue theorem) Let the function $f(z)$ be analytic everywhere inside and on a closed elementary contour C, except at a finite number of isolated singularities inside C. Then $\int_C f(z)\,dz = 2\pi i \sum \text{Res}\,(z_k)$ where $\sum \text{Res}\,(z_k)$ denotes the sum of the residues at the singularities lying inside C.

Proof From the Laurent expansion, the residue b_{1k} at the singularity z_k is given by $b_{1k} = (1/2\pi i) \int_{C_k} f(z)\,dz$ where C_k is a closed circle centred on z_k and including no other singularity of $f(z)$.

Let all the singularities $z_1, \ldots, z_n$ inside C be surrounded by such circles C_k, $k = 1, 2, \ldots, n$, with their radii taken so small that they lie completely within C and do not intersect one another. It now follows from Cauchy's theorem and a simple extension of Exercise 3.7 that

$$\int_C f(z)\,dz = \sum_{k=1}^{n} \int_{C_k} f(z)\,dz$$

$$= 2\pi i \sum_{k=1}^{n} b_{1k}$$

proving the theorem.

The residue of the function $f(z)$ at infinity is defined as the negative of the coefficient of z in the expansion of $f(1/z)$ about the origin. Thus the residue at infinity is given by $(1/2\pi i) \int_C f(z)\,dz$ where C is a circle large enough to contain all the finite singularities of $f(z)$ and is described in the negative (clockwise) sense. By applying the transformation $z = 1/w$ it is easily seen that the residue of $f(z)$ at infinity is given by $\lim_{z\to\infty} \{-zf(z)\}$ when this limit exists. It is important to appreciate that although a function is analytic at infinity it can still have a residue there. For example the rational function of Example 4.5 has residue -1 at infinity.

EXERCISE 5.1 The sum of the residues at the finite and infinite singularities of the rational function of Example 4.5 is zero; prove that this is so for all rational functions.

Let $f(z)$ be analytic inside and on a closed contour C except at a finite number of poles inside C, and assume that none of the zeros of $f(z)$ lies on C. If z_0 is a zero of $f(z)$ of order m then $f(z) = (z - z_0)^m g(z)$ where $g(z_0) \neq 0$ and $g(z)$ is analytic at $z = z_0$. Hence

$$\frac{f'(z)}{f(z)} = \frac{m}{z - z_0} + \frac{g'(z)}{g(z)}$$

and it follows that $f'(z)/f(z)$ has a simple pole at z_0 with residue m. Similarly, $f'(z)/f(z)$ has a simple pole of residue $-m$ at a pole of order m of $f(z)$. Therefore a straightforward application of Theorem 5.1 gives the following theorem.

Theorem 5.2 If $f(z)$ is analytic in and on a closed elementary contour except at a finite number of poles inside C, and $f(z)$ does not vanish on C, then

$$\frac{1}{2\pi i}\int_C \frac{f'(z)}{f(z)}\,dz = N - P$$

where N is the number of zeros inside C, and P the number of poles (each pole or zero of order m being counted m times).

Example 5.1 Consider the integral

$$I = \int_0^{2\pi} \frac{a\,d\theta}{a^2 + \sin^2\theta}, \quad a > 0$$

Write $z = \exp(i\theta)$ so that $\sin\theta = (1/2i)(z - 1/z)$ and $d\theta/dz = 1/iz$, giving

$$I = \int_C \frac{-4aiz\,dz}{4a^2z^2 - (z^2 - 1)^2}$$

where C is the unit circle. Now

$$\frac{z}{4a^2z^2 - (z^2 - 1)^2} = \frac{-z}{(z^2 - 2az - 1)(z^2 + 2az - 1)}$$

Let

$$(z^2 - 2az - 1) = (z - \alpha)(z - \beta)$$

and

$$(z^2 + 2az - 1) = (z - \alpha')(z - \beta')$$

where

$$\alpha, \beta = a \pm \sqrt{(a^2 + 1)}$$
$$a', \beta' = -a \pm \sqrt{(a^2 + 1)}$$

Then the above rational function has simple poles at β and α' inside the unit circle (since $a > 0$ the other two poles are outside C).

The residue at β is given by

$$\frac{-\beta}{(\beta - \alpha)(\beta - \alpha')(\beta - \beta')}$$

and that at α' by

$$\frac{-\alpha'}{(\alpha' - \alpha)(\alpha' - \beta)(\alpha' - \beta')}$$

But

$$\begin{aligned}
\beta - \alpha &= -2\sqrt{(a^2 + 1)} \\
\beta - \alpha' &= 2a - 2\sqrt{(a^2 + 1)} \\
\beta - \beta' &= 2a \\
\alpha' - \alpha &= -2a \\
\alpha' - \beta &= -2a + 2\sqrt{(a^2 + 1)} \\
\alpha' - \beta' &= 2\sqrt{(a^2 + 1)}
\end{aligned}$$

so that the sum of the residues inside C is just

$$\frac{-1}{(\beta - \alpha)(\beta - \beta')} = \frac{1}{4a\sqrt{(a^2 + 1)}}$$

Theorem 5.1 then gives

$$\int_0^{2\pi} \frac{a\,d\theta}{a^2 + \sin^2\theta} = \frac{2\pi}{\sqrt{(a^2 + 1)}}$$

Example 5.2 Consider the integral $I = \int_C e^{imz} f(z)\,dz$, where the only singularities of $f(z)$ for finite z are poles (that is $f(z)$ is meromorphic) and C is a semicircle in the upper half-plane of radius R and centre the origin. Let m be positive and assume that $f(z)$ tends uniformly to zero as $|z| \to \infty$.

By this last condition, for any given $\epsilon > 0$, R can be chosen large enough to ensure that, for all z on C, $|f(z)| < \epsilon$. On putting $z = R(\cos\theta + i\sin\theta) = R\exp(i\theta)$ there results

$$\begin{aligned}
\left|\int_C f(z)\exp(imz)\,dz\right| &= \left|\int_0^{\pi} f(z)\exp(imz)R\exp(i\theta)i\,d\theta\right| \\
&< \epsilon\int_0^{\pi}\exp(-mR\sin\theta)R\,d\theta \\
&= 2\epsilon R\int_0^{\pi/2}\exp(-mR\sin\theta)\,d\theta
\end{aligned}$$

because $|\exp imz| = \exp(-mR\sin\theta)$. But $(\sin\theta)/\theta$ decreases steadily from 1 to $2/\pi$ as θ increases from 0 to $\pi/2$ so that for $0 < \theta < \pi/2$, $(\sin\theta)/\theta > 2/\pi$. Therefore

$$|I| < 2R\epsilon\int_0^{\pi/2}\exp\left(-\frac{2mR\theta}{\pi}\right)d\theta = \frac{\pi\epsilon}{m}(1 - e^{-mR})$$

showing that the integral $I \to 0$ as $R \to \infty$. This result is known as *Jordan's lemma*.

To illustrate the use of Jordan's lemma let $f(z) = z/(z^2 + 1)$ which satisfies the above conditions. Integrate $f(z) \exp(iz)$ around a contour C consisting of the above semicircle and the real axis from $-R$ to $+R$, and let R tend to infinity. The result is

$$\int_{-\infty}^{+\infty} \frac{x \exp(ix)}{x^2 + 1} \, dx = \frac{\pi i}{e}$$

because the only pole of $f(z)$ in the upper half-plane is at i and has residue $1/2e$. The integral on the semicircle is zero by the lemma.

Now on taking imaginary parts in the above integral the result is

$$\int_{-\infty}^{\infty} \frac{x \sin x}{x^2 + 1} \, dx = \frac{\pi}{e}$$

The integrand is an even function so that

$$\int_{0}^{\infty} \frac{x \sin x}{x^2 + 1} \, dx = \frac{\pi}{2e}$$

Example 5.3 The following result is useful in the inversion of Laplace transforms (see Chapter 8). Suppose that $f(z)$ is a meromorphic function such that $|f(z)| < M/|z|^k$ for all $|z| > K$, where M, k, and K are positive constants. Consider the closed contour C shown in Fig. 4 consisting of part of a circle of radius

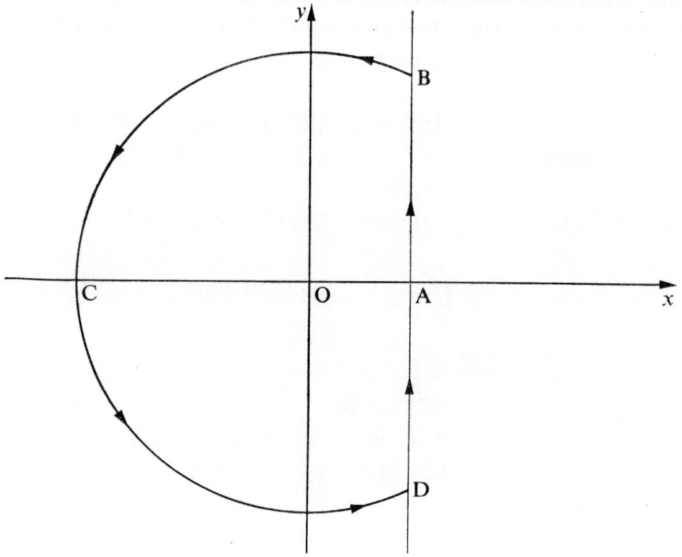

FIG. 4

R and part of the straight line Re $z = \alpha$. Assume that no poles of f lie on BD and that R is sufficiently large to include all the poles of f lying to the left of this line. Assume also that $R > K$. Then by Theorem 5.1,

$$I = \int_C e^{zt} f(z)\, dz = 2\pi i \sum \text{Res}\, [e^{zt} f(z)]$$

where t is real and positive. On considering the above integral on the circular arcs BC and CD, an argument similar to that used in Jordan's lemma shows that these parts of I vanish as $R \to \infty$. It follows that

$$\int_{\alpha - i\infty}^{\alpha + i\infty} e^{zt} f(z)\, dz = 2\pi i \sum \text{Res}\, [e^{zt} f(z)]$$

where the sum is taken over all the poles to the left of Re $z = \alpha$ and the integral is the Cauchy principal value.

EXERCISE 5.2 Prove that

$$\frac{1}{2\pi i} \int_{\alpha - i\infty}^{\alpha + i\infty} z^{-1} e^{zt}\, dz = \begin{cases} 0 & t < 0 \\ \frac{1}{2} & t = 0 \\ 1 & t > 0 \end{cases}$$

where the Cauchy principal value of the integral is understood and $\alpha > 0$. [Use the part of a circle to the right of Re $z = \alpha$ for $t < 0$, and the part to the left of Re $z = \alpha$ for $t > 0$. For $t = 0$ use a contour consisting of the line Re $z = \alpha$, the line Re $z = 0$ indented to the right at $z = 0$, and the two connecting circular arcs.]

EXERCISE 5.3 Prove that

$$\frac{1}{2\pi} \int_0^{2\pi} (x^2 + y^2 + 2xy \cos \theta)^n\, d\theta = \sum_{r=0}^{n} \binom{n}{r}^2 x^{2r} y^{2(n-r)}$$

These examples will give some idea of the techniques used in the calculus of residues. Further details and other applications including the expansion of meromorphic functions, the summation of series, and partial fractions can be found in the references at the end of this chapter.

6 Analytic continuation

This chapter is concluded with a brief account of analytic continuation. If a function $f_1(z)$ is analytic on the region S_1 within a given closed contour C_1, then it can be represented as a Taylor series in a neighbourhood of any point $z_0 \in S_1$ with radius of convergence at least the minimum distance from z_0 to C_1. Should this Taylor series converge in the interior S_2 of a larger circle, C_2 say, then an analytic function $f_2(z)$ is defined on S_2 which coincides with $f(z)$ on $S_1 \cap S_2$: $f_2(z)$ is called the *analytic continuation* of $f_1(z)$ into S_2. Conversely f_1 can be said to be the analytic continuation of f_2 into S_1. Obviously, this process can be repeated for any point close to C_2 and so on. This particular method of continuation is known as the *method of power series.*

More generally, if the two functions f_1 and f_2 are analytic within the regions S_1 and S_2 interior to the closed contours C_1 and C_2, and $f_1 = f_2$ on $S_1 \cap S_2$, then f_2 is said to be the analytic continuation of f_1 into S_2. It is then possible to think of the two functions f_1 and f_2 as representing a single analytic function on $S_1 \cup S_2$. If f_2 and g_2 are both analytic continuations of f_1 from S_1 into S_2 then $f_2 = g_2$.

A *complete analytic function* can then be defined as the original function together with the totality of its direct analytic continuations, together with the continuation of these continuations, and so on.

To study analytic continuation in any depth is beyond the scope of this book and the interested reader is referred, for example, to Ahlfors (1966) for greater detail.

Example 6.1 Consider the series

$$f(z) = 1 + z^2 + z^4 + z^8 + \cdots + z^{2^n} + \cdots$$

which represents an analytic function for $|z| < 1$. The sets of roots of the equations $z^{2^n} = 1$ for $n = 1, 2, \ldots$ are all singularities of $f(z)$, for when z equals such a root all terms after the nth in the series equal 1. But since there are infinitely many such roots on any arc of the circle $|z| = 1$ it is impossible to continue $f(z)$ beyond this circle. In this case $|z| = 1$ is said to form the *natural boundary* of $f(z)$.

EXERCISE 6.1 The largest region in which the function $f(z) = \sum_{n=0}^{\infty} z^n$ is analytic is $|z| < 1$. Find a point inside this region about which the Taylor expansion for $f(z)$ would give an analytical continuation outside the original region, and also a point from which no continuation is possible.

Further reading

The whole of the material of this chapter is dealt with in greater detail in the classical texts by Copson (1935) and Titchmarsh (1939). Many interesting historical details regarding Cauchy's theorem and examples of contour integration can be found in Watson (1914). A useful short account of the subject is that of Phillips (1951).

More recent introductory texts are Franklin (1959), Goodstein (1965), and Ahlfors (1966), the last of these is presented in a highly individual way and contains a considerable background in modern analysis. A very comprehensive and modern treatment is to be found in Markushevich (1965).

Finally, on the applied aspects of the subject the reader is referred to Dettman (1965) and Carrier (1966).

References

AHLFORS, L. V., 1966, *Complex analysis*, McGraw-Hill, New York.

APOSTOL, T. M., 1957, *Mathematical analysis*, Addison-Wesley, Reading, Mass.

CARRIER, G. F., M. KROOK, and C. E. PEARSON, 1966, *Functions of a complex variable; theory and technique*, McGraw-Hill, New York.

COPSON, T., 1935, *An introduction to the theory of functions of a complex variable*, Oxford University Press, London.

DETTMAN, J. W., 1965, *Applied complex variables*, Macmillan, New York.

FRANKLIN, P., 1959, *Functions of complex variables*, Pitman, London.

GOODSTEIN, R. L., 1965, *Complex functions*, McGraw-Hill, New York.

MARKUSHEVICH, A. I. (Trans. R. A. Silverman), 1965, *Theory of functions of a complex variable* (Vols. I, II, and III), Prentice-Hall, Englewood Cliffs, N.J.

PHILLIPS, E. G., 1951, *Functions of a complex variable*, Oliver and Boyd, Edinburgh.

TITCHMARSH, E. C., 1939, *The theory of functions* (2nd ed.), Oxford University Press, London.

WATSON, G. N., 1914, *Complex integration and Cauchy's theorem*, Hafner, New York.

Problems

1 Show that if n is an odd integer

$$\cos^n \theta = \frac{1}{2^{n-1}}\left\{\cos n\theta + \sum \binom{n}{r} \cos (n - 2r)\theta\right\}$$

where $r = 1, 2, \ldots, \frac{1}{2}(n - 1)$.

2 Show that the function $w = f(z) = z^{1/2}$ maps the entire z plane with the exception of zero and the negative real axis, into the right half-plane of w. Prove also that the mappings of two curves will intersect at the same angle as the original curves.

3 If C is the unit circle centred at the origin show that

$$\int_C z^n \, dz = 0$$

where n is an integer other than -1.

4 Expand cosec z as a Laurent series about $z = 0$ and show that it has a simple pole there with residue 1. Show also that cosec $(1/z)$ has a non-isolated essential singularity at $z = 0$.

5 Prove that if the two functions $f(z)$ and $g(z)$ are analytic inside and on the closed contour C, $f(z) \neq 0$ on C and $|g(z)| < |f(z)|$ on C, then $f(z)$ and $g(z) + f(z)$ have the same number of zeros inside C. (This result is known as *Rouché's theorem.*)

6 Prove that if $f(x, y)$ is a rational function of x and y then

$$\int_0^{2\pi} f(\cos \theta, \sin \theta) \, d\theta = 2\pi i \sum \operatorname{Res}\left\{\frac{1}{iz} f\left[\frac{1}{2}\left(z + \frac{1}{z}\right), \frac{1}{2i}\left(z - \frac{1}{z}\right)\right]\right\}$$

where the sum is taken over the poles of f inside the circle $|z| = 1$.

7 Prove that a polynomial of degree n has n roots.

8 If the function $f(z)$ is analytic on the closed disk $|z| \leqslant R$, show that

$$u(r, \theta) = \frac{1}{2\pi} \int_0^{2\pi} \frac{R^2 - r^2}{R^2 - 2Rr\cos(\theta - \phi) + r^2}\, u(R, \phi)\, d\phi$$

where u is the real part of f and $0 \leqslant r < R$. Give a similar result for the imaginary part of f. [This result is known as *Poisson's formula.*]

9 Use contour integration to prove the following results:

(a) $\displaystyle\int_0^\infty \frac{\operatorname{sech}^2 x\, dx}{4x^2 + \pi^2} = \frac{1}{12}$

(b) $\displaystyle\int_0^\infty \frac{\cos \pi x}{1 + x^2 + x^4}\, dx = \frac{\pi}{2} \exp\left(-\frac{\sqrt{3}}{2}\pi\right)$

10 Find the Laurent expansions of $1/(z - 1)(z - 2)$ valid in the regions $1 < |z| < 2$, $|z| < 1$, and $|z| > 2$.

8 Laplace transform and z-transform

1 Introduction

The use of the Laplace transform in engineering grew out of the operational methods initiated by Heaviside. Given the equation

$$\ddot{x} + 3\dot{x} + 2x = 1; \qquad \dot{x}(0) = x(0) = 0 \tag{1.1}$$

Heaviside wrote

$$(p^2 + 3p + 2)x = 1 \tag{1.2}$$

where the 'operator' p represents differentiation. From (1.2) he obtained

$$x = \frac{1}{p^2 + 3p + 2} = \frac{1}{p + 1} - \frac{1}{p + 2} \tag{1.3}$$

and if the replacement of (1.1) by (1.2) is valid, then equally $1/(p + 1)$ should be the solution of

$$\dot{y} + y = 1; \qquad y(0) = 0 \tag{1.4}$$

which is

$$y(t) = 1 - e^{-t} \tag{1.5}$$

Interpreting $1/(p + 2)$ similarly, (1.3) leads to

$$x(t) = \tfrac{1}{2} - e^{-t} + \tfrac{1}{2}e^{-2t} \tag{1.6}$$

which will be found to satisfy (1.1).

This procedure is not difficult to justify, but the justification is less easy when similar methods are applied to partial differential equations. As a result of work by mathematicians after Heaviside, the Laplace transform was made the basis of a procedure which for problems such as (1.1) leads to exactly parallel results. This procedure based on the Laplace transform has some small disadvantages compared with Heaviside's method when applied to (1.1) (see Section 5.4). For partial differential equations, however, the Laplace transform is greatly superior, and it has therefore displaced the older method.

The Laplace transform is closely related to the Fourier transform, and it can be interpreted in terms of vector spaces. Let f be a function mapping E_1 into E_1: such functions define a vector space (Chapter 3). Let K be a function mapping $C_1 \times E_1$ into C_1, where C_1 is the space of complex numbers. If f and K satisfy suitable conditions the integral

$$\phi(s) = \int_{-\infty}^{\infty} K(s, t)f(t)\, dt \tag{1.7}$$

will exist, and will define a mapping of the space of functions f into the space of functions ϕ. With suitable further conditions on K and f, this mapping is one-to-one. Notice the analogy between (1.7) and the product of a matrix (k_{ij}) with a vector (f_j).

The particular choice $K(s, t) = e^{-st}$ in (1.7) gives the two-sided Laplace transformation. Its importance in the solution of linear time-invariant differential equations lies in the fact that if ϕ is the Laplace transform of f, and ψ is the Laplace transform of Df, then subject again to suitable conditions

$$\psi(s) = s\phi(s) \tag{1.8}$$

The remainder of this chapter is devoted to proving (1.8) and the subsidiary results needed for the solution of (1.1), or for other applications to dynamical systems.

2 The Laplace transform

Let $s = \sigma + i\omega$ be a complex number, and $f \in D_R^1$ be a function with domain E_1 and range $R \subseteq E_1$. Then

$$e^{-st} = e^{-\sigma t}(\cos \omega t - i \sin \omega t) \tag{2.1}$$

and so

$$\begin{aligned} \phi(a, b, s) &\triangleq \int_a^b e^{-st} f(t)\, dt \\ &= \int_a^b f(t)e^{-\sigma t} \cos \omega t\, dt - i \int_a^b f(t)e^{-\sigma t} \sin \omega t\, dt \end{aligned} \tag{2.2}$$

Each of the integrands, $f(t)e^{-\sigma t} \cos \omega t$ and $f(t)e^{-\sigma t} \sin \omega t$, is in D_R^1 and so by Section 4 of Chapter 5 both integrals in (2.2) exist and ϕ is defined as a complex number depending on a, b, and s.

Now suppose that for s in some given region S of the complex plane, $\phi(-\infty, \infty, s)$ exists as an improper integral. The function ϕ having domain S and range $R \subseteq C_1$ and having its value at $s \in S$ defined by (2.2) with $a = -\infty$, $b = \infty$ is said to be the *two-sided* (or *bilateral*) *Laplace transform* of f. Various notations are in use for the Laplace transform, and all aim to stress the connection between the transform and the function f from which it arose: common notations are $\mathscr{L}(f)$, F, $\bar{f}$. Usually, however, there is no confusion if the same symbol f is used for the function and its transform, and they are distinguished by their argument, $f(t)$ and $f(s)$: this is the commonest practice.

In this book the bar notation will generally be used, $\bar{f}$. This permits the distinction between the functions f and $\bar{f}$ and their values $f(t)$ and $\bar{f}(s)$ (Chapter 1) which is lost when f is used both for the function and its transform. The notation $\mathscr{L}(f)$ will also be used occasionally, and notational difficulties will sometimes constrain us to write $\mathscr{L}[f(t)] = \bar{f}(s)$ rather than $\mathscr{L}(f) = \bar{f}$. For example, the only way to avoid $\mathscr{L}[U(t)] = s^{-1}$ is to write, somewhat clumsily, '$\mathscr{L}(\phi) = \psi$, where ϕ is defined by $\phi(t) = U(t)$ and ψ by $\psi(s) = s^{-1}$'.

So far, $\bar{f}(s)$ has been assumed to exist for s in some region S of the complex plane. To ensure this, it is convenient to add the following restriction on the function $f \in D^1_R$. Let there exist constants M_1, M_2, α_1, α_2, such that $\alpha_1 < \alpha_2$ and

$$|f(t)| < M_1 e^{\alpha_1 t}, \quad t > 0 \tag{2.3}$$

$$|f(t)| < M_2 e^{\alpha_2 t}, \quad t < 0 \tag{2.4}$$

Then $\bar{f}(s)$ exists for all $s = \sigma + i\omega$ having $\alpha_1 < \sigma < \alpha_2$. In other words, $\bar{f}(s)$ is defined as a complex function of s in the strip $\alpha_1 < \sigma < \alpha_2$. This result is proved in the following way.

Consider the first integral in (2.2); the second integral is treated in the same way. Write

$$\int_a^b f(t)e^{-\sigma t}\cos\omega t\,dt = \int_a^0 f(t)e^{-\sigma t}\cos\omega t\,dt + \int_0^b f(t)e^{-\sigma t}\cos\omega t\,dt \tag{2.5}$$

Then for $0 < c < b$ and $\sigma > \alpha_1$,

$$\int_0^b f(t)e^{-\sigma t}\cos\omega t\,dt = \left(\int_0^c + \int_c^b\right) f(t)e^{-\sigma t}\cos\omega t\,dt \tag{2.6}$$

and

$$\left|\int_c^b f(t)e^{-\sigma t}\cos\omega t\,dt\right| \leqslant \int_c^b |f(t)|e^{-\sigma t}\,dt \tag{2.7}$$

$$< \int_c^b M_1 e^{-(\sigma-\alpha_1)t}\,dt \tag{2.8}$$

$$= \frac{M_1}{\sigma-\alpha_1}\left[e^{-(\sigma-\alpha_1)c} - e^{-(\sigma-\alpha_1)b}\right] \tag{2.9}$$

$$< \frac{M_1}{\sigma-\alpha_1} e^{-(\sigma-\alpha_1)c} \tag{2.10}$$

Therefore, by Cauchy's condition, the integral on the left of (2.6) tends to a finite limit as $b \to \infty$. In the same way, the first integral on the right of (2.5) tends to a finite limit as $a \to -\infty$ provided that $\sigma < \alpha_2$. Hence the result.

Given a function f satisfying (2.3), either (2.3) remains true for some finite $M_1(\alpha_1)$ and for all values of α_1 however low, or there is an infimum α to the values of α_1 for which (2.3) is true. Similarly, α_2 in (2.4) has a supremum β if (2.4) is not true for all values of α_2 however high. The strip in the complex plane defined by $\alpha < \sigma < \beta$ (where α or β or both may be replaced by infinity in appropriate cases) will be called the *strip of exponential convergence of* $\bar{f}$. It follows from what has been proved that $\bar{f}(s)$ exists and is finite for all s in this strip. Also, the method of proof shows that the defining integral for $\bar{f}(s)$ remains convergent in the strip of exponential convergence if the integrand is replaced by its modulus; that is, the infinite integral is *absolutely convergent* in this strip.

The defining integral for $\bar{f}(s)$ may remain convergent, or even absolutely convergent, in a wider strip than that just defined. It is possible, therefore, to define also a *strip of convergence* and a *strip of absolute convergence*. The strip of exponential convergence will, however, be sufficient for our purposes. It will be written (α, β).

Example 2.1 For the function defined by

$$f(t) = U(t) - U(t-1) \tag{2.11}$$

the strip of exponential convergence is the entire s-plane. The Laplace transform of f is given by

$$\bar{f}(s) = \int_{-\infty}^{\infty} f(t)e^{-st}\,dt \tag{2.12}$$

$$= \int_0^1 e^{-st}\,dt \tag{2.13}$$

$$= \frac{1-e^{-s}}{s} \tag{2.14}$$

Example 2.2 Let f be defined by

$$f(t) = U(t)e^{-at} \tag{2.15}$$

Then the strip of exponential convergence is $(-a, \infty)$. The Laplace transform is defined by

$$\bar{f}(s) = \int_0^{\infty} e^{-(s+a)t}\,dt \tag{2.16}$$

$$= \frac{1}{s+a} \tag{2.17}$$

Direct examination of (2.16) shows that the integral does not converge at the point $s = -a$.

Example 2.3 Let f be defined by

$$f(t) = -U(-t)e^{-at} \tag{2.18}$$

The strip of exponential convergence is $(-\infty, -a)$. The Laplace transform is given by

$$\bar{f}(s) = -\int_{-\infty}^{0} e^{-(s+a)t}\, dt = \frac{1}{s+a} \tag{2.19}$$

where the integral fails to converge at $s = -a$. Comparison with (2.17) shows that two different functions may have the same Laplace transform, though the strips of convergence have no point in common in this example (compare Section 4 below).

Example 2.4 Consider the function defined by

$$f(t) = \sum_{r=1}^{\infty} \phi_r(t) \tag{2.20}$$

where

$$\phi_r(t) = U(t - r + e^{-r^2}) - U(t - r - e^{-r^2}) \tag{2.21}$$

When $t = n$, a positive integer, $f(t) = 1$; and $f(t)$ is everywhere either 1 or 0. It follows that the strip of exponential convergence is $(0, \infty)$. On the other hand

$$\int_{-\infty}^{\infty} f(t)e^{-st}\, dt = \sum_{r=1}^{\infty} \int_{r-e^{-r^2}}^{r+e^{-r^2}} e^{-\sigma t}(\cos \omega t - i \sin \omega t)\, dt \tag{2.22}$$

If $\sigma \leqslant 0$, each term in the series has its real and its imaginary part less in modulus than

$$\exp[-\sigma(r+1)].2\exp(-r^2) = 2\exp\left(-\sigma + \frac{\sigma^2}{4}\right)\exp -\left(r + \frac{\sigma}{2}\right)^2 \tag{2.23}$$

$$< 2\exp\left(-\sigma + \frac{\sigma^2}{4}\right)\exp - (r/2)^2, \quad r > -\sigma \tag{2.24}$$

The terms in the series $\sum_{r=1}^{\infty} \exp -(r/2)^2$ are among those in the series $\sum_{r=1}^{\infty} \exp(-r/4)$: this last series has positive terms and is convergent, hence the first series is convergent. It follows that the defining integral for $\bar{f}(s)$ is absolutely convergent for all s. That is to say, the strip of absolute convergence is $(-\infty, \infty)$, and is wider than the strip of exponential convergence.

EXERCISE 2.1 Let f and g be two functions in D_R^1, and let the corresponding strips of exponential convergence be (α_f, β_f), (α_g, β_g). Show that if these strips have a non-null intersection (α, β) then the function $h = c_1 f + c_2 g$ has Laplace transform $\bar{h} = c_1\bar{f} + c_2\bar{g}$ with exponential convergence in (α, β). Here c_1 and c_2 are arbitrary real constants.

2.1 Special cases

There are two important special cases of the two-sided Laplace transform. Suppose first that the strip of exponential convergence corresponding to a function f includes the imaginary axis, Re $s = 0$. Then the two-sided Laplace transform exists on this axis, and the restricted transform $\bar{f}(0 + i\omega)$ is known as the *exponential Fourier transform* of f:

$$\bar{f}(0 + i\omega) = \int_{-\infty}^{\infty} f(t)e^{-i\omega t}\, dt \tag{2.25}$$

$$= \int_{-\infty}^{\infty} f(t) \cos \omega t\, dt - i \int_{-\infty}^{\infty} f(t) \sin \omega t\, dt \tag{2.26}$$

The two integrals in (2.26) are known as the *Fourier cosine transform* and the *Fourier sine transform*.

Note that there is some divergence in the literature concerning the definition of the Fourier transform. For example the definition given in (2.25) is sometimes replaced by

$$\frac{1}{\sqrt{(2\pi)}} \int_{-\infty}^{\infty} f(t)e^{-i\omega t}\, dt \tag{2.27}$$

Also, the integral in (2.25) may converge even when the imaginary axis is not within the strip of exponential convergence (compare Example 2.4).

A second particular case of the two-sided Laplace transform arises when the functions f satisfy $f(t) = 0$, $t \leqslant 0$. This condition, which arises naturally when non-anticipative systems are disturbed from rest, has already been used in Chapter 5. The two-sided Laplace transform is $\bar{f}$ defined by

$$\bar{f}(s) = \int_{-\infty}^{\infty} e^{-st} f(t)\, dt$$

$$= \int_{0}^{\infty} e^{-st} f(t)\, dt \tag{2.28}$$

The last integral in (2.28) defines the *one-sided* (or *unilateral*) *Laplace transform*. The strip of exponential convergence now extends to infinity on the right; that is, if there is a strip of exponential convergence it has the form (α, ∞).

The one-sided Laplace transform is the most natural form to use for the analysis of linear time-invariant systems. In the remainder of this book, 'Laplace transform' used without qualification will mean the one-sided Laplace transform. It is naturally implied that the functions f which are transformed satisfy $f(t) = 0$, $t \leqslant 0$.

Example 2.5 The Laplace transform of the unit step $U(t - \tau)$ with $\tau \geqslant 0$ is

$$\mathscr{L}\{U(t - \tau)\} = \int_{\tau}^{\infty} e^{-st}\, dt = e^{-s\tau}/s \tag{2.29}$$

with exponential convergence in $(0, \infty)$.

Example 2.6 The Laplace transform of the delta function $\delta(t - \tau)$ with $\tau \geqslant 0$ is defined formally by

$$\mathscr{L}\{\delta(t - \tau)\} = \int_0^\infty \delta(t - \tau)e^{-st}\, dt \tag{2.30}$$

which by Section 8 of Chapter 5 can be interpreted to mean

$$\mathscr{L}\{\delta(t - \tau)\} = \int_0^\infty e^{-st}\, dU(t - \tau) \tag{2.31}$$

$$= e^{-s\tau} \tag{2.32}$$

with exponential convergence in $(-\infty, \infty)$.

EXERCISE 2.2 Show that if $\mathscr{L}\{f(t)\} = \bar{f}(s)$, with exponential convergence in (α, ∞), then $\mathscr{L}\{e^{-at} f(t)\} = \bar{f}(s + a)$ with exponential convergence in $(\alpha - a, \infty)$. Here a is a real constant.

EXERCISE 2.3 Show that if $\mathscr{L}\{f(t)\} = \bar{f}(s)$, with exponential convergence in (α, ∞), then $\mathscr{L}\{f(t - \tau)\} = e^{-s\tau}\bar{f}(s)$, $\tau > 0$, with the same strip of exponential convergence. [Make a change of variable in the defining integral for $\mathscr{L}\{f(t - \tau)\}$.]

EXERCISE 2.4 Verify the Laplace transforms and strips of exponential convergence in the accompanying table. [Use the exponential equivalents of the circular functions, the result of Exercise 2.2, and integration by parts.]

$f(t)$	$\bar{f}(s)$	*Strip of exponential convergence*
$U(t)e^{-at}$	$\dfrac{1}{s + a}$	$(-a, \infty)$
$U(t) \sin \omega t$	$\dfrac{\omega}{s^2 + \omega^2}$	$(0, \infty)$
$U(t) \cos \omega t$	$\dfrac{s}{s^2 + \omega^2}$	$(0, \infty)$
$U(t)\, e^{-at} \sin \omega t$	$\dfrac{\omega}{(s + a)^2 + \omega^2}$	$(-a, \infty)$
$U(t)\, e^{-at} \cos \omega t$	$\dfrac{s + a}{(s + a)^2 + \omega^2}$	$(-a, \infty)$
$U(t)e^{-at}\left\{\cos \omega t + \dfrac{b - a}{\omega} \sin \omega t\right\}$, $a = \frac{1}{2}c_1$, $\omega^2 = c_0 - \frac{1}{4}c_1^2$	$\dfrac{s + b}{s^2 + c_1 s + c_0}$	$(-a, \infty)$

$f(t)$	$\bar{f}(s)$	*Strip of exponential convergence*
$U(t)t^r e^{-at}$	$\dfrac{r!}{(s+a)^{r+1}}$	$(-a, \infty)$
$U(t)t^r \sin \omega t$	$\dfrac{r!(r+1)!}{(s^2+\omega^2)^{r+1}} \times \left\{\dfrac{s^r\omega}{r!1!} - \dfrac{s^{r-2}\omega^3}{(r-2)!3!} + \cdots\right\}$	$(0, \infty)$
$U(t)t^r \cos \omega t$	$\dfrac{r!(r+1)!}{(s^2+\omega^2)^{r+1}} \times \left\{\dfrac{s^{r+1}}{(r+1)!} - \dfrac{s^{r-1}\omega^2}{(r-1)!2!} + \cdots\right\}$	$(0, \infty)$

2.2 Laplace transform of a vector or matrix

So far, it has been assumed that f has domain E_1 and range $R \subseteq E_1$. If instead $R \subseteq E_n$, so that f is an n-vector, suppose that each element f_i of f satisfies

$$f_i(t) = 0, \quad t \leqslant 0; \qquad |f_i(t)| < Me^{\alpha t}, \quad t > 0 \tag{2.33}$$

Then each element f_i has Laplace transform $\bar{f}_i$, exponentially convergent for $\operatorname{Re} s > \alpha$. The vector $\bar{f}$ with elements $\bar{f}_i$ is called the Laplace transform of f. In the same way, if F is a matrix with its elements f_{ij} satisfying conditions similar to (2.33), the matrix $\bar{F}$ with elements $\bar{f}_{ij}$ is the Laplace transform of F.

Example 2.7 The Laplace transform of F defined by

$$F(t) = \begin{pmatrix} U(t) & U(t)e^{-at} \\ U(t)\cos \omega t & U(t)\sin \omega t \end{pmatrix} \tag{2.34}$$

where $a \geqslant 0$, is $\bar{F}$ defined by

$$\bar{F}(s) = \begin{pmatrix} \dfrac{1}{s} & \dfrac{1}{s+a} \\ \dfrac{s}{s^2+\omega^2} & \dfrac{\omega}{s^2+\omega^2} \end{pmatrix} \tag{2.35}$$

with strip of exponential convergence $(0, \infty)$.

3 Laplace transform of the derivative

Let $f \in D_R^1$ satisfy $f(t) = 0$, $t \leqslant 0$, and let the strip of exponential convergence for $\bar{f}(s)$ be (α, ∞). Then f has a left-hand derivative D_-f at every point. The derivative Df does not exist in the ordinary sense where f is discontinuous, but if f is written

$$f(t) = \phi(t) + \sum_i f_i U(t - p_i) \tag{3.1}$$

then with the convention of Chapter 5, Section 8, Df may be defined by

$$Df(t) = D_-\phi(t) + \sum_i f_i \,\delta(t - p_i) \tag{3.2}$$

The question we wish to answer is, what is the relation between $\bar{f}$ and the Laplace transform of Df, assuming the latter to exist in the ordinary sense or in the sense of equation (3.2)?

By equation (7.25) of Chapter 5

$$\int_0^b e^{-st}\, D_-f(t)\, dt - \int_0^b se^{-st} f(t)\, dt = f(b)e^{-sb} - \sum_i f_i e^{-sp_i} \tag{3.3}$$

because $f(0) = 0$ and e^{-st} has no discontinuities. The sum over i includes all discontinuities of f in $[0, b]$: we need not insist that the discontinuities of f are finite in number, but only that there is a finite number in any finite interval.

Now by definition,

$$\bar{f}(s) = \lim_{b\to\infty} \int_0^b e^{-st} f(t)\, dt \tag{3.4}$$

and the limit exists for s in the strip of exponential convergence. So from (3.3)

$$\zeta(s) \triangleq \lim_{b\to\infty} \int_0^b e^{-st}\, D_-f(t)\, dt \tag{3.5}$$

$$= s\bar{f}(s) - \lim_{b\to\infty} \sum_i f_i e^{-sp_i}, \quad \text{Re}\, s > \alpha \tag{3.6}$$

provided that the limits exist, because

$$\lim_{b\to\infty} f(b)e^{-sb} = 0, \quad \text{Re}\, s > \alpha \tag{3.7}$$

In (3.5), $\zeta(s)$ is by definition the Laplace transform of $D_-f(t)$. It follows from (3.3) and (3.7) that if the sum in (3.6) is finite, the integral in (3.5) converges for all s in the strip (α, ∞). If the sum in (3.6) is infinite, and if it converges for $\text{Re}\, s > \alpha_1 \geqslant \alpha$, then the integral in (3.5) converges in the strip (α_1, ∞).

Exercise 3.1 If $f \in D_R^1$, it has bounded variation on any finite interval and can therefore be expressed as the difference of two non-decreasing functions, $f(t) = g(t) - h(t)$, where $g(t)$ and $h(t)$ can be taken to have the value zero for $t \leqslant 0$. If the strips of exponential convergence for g and h have a non-null intersection (α_1, ∞), show that the defining integral for $\mathscr{L}\{D_-f(t)\}$ is absolutely convergent for s in the strip (α_1, ∞). [For the non-decreasing positive function

$g \in D^1_R$, $\sum_{i=1}^n g_i \leqslant g(p_n +) < Me^{\alpha_1 p_n}$ and so $g_n < Me^{\alpha_1 p_n}$, whence $\sum_i g_i\, e^{-\sigma p_i}$ is convergent for $\sigma > \alpha_1$. Then use $|D_- f(t)| \leqslant D_- g(t) + D_- h(t)$.]

If now f has a derivative in the usual sense at every point, so that $Df = D_- f$ and f is necessarily continuous, (3.6) gives

$$\mathscr{L}\{Df(t)\} = \zeta(s) = s\bar{f}(s) \tag{3.8}$$

Secondly, if f is continuous but its derivative Df does not exist at certain points where it is discontinuous, (3.6) gives for the Laplace transform of $D_- f(t)$

$$\mathscr{L}\{D_- f(t)\} = s\bar{f}(s) \tag{3.9}$$

Thirdly, if f is discontinuous at the points p_i, (3.6) gives

$$\mathscr{L}\{D_- f(t)\} = s\bar{f}(s) - \sum_i f_i e^{-sp_i} \tag{3.10}$$

These results, (3.8), (3.9), and (3.10), are all firmly based on the analysis given in Sections 2–5 of Chapter 5, and make no use of delta functions. If, however, f is discontinuous and Df is defined by (3.2), then

$$\mathscr{L}\{Df(t)\} = \mathscr{L}\{D_- f(t)\} + \mathscr{L}\{\sum_i f_i\, \delta(t - p_i)\} \tag{3.11}$$

On using (3.6) and (2.32) this becomes

$$\mathscr{L}\{Df(t)\} = s\bar{f}(s) \tag{3.12}$$

showing the simplification of formulae attained in the theory of distributions or its equivalents (compare Sections 8 and 9 of Chapter 4). Equation (3.12) will not be used in any essential way in subsequent developments.

Example 3.1 If f is the unit step $U(t)$, $D_- f(t) = 0$, all t. By equation (2.29), $\bar{f}(s) = 1/s$ with exponential convergence in $(0, \infty)$. The only discontinuity of f is $f_1 = 1$ at $t = 0$. Therefore (3.10) becomes

$$\mathscr{L}\{D_- f(t)\} = 0 = s(1/s) - 1 \tag{3.13}$$

which is clearly correct. Equation (3.12) gives

$$\mathscr{L}\{Df(t)\} = 1 \tag{3.14}$$

absolutely convergent at least in $(0, \infty)$ by Exercise 3.1. Compare this with (2.32).

Example 3.2 Let f be defined by

$$f(t) = U(t)e^{-at} \tag{3.15}$$

so that (2.17) gives $\bar{f}(s) = 1/(s + a)$ with exponential convergence in $(-a, \infty)$. Then $D_- f(t) = -aU(t)e^{-at}$ and (3.10) gives

$$\mathscr{L}\{-aU(t)e^{-at}\} = \frac{s}{s + a} - 1 = \frac{-a}{s + a} \tag{3.16}$$

which is correct. Equation (3.12) gives

$$\mathscr{L}\{DU(t)e^{-at}\} = \frac{s}{s+a} \tag{3.17}$$

which is readily confirmed. Note that Exercise 3.1 for this example assures absolute convergence only in $(0, \infty)$, whereas absolute convergence holds in fact in the wider strip $(-a, \infty)$.

It is perhaps worth noticing that in the usual treatments, in which delta functions can occur at $t = 0$, the defining integral (2.28) is replaced by

$$\int_{0-}^{\infty} s^{-st} f(t)\, dt \tag{3.18}$$

in order to 'include the delta function in the range of integration'. Alternatively, the lower limit of integration may be made $0+$, and the effect of a delta function at the origin included separately. The defining integral is, in either case, improper in regard to both limits. (Compare Zadeh and Desoer, 1963, Appendix B.)

4 Uniqueness

It is clear from (2.28) that to any function $f \in D_R^1$ satisfying the conditions

$$f(t) = 0, \quad t \leqslant 0 \tag{4.1}$$

$$|f(t)| < Me^{\alpha t}, \quad t > 0 \tag{4.2}$$

there corresponds a unique Laplace transform $\bar{f}$ having exponential convergence in the strip (α, ∞). We are interested in the converse question: under what conditions does there exist a unique function f corresponding to a given $\bar{f}$ having exponential convergence in the strip (α, ∞)? This is answered by the following result.

Theorem 4.1 Let $f_1 \in D_R^1$ and $f_2 \in D_R^1$ be two functions, satisfying

$$f_1(t) = f_2(t) = 0, \quad t \leqslant 0 \tag{4.3}$$

and, for some M and α,

$$|f_1(t)| < Me^{\alpha t}, \quad t > 0 \tag{4.4}$$

$$|f_2(t)| < Me^{\alpha t}, \quad t > 0 \tag{4.5}$$

and let the Laplace transforms of f_1 and f_2 be equal in the strip (α, ∞). Then $f_1(t) = f_2(t)$, all t.

†*Proof* Write $g = f_1 - f_2$, so that

$$g(t) = 0, \quad t \leqslant 0 \tag{4.6}$$

$$|g(t)| < 2Me^{\alpha t}, \quad t > 0 \tag{4.7}$$

†May be omitted at first reading.

The first stage in the proof is to show that for any real $\sigma < \alpha$, all the moments $\int_0^\infty t^r e^{-\sigma t} g(t)\,dt$, $r = 0, 1, 2, \ldots$, are zero. From (4.7), if $\alpha_1 > \alpha$ and $r \geqslant 1$,

$$|t^r g(t)| < 2t^r e^{-(\alpha_1 - \alpha)t} M e^{\alpha_1 t} < R e^{\alpha_1 t}, \quad t > 0 \tag{4.8}$$

because $t^r e^{-(\alpha_1-\alpha)t}$ is bounded for $t > 0$. In (4.8), R depends upon r and α_1, but α_1 may be chosen arbitrarily subject to $\alpha_1 > \alpha$. Also, by the hypotheses of the theorem, g has Laplace transform $\bar{g}$ satisfying

$$\bar{g}(s) = 0, \quad \text{Re}\, s > \alpha \tag{4.9}$$

Then if s_0 is some value of s, with Re $s_0 > \alpha$, $\bar{g}$ has a derivative at s_0 (see Chapter 7, Section 2.1) and

$$\bar{g}'(s_0) = 0 \tag{4.10}$$

Now let h be real and satisfy $|h| < \text{Re}\, s_0 - \alpha$, and consider

$$\bar{g}(s_0 + h) - \bar{g}(s_0) = \int_0^\infty e^{-(s_0+h)t} g(t)\,dt - \int_0^\infty e^{-s_0 t} g(t)\,dt \tag{4.11}$$

$$= \lim_{a\to\infty} \int_0^a e^{-(s_0+h)t} g(t)\,dt - \lim_{a\to\infty} \int_0^a e^{-s_0 t} g(t)\,dt$$

$$= \lim_{a\to\infty} \left\{ \int_0^a e^{-s_0 t}(e^{-ht} - 1) g(t)\,dt \right\} \tag{4.12}$$

$$= \lim_{a\to\infty} \left\{ \int_0^a [-hte^{-s_0 t} g(t) + (e^{-ht} - 1 + ht) e^{-s_0 t} g(t)]\,dt \right\} \tag{4.13}$$

$$= -h \lim_{a\to\infty} \int_0^a e^{-s_0 t} t g(t)\,dt + \lim_{a\to\infty} \int_0^a (e^{-ht} - 1 + ht) e^{-s_0 t} g(t)\,dt \tag{4.14}$$

where properties of limits and of the integral have been used. In (4.14) note that the first term on the right defines the Laplace transform of $tg(t)$ at s_0 which certainly exists because, by (4.8), $|tg(t)| < M_1 e^{\alpha_1 t}$, $t > 0$, where $\alpha < \alpha_1 < \text{Re}\, s_0$. Consequently the limit of the second integral also exists, and the passage from (4.13) to (4.14) is justified.

Next it will be shown that

$$\left| \lim_{a\to\infty} \int_0^a (e^{-ht} - 1 + ht) e^{-s_0 t} g(t)\,dt \right| < M_2 h^2 \tag{4.15}$$

Direct evaluation shows that

$$e^{-ht} - 1 + ht \triangleq \phi(t) = h^2 \int_0^t \left\{ \int_0^x e^{-hy}\,dy \right\} dx \tag{4.16}$$

and it follows from the fact that $0 < e^{-hy} \leqslant 1$, $y \geqslant 0$, that

$$0 \leqslant \phi(t) \leqslant h^2 \int_0^t \left\{\int_0^x dy\right\} dx = h^2t^2/2 \tag{4.17}$$

From (4.16) and (4.17), and equation (4.34) of Chapter 5,

$$\left|\int_0^a \phi(t)e^{-s_0t}g(t)\,dt\right| \leqslant \frac{1}{2}h^2 \int_0^a e^{-\sigma t}\,|t^2g(t)|\,dt \tag{4.18}$$

where $\sigma = \text{Re}\, s_0$, and because the 'modulus' function is continuous

$$\left|\lim_{a\to\infty} \int_0^a \phi(t)e^{-s_0t}g(t)\,dt\right| = \lim_{a\to\infty}\left|\int_0^a \phi(t)e^{-s_0t}g(t)\,dt\right|$$

$$\leqslant \tfrac{1}{2}h^2 \lim_{a\to\infty} \int_0^a e^{-\sigma t}\,|t^2g(t)|\,dt \tag{4.19}$$

$$< h^2M_2 \tag{4.20}$$

where M_2 is the Laplace transform of $|t^2g(t)|$ evaluated at the real value σ of s: this transform exists because, by (4.8), $|t^2g(t)| < M_3\,e^{\alpha_1 t}$, $t > 0$, where $\alpha < \alpha_1 < \sigma$.

On using (4.20) in (4.14), for $\text{Re}\, s_0 > \alpha$,

$$0 = \bar{g}'(s_0) = \lim_{h\to 0} \frac{\bar{g}(s_0 + h) - \bar{g}(s_0)}{h}$$

$$= -\int_0^\infty e^{-s_0t}tg(t)\,dt = -\,\bar{h}(s_0) \tag{4.21}$$

where $\bar{h}(s_0)$ is the transform of $h(t) = tg(t)$ evaluated at $s = s_0$. Now $\bar{h}(s)$ can be substituted for $\bar{g}(s)$ in (4.9), and the argument can be repeated to give, for $\text{Re}\, s_0 > \alpha$,

$$0 = \bar{g}''(s_0) = \int_0^\infty e^{-s_0t}\,t^2g(t)\,dt \tag{4.22}$$

$$\vdots$$

$$0 = \bar{g}^{(r)}(s_0) = (-)^r \int_0^\infty e^{-s_0t}\,t^rg(t)\,dt \tag{4.23}$$

In particular, equations (4.21), (4.22), and (4.23) are true if s_0 is given the real value $\sigma > \alpha$. This shows that all the moments of $e^{-\sigma t}g(t)$ are zero. The remaining part of the proof, which is somewhat tedious, is to show that this implies $g(t) = 0$ for all t.

Suppose, contrary to what is to be proved, that $g(t_0) = 2\eta > 0$ for some $t_0 > 0$; a similar argument is used when $g(t_0) < 0$. Then because $g \in D_R^{\frac{1}{R}}$ there is an $\epsilon_1 > 0$ such that g is continuous on $[t_0 - \epsilon_1, t_0]$. Moreover, there is an $\epsilon_2 > 0$ such that g has continuous first derivative on $(t_0 - \epsilon_2, t_0)$, satisfying $|Dg| < M_4$. If $\epsilon_3 = \eta/M_4$ and $\epsilon = \min(\epsilon_1, \epsilon_2, \epsilon_3, \tfrac{1}{2}t_0)$, it readily follows that

$$g(t) > \eta, \quad t_0 - \epsilon \leqslant t \leqslant t_0 \tag{4.24}$$

It will now be shown that (4.23) and (4.24) imply a contradiction. First, put $q = t_0 - \frac{1}{2}\epsilon > 0$, and make the change of variable $qu = t$ in (4.23). Choose $s_0 = \beta + r/q$, where $\beta > \max\{\alpha_1, 0\}$, so that after multiplication by $r^{r+1}/r!q^{r+1}$, (4.23) gives

$$\frac{r^{r+1}}{r!}\int_0^\infty e^{-ru}u^r g(qu)e^{-\beta qu}\,du = 0 \tag{4.25}$$

while (4.24) gives

$$g(qu)e^{-\beta qu} > \eta_1, \qquad 1 - \delta \leqslant u \leqslant 1 + \delta \tag{4.26}$$

where

$$\eta_1 = \eta e^{-\beta q(1+\delta)} > 0, \qquad \delta = \epsilon/2q > 0 \tag{4.27}$$

Also by (4.7)

$$|g(qu)e^{-\beta qu}| < 2Me^{-(\beta-\alpha)qu} < 2M, \quad u > 0 \tag{4.28}$$

because $\beta > \alpha_1 > \alpha$.

In the lemma which follows this proof, the following results are obtained.

$$\frac{r^{r+1}}{r!}\int_0^\infty e^{-ru}u^r\,du = 1, \quad \text{all } r \tag{4.29}$$

$$\lim_{r\to\infty}\left\{\frac{r^{r+1}}{r!}\int_0^{1-\delta} e^{-ru}u^r\,du\right\} = 0 \tag{4.30}$$

$$\lim_{r\to\infty}\left\{\frac{r^{r+1}}{r!}\int_{1+\delta}^\infty e^{-ru}u^r\,du\right\} = 0 \tag{4.31}$$

These results imply that $(r^{r+1}/r!)e^{-ru}u^r$ behaves, as $r \to \infty$, more and more like a delta function at $u = 1$. This is the idea underlying the remaining part of the proof.

From (4.30) and (4.31) it follows that there exists an r_0 such that for all $r > r_0$ the expressions within braces in those two equations (which are positive) are each less than $\frac{1}{5}$. Hence for $r > r_0$,

$$\begin{aligned}\frac{r^{r+1}}{r!}\int_{1-\delta}^{1+\delta} e^{-ru}u^r\,du &= \frac{r^{r+1}}{r!}\left\{\int_0^\infty - \int_0^{1-\delta} - \int_{1+\delta}^\infty\right\}e^{-ru}u^r\,du \\ &> \frac{3}{5}\end{aligned} \tag{4.32}$$

and by (4.26)

$$\frac{r^{r+1}}{r!}\int_{1-\delta}^{1+\delta} e^{-ru}u^r g(qu)e^{-\beta qu}\,du > \frac{3}{5}\eta_1 > 0 \tag{4.33}$$

On the other hand, by (4.28), (4.30), and (4.31) there exists an r_1 such that for all $r > r_1$

$$\left| \frac{r^{r+1}}{r!} \int_0^{1-\delta} e^{-ru} u^r g(qu) e^{-\beta q u}\, du \right| < \frac{1}{5} \eta_1 \tag{4.34}$$

and there is an r_2 such that for all $r > r_2$

$$\left| \frac{r^{r+1}}{r!} \int_{1+\delta}^{\infty} e^{-ru} u^r g(qu) e^{-\beta q u}\, du \right| < \frac{1}{5} \eta_1 \tag{4.35}$$

From (4.33), (4.34), and (4.35), for $r > \max\{r_0, r_1, r_2\}$,

$$\frac{r^{r+1}}{r!} \int_0^{\infty} e^{-ru} u^r g(qu) e^{-\beta q u}\, du > \frac{1}{5} \eta_1 > 0 \tag{4.36}$$

which contradicts (4.25). This completes the proof.

Lemma Integrate by parts in (4.29) to obtain

$$\frac{r^{r+1}}{r!} \left\{ \left[-\frac{e^{-ru}}{r} u^r \right]_0^{\infty} + \int_0^{\infty} \frac{e^{-ru}}{r} r u^{r-1}\, du \right\} \tag{4.37}$$

and note that $u^r e^{-ru} \to 0$ as $u \to \infty$, so that the integrated part in (4.37) is zero. Proceed in this way to obtain

$$\frac{r^{r+1}}{r!} \left\{ \frac{r}{r} \times \frac{r-1}{r} \times \cdots \times \frac{1}{r} \int_0^{\infty} e^{-ru}\, du \right\} = 1 \tag{4.38}$$

Stirling's series (Bromwich, 1926, p. 330) will be assumed here as a known result. It shows that

$$\log(r!) > (r + \tfrac{1}{2}) \log r - r \tag{4.39}$$

Hence

$$\phi(r, u) \triangleq \log \left[\frac{r^{r+1}}{r!} e^{-ru} u^r \right] < \tfrac{1}{2} \log r + r[1 - u + \log u] \tag{4.40}$$

Put $u = 1 + x$ and use the series expansion of $\log(1 + x)$ (Chapter 7, Example 3.2),

$$\log(1 + x) = x - \tfrac{1}{2}x + \tfrac{1}{3}x^3 - \cdots \tag{4.41}$$

convergent for $|x| < 1$, to obtain

$$\phi(r, 1 + x) < \tfrac{1}{2} \log r - [\tfrac{1}{2}x^2 - \tfrac{1}{3}x^3 + \cdots] r \tag{4.42}$$

Now the series multiplying r in (4.42) can be written

$$\begin{aligned} (\tfrac{1}{2}x^2 - \tfrac{1}{3}x^3) + (\tfrac{1}{4}x^4 - \tfrac{1}{5}x^5) + \cdots & \\ > \tfrac{1}{2}x^2 - \tfrac{1}{3}x^3, \quad |x| < 1 & \\ > \tfrac{1}{6}x^2, \quad |x| < 1 & \end{aligned} \tag{4.43}$$

and so, because the exponential function is monotonic increasing,

$$\exp \phi(r, u) < r^{1/2} \exp\left(-\frac{rx^2}{6}\right), \quad |x| < 1, \quad r \geqslant 1 \tag{4.44}$$

which by Exercise 12.1 of Chapter 4 shows that

$$\frac{r^{r+1}}{r!} e^{-ru} u^r \to 0 \quad \text{as } r \to \infty \tag{4.45}$$

when $u = 1 - \delta$, $0 < \delta < 1$. The last condition is satisfied by the δ defined in (4.27) because $q = t_0 - \frac{1}{2}\epsilon$, $0 < \epsilon \leqslant \frac{1}{2}t_0$, $0 < \delta = \epsilon/2q \leqslant \frac{1}{3}$.

Now for $r \geqslant 1$

$$\frac{\partial}{\partial u} e^{-ru} u^r = r u^{r-1} e^{-ru}(1 - u) \tag{4.46}$$

$$> 0, \quad 0 < u < 1 \tag{4.47}$$

Given any $\epsilon_4 > 0$, there exists by (4.45) an $r_3 > 1$ such that for all $r > r_3$,

$$\frac{r^{r+1}}{r!} e^{-ru} u^r < \epsilon_4, \quad u = 1 - \delta \tag{4.48}$$

By (4.47) and the mean-value theorem, and the fact that $e^{-ru}u^r \geqslant 0$, $u \geqslant 0$, the inequality in (4.48) holds for all u in $[0, 1 - \delta]$. It follows that

$$\int_0^{1-\delta} \frac{r^{r+1}}{r!} e^{-ru} u^r \, du < (1 - \delta)\epsilon_4 \tag{4.49}$$

which shows that (4.30) is true.

Finally, for $r \geqslant 1$, (4.40) gives

$$0 < \frac{r^{r+1}}{r!} e^{-ru} u^r < r^{1/2} u^r e^{r(1-u)} \tag{4.50}$$

whence, integrating by parts,

$$\frac{r^{r+1}}{r!} \int_{1+\delta}^{\infty} e^{-ru} u^r \, du < e^r r^{1/2} \left\{ -\left[\frac{e^{-ru}}{r} u^r \right]_{1+\delta}^{\infty} + \int_{1+\delta}^{\infty} \frac{e^{-ru}}{r} r u^{r-1} \, du \right\} \tag{4.51}$$

$$= e^r r^{1/2} e^{r(1+\delta)} \left\{ \frac{(1+\delta)^r}{r} + \frac{r(1+\delta)^{r-1}}{r^2} + \frac{r(r-1)(1+\delta)^{r-2}}{r^3} + \cdots + \frac{r!}{r^{r+1}} \right\} \tag{4.52}$$

$$< e^{-r\delta} r^{1/2} (r + 1) \frac{(1+\delta)^r}{r} \tag{4.53}$$

$$\leqslant 2r^{1/2} \exp r[-\delta + \log(1 + \delta)] \tag{4.54}$$

For $|\delta| < 1$,

$$-\delta + \log(1 + \delta) = -\delta + \delta - \frac{\delta^2}{2} + \frac{\delta^3}{3} - \cdots \tag{4.55}$$

$$= -\left(\frac{\delta^2}{2} - \frac{\delta^3}{3}\right) - \left(\frac{\delta^4}{4} - \frac{\delta^5}{5}\right) - \cdots \tag{4.56}$$

$$< 0$$

and with (4.54) and Exercise 12.1 of Chapter 4 this shows that (4.31) is true. This concludes the lemma.

EXERCISE 4.1 Plot the function ψ defined by

$$\psi(r, u) = \frac{r^{r+1}}{r!} e^{-ru} u^r \tag{4.57}$$

for $r = 2, 4, 6$ and for $0 \leqslant u \leqslant 2$.

5 Solution of linear time-invariant differential equations

It was shown in Section 2 of Chapter 6 that if ξ is an n-vector and A a constant $n \times n$ matrix, the equation

$$\dot{\xi}(t) = A\xi(t); \quad \xi(0) = c \tag{5.1}$$

has a unique solution which exists for all t and can be written explicitly

$$\xi(t) = e^{At}c \tag{5.2}$$

Though (5.2) expresses the solution of (5.1) in a compact form, it does not allow this solution to be written easily in terms of tabulated functions. For this purpose the Laplace transform can be used.

Since the one-sided Laplace transform demands that the functions considered should be zero for $t \leqslant 0$, let x be the function defined by

$$x(t) = \xi(t)U(t) \tag{5.3}$$

where $\xi(t)$ satisfies (5.1). It is clear from (5.2) and (5.3) that $x \in D_R^1$. Also by Theorem 13.5 of Chapter 4 there exists a constant $M(\alpha)$ such that

$$|\xi_i(t)| < Me^{\alpha t}, \quad t > 0, \quad \text{all } i \tag{5.4}$$

where α is any constant greater than ρ, and ρ is the greatest real part of an eigenvalue of A. It follows that x has a Laplace transform with strip of exponential convergence at least (ρ, ∞). The strip may in fact be larger than this if the solution of (5.1) contains no term with exponent depending on the eigenvalue(s) with real part ρ.

From (5.3), $D_-x(t)$ exists for all t, and

$$\left.\begin{aligned} D_-x(t) &= D_-\xi(t) = D\xi(t) = A\xi(t) = Ax(t), \quad t > 0 \\ D_-x(t) &= Ax(t) = 0, \quad t \leqslant 0 \end{aligned}\right\} \tag{5.5}$$

From (5.5) it follows that D_-x has a Laplace transform with strip of exponential convergence at least (ρ, ∞). Equation (3.10) gives

$$\mathscr{L}\{D_-x(t)\} = s\bar{x}(s) - c \tag{5.6}$$

where c is the jump of x at its discontinuity at $t = 0$. Equations (5.5) give

$$\mathscr{L}\{D_-x(t)\} = A\int_0^\infty e^{-st}x(t)\,dt \tag{5.7}$$

$$= A\bar{x}(s) \tag{5.8}$$

Combining (5.6) and (5.8),

$$(sI - A)\bar{x}(s) = c \tag{5.9}$$

$$\bar{x}(s) = (sI - A)^{-1}c \tag{5.10}$$

The elements of the vector $(sI - A)^{-1}c$ are rational functions of s. These may be put into partial fractions, and the fractions may then be found in a table of Laplace transforms. The function of time so obtained is in D_R^1 and is exponentially bounded, and has Laplace transform $\bar{x}$ satisfying (5.9) which is a necessary condition for the solution $x(t) = \xi(t)U(t)$ of (5.1) for $t > 0$. It was shown in Section 4 that there is no other exponentially bounded $\phi \in D_R^1$ having the same Laplace transform $\bar{\phi} = \bar{x}$. It follows that the function of time obtained by means of the table of transforms is in fact the required solution defined by $x(t) = \xi(t)U(t)$.

Example 5.1 Let ξ satisfy

$$\dot{\xi}_1 = \xi_2; \qquad \xi_1(0) = c_1 \tag{5.11}$$

$$\dot{\xi}_2 = -a_0\xi_1 - a_1\xi_2; \qquad \xi_2(0) = c_2 \tag{5.12}$$

or, what is the same thing,

$$\ddot{\xi}_1 + a_1\dot{\xi}_1 + a_0\xi_1 = 0; \qquad \xi_1(0) = c_1, \qquad \dot{\xi}_1(0) = c_2 \tag{5.13}$$

Then

$$A = \begin{pmatrix} 0 & 1 \\ -a_0 & -a_1 \end{pmatrix} \tag{5.14}$$

and

$$\bar{x}(s) = \begin{pmatrix} s & -1 \\ a_0 & s + a_1 \end{pmatrix}^{-1} \begin{pmatrix} c_1 \\ c_2 \end{pmatrix} \tag{5.15}$$

$$= \frac{1}{s^2 + a_1 s + a_0} \begin{pmatrix} s + a_1 & 1 \\ -a_0 & s \end{pmatrix} \begin{pmatrix} c_1 \\ c_2 \end{pmatrix} \tag{5.16}$$

$$= \frac{1}{s^2 + a_1 s + a_0} \begin{pmatrix} c_1 s + a_1 c_1 + c_2 \\ c_2 s - a_0 c_1 \end{pmatrix} \tag{5.17}$$

From the table of transforms

$$x(t) = U(t)e^{-\sigma t} \begin{pmatrix} c_1 \cos \omega t + \dfrac{c_2 + \sigma c_1}{\omega} \sin \omega t \\ c_2 \cos \omega t - \dfrac{a_0 c_1 + \sigma c_2}{\omega} \sin \omega t \end{pmatrix} \tag{5.18}$$

where

$$\sigma = \tfrac{1}{2} a_1 \tag{5.19}$$

$$\omega^2 = a_0 - \tfrac{1}{4} a_1^2 \tag{5.20}$$

It is easily confirmed that in (5.18)

$$\dot{x}_1(t) = x_2(t), \quad t > 0 \tag{5.21}$$

If the problem is stated as in (5.13), it will usually be sufficient to obtain x_1: see Exercise 5.1.

EXERCISE 5.1. Show that if A is the companion matrix

$$A = \begin{pmatrix} 0 & 1 & 0 & \cdots & 0 \\ 0 & 0 & 1 & \cdots & 0 \\ \vdots & \vdots & \vdots & & \vdots \\ -a_0 & -a_1 & -a_2 & \cdots & -a_{n-1} \end{pmatrix} \tag{5.22}$$

then the element in position $(1, r)$ in $(sI - A)^{-1}$ is

$$\frac{s^{n-r} + a_{n-1}s^{n-r-1} + \cdots + a_r}{s^n + a_{n-1}s^{n-1} + \cdots + a_0}, \quad r = 1, 2, \ldots, n-1 \tag{5.23}$$

while when $r = n$ the numerator in (5.23) is 1. Hence show that if

$$\xi^{(n)} + a_{n-1}\xi^{(n-1)} + \cdots + a_1\xi + a_0\xi = 0 \tag{5.24}$$

$$\xi^{(n-1)}(0) = c_{n-1}, \quad \xi^{(n-2)}(0) = c_{n-2}, \quad \ldots, \quad \xi(0) = c_0 \tag{5.25}$$

the solution $x(t) = U(t)\xi(t)$ is given by solving for $\bar{x}(s)$ in (5.26) and finding the corresponding time function:

$$\begin{aligned} &[s^n\bar{x}(s) - s^{n-1}c_0 - \cdots - sc_{n-2} - c_{n-1}] \\ &+ a_{n-1}[s^{n-1}\bar{x}(s) - s^{n-2}c_0 - \cdots - c_{n-2}] \\ &+ \cdots \\ &+ a_1[s\bar{x}(s) - c_0] \\ &+ a_0\bar{x}(s) = 0 \end{aligned} \tag{5.26}$$

Deduce the usual rule for obtaining (5.26) from (5.24) and (5.25), namely that $D^r\xi$ in (5.24) is to be replaced in (5.26) by $s^r\bar{x}(s) - s^{r-1}c_0 - \cdots - sc_{r-2} - c_{r-1}$.

EXERCISE 5.2 If $x(t) = U(t)\xi(t)$, as in Example 5.1, show that the approach of Section 8, Chapter 5 leads formally to

$$\left.\begin{aligned} Dx(t) &= U(t)D\xi(t) + c_0\delta(t) \\ D^2x(t) &= U(t)D^2\xi(t) + c_0D\delta(t) + c_1\delta(t) \\ &\vdots \\ D^rx(t) &= U(t)D^r\delta(t) + c_0D^{r-1}\delta(t) + \cdots + c_{r-1}\delta(t) \end{aligned}\right\} \quad (5.27)$$

Using the results of Section 3 in a formal way, show that (5.27) leads to

$$\left.\begin{aligned} \mathscr{L}\{U(t)D\xi(t)\} &= s\bar{x}(s) - c_0 \\ \mathscr{L}\{U(t)D^2\xi(t)\} &= s^2\bar{x}(s) - c_0s - c_1 \\ &\vdots \\ \mathscr{L}\{U(t)D^r\xi(t)\} &= s^r\bar{x}(s) - c_0s^{r-1} - \cdots - c_{r-1} \end{aligned}\right\} \quad (5.28)$$

From (5.24) obtain

$$\int_0^\infty e^{-st}U(t)\{D^n\xi(t) + a_{n-1}D^{n-1}\xi(t) + \cdots + a_0\xi(t)\}\,dt = 0 \quad (5.29)$$

Use (5.28) in (5.29) and compare with (5.26). [As presented here, the approach in this Exercise is purely formal. It becomes legitimate in the theory of distributions. In comparing equations such as (5.28) with alternative treatments it should be remembered that the usual approach is to allow delta functions at $t = 0$, rather than at $t = 0+$ as (in effect) is done here, and to make the lower limit of the defining integral for the Laplace transform either $0-$ or $0+$ (rather than 0). To reconcile these differences, the typical equation in (5.28) is replaced, with a self-evident notation, by

$$\begin{aligned} \mathscr{L}_{0\pm}\{U(t)\,D^r\xi(t)\} &= \mathscr{L}_{0\pm}\{D^rx(t) - c_0\,D^{r-1}\,\delta(t) - \cdots - c_{r-1}\,\delta(t)\} \\ &\quad - s^{r-1}x(0\pm) - s^{r-2}\,Dx(0\pm) - \cdots - D^{r-1}x(0\pm) \end{aligned} \quad (5.30)$$

When the lower limit of the defining integral for the Laplace transform is $0-$, the delta functions in (5.30) contribute to the transform and $x(0-) = Dx(0-) = \ldots = D^{r-1}x(0-) = 0$. When the lower limit is $0+$, the delta functions do not contribute but $x(0+) = c_0$, $Dx(0+) = c_1, \ldots, D^{r-1}x(0+) = c_{r-1}$.]

5.1 Non-homogeneous equations

In Section 5 the homogeneous equation (5.1) was considered. We now turn to the non-homogeneous equation with zero initial conditions:

$$\dot{\xi}(t) = A\xi(t) + Bu(t); \qquad \xi(0) = 0 \quad (5.31)$$

where u is an l-vector of input functions. It is assumed that $u \in D_R^1$, $u(t) = 0$, $t \leqslant 0$, and $|u_i(t)| < M_1 e^{\alpha_1 t}$ for $t > 0$ and all i. Consequently, u will have a Laplace transform $\bar{u}$ with exponential convergence in the strip (α_1, ∞). Let x be defined by

$$x(t) = \begin{cases} 0, & t \leqslant 0 \\ \displaystyle\int_0^t e^{(t-\tau)A} Bu(\tau)\, d\tau, & t > 0 \end{cases} \tag{5.32}$$

Then it was shown in Section 2 of Chapter 6 that when $u \in D_R^1$ (and therefore $u \in D_R$) and $u(t) = 0$, $t \leqslant 0$, the function x given by (5.32) is continuous, belongs to D_R^1, and is the unique continuous function which satisfies the differential equation (5.31) at every point where u is continuous. Where u is discontinuous, $Dx(t)$ does not exist, but $D_-x(t)$ exists and is equal to $Ax(t) + Bu(t)$. With this understanding [which agrees with the usual physical basis of (5.31)] x may be equated to the solution ξ of (5.31).

Example 5.2 Let an input u defined by $u(t) = U(t - 1)$ be applied to the analogue circuit shown in Fig. 1, with zero initial conditions. The equation governing the system is usually written

$$\dot{x} = -ax + bu; \qquad x(0) = 0 \tag{5.33}$$

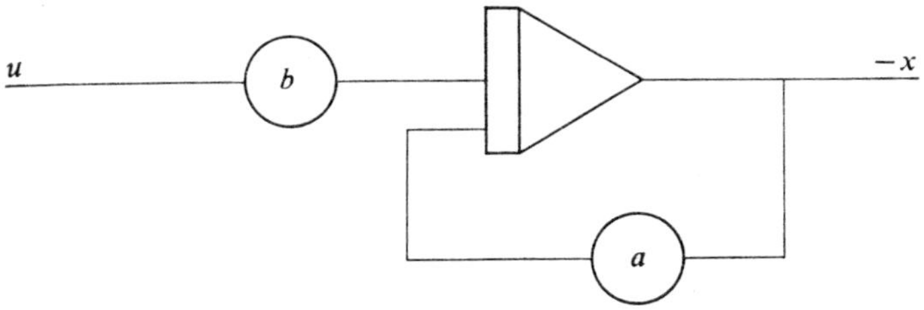

FIG. 1

and the output for the given u is, from (5.32),

$$x(t) = \begin{cases} 0, & t \leqslant 1 \\ \dfrac{b}{a}[1 - e^{-a(t-1)}], & t > 1 \end{cases} \tag{5.34}$$

It is easily verified that x satisfies (5.33) except at $t = 1$, where $D_-x(0) = 0$, $D_+x(0) = b$ and $Dx(0)$ does not exist. Physical consideration of the circuit shows that (5.33) should in fact be replaced, at a point t where u has a simple discontinuity on the right, by $D_-x(t) = -ax(t) + bu(t)$, x continuous at t. This reflects the non-anticipative nature of the system.

If $\Phi(t) = e^{At}$, then every element ϕ_{ij} of Φ satisfies

$$|\phi_{ij}(t)| < M_2 e^{\alpha_2 t}, \quad t > 0 \tag{5.35}$$

where M_2 is a constant and α_2 is greater than the greatest real part ρ of an eigenvalue of A. Consequently (5.32) gives, if $B = (b_{ij})$,

$$|x_i(t)| \leqslant \int_0^t \left| \sum_{j,k} \phi_{ij}(t - \tau) b_{jk} u_k(\tau) \right| d\tau \tag{5.36}$$

$$< \int_0^t nlbM_2 e^{\alpha_2(t-\tau)} M_1 e^{\alpha_1 \tau}\, d\tau$$

$$= M_3 e^{\alpha_2 t}\, [e^{(\alpha_1 - \alpha_2)t} - 1] \tag{5.37}$$

where $b = \max_{j,k} |b_{jk}|$. From (5.37)

$$|x_i(t)| < Me^{\alpha t}, \quad t > 0 \tag{5.38}$$

where $\alpha = \max\{\alpha_1, \alpha_2\}$. Hence $x \in D_R^1$, satisfying (5.38) and $x(t) = 0$, $t \leqslant 0$, has a Laplace transform with exponential convergence in a strip (α, ∞).

It follows from what was said earlier that

$$D_- x(t) = Ax(t) + Bu(t), \quad \text{all } t \tag{5.39}$$

Equation (3.9) gives

$$s\bar{x}(s) = A\bar{x}(s) + B\bar{u}(s) \tag{5.40}$$

valid for $\operatorname{Re} s > \alpha$, or

$$\bar{x}(s) = (sI - A)^{-1} B\bar{u}(s) \tag{5.41}$$

Equation (5.41) can be made the basis of a procedure for finding x in terms of elementary functions. This procedure relies on the fact, proved in Section 4, that given an $\bar{x}$ valid for $\operatorname{Re} s > \alpha$, and an exponentially bounded function $x \in D_R^1$ such that $\mathscr{L}\{x\} = \bar{x}$, then if $y \in D_R^1$ is exponentially bounded and $\mathscr{L}\{y\} = \bar{x}$ for $\operatorname{Re} s > \alpha_1$, $y = x$.

Example 5.3 Let x satisfy

$$\dot{x}_1 = x_2; \qquad x_1(0) = 0 \tag{5.42}$$

$$\dot{x}_2 = -a_0 x_1 - a_1 x_2 + bu; \quad x_2(0) = 0 \tag{5.43}$$

$$u(t) = tU(t - 1) \tag{5.44}$$

or, what is the same thing,

$$\ddot{x}_1(t) + a_1 \dot{x}_1(t) + a_0 x_1(t) = btU(t - 1); \qquad \dot{x}_1(0) = x_1(0) = 0 \tag{5.45}$$

Note that (5.44) and (5.45) must be replaced by an appropriate condition where $u(t)$ is discontinuous, as discussed above. First

$$\begin{aligned}\bar{u}(s) &= \mathscr{L}\{(t-1)U(t-1) + U(t-1)\} \\ &= e^{-s}\mathscr{L}\{tU(t) + U(t)\} \\ &= e^{-s}(1+s)/s^2 \qquad (5.46)\end{aligned}$$

where the results of Exercises 2.3 and 2.4 have been used. Then (5.41) gives, with (5.46),

$$\bar{x}(s) = \frac{e^{-s}(1+s)}{s^2(s^2+a_1s+a_0)}\begin{pmatrix} s+a_1 & 1 \\ -a_0 & s\end{pmatrix}\begin{pmatrix}0 \\ b\end{pmatrix} \qquad (5.47)$$

$$\begin{aligned}\bar{x}_1(s) &= \frac{be^{-s}(1+s)}{s^2(s^2+a_1s+a_0)} \\ &= be^{-s}\left\{\frac{-a_0 - a_0a_1 + a_1^2 - (a_0 - a_1)s}{a_0^2(s^2+a_1s+a_0)} + \frac{a_0 + (a_0-a_1)s}{a_0^2s^2}\right\} \qquad (5.48)\end{aligned}$$

where it is assumed that $s = 0$ is not a root of $s^2 + a_1s + a_0 = 0$. On again using results from Exercises 2.3 and 2.4,

$$\begin{aligned}x_1(t) = AU(t-1)e^{-\sigma(t-1)}&\left\{\cos\omega(t-1) + \frac{\beta-\sigma}{\omega}\sin\omega(t-1)\right\} \\ &+ B(t-1)U(t-1) + CU(t-1) \qquad (5.49)\end{aligned}$$

where

$$\begin{aligned}\sigma &= \tfrac{1}{2}a_1 \\ \omega^2 &= a_0 - \tfrac{1}{4}a_1^2 \\ A &= -b(a_0 - a_1)/a_0^2 \\ \beta &= (a_0 + a_0a_1 - a_1^2)/(a_0 - a_1) \\ B &= b/a_0 \\ C &= b(a_0 - a_1)/a_0^2 = -A\end{aligned}$$

Exercise 5.3 By comparison with Exercise 5.1, show that if

$$x^{(n)} + a_{n-1}x^{(n-1)} + \cdots + a_1\dot{x} + a_0x = u \qquad (5.50)$$

$$x^{(n-1)}(0) = x^{(n-2)}(0) = \cdots = x(0) = 0 \qquad (5.51)$$

and $u \in D_R^1$; $u(t) = 0, t \leqslant 0$; $|u(t)| < Me^{\alpha t}, t > 0$; then the Laplace transform of x satisfying (5.50) and (5.51) is

$$\bar{x}(s) = \bar{u}(s)/(s^n + a_{n-1}s^{n-1} + \cdots + a_1s + a_0) \qquad (5.52)$$

As usual, (5.50) is satisfied only where u is continuous.

5.2 Transfer function matrix

Let a dynamical system be described by the equations

$$\dot{x} = Ax + Bu \tag{5.53}$$

$$y = Cx + Du \tag{5.54}$$

where x is an n-vector (the *state* vector), u is an l-vector of inputs, and y is an m-vector of outputs. Let u satisfy the conditions assumed in Section 5.1, namely

$$u \in D_R^1; \qquad u(t) = 0, \quad t \leqslant 0; \qquad |u_i(t)| < Me^{\alpha t}, \quad t > 0, \quad \text{all } i \tag{5.55}$$

and let the initial condition for x be

$$x(0) = 0 \tag{5.56}$$

By Section 5.1, if x is the solution of (5.53) in the usual sense (x satisfies (5.53) where u is continuous; x is continuous) then it has Laplace transform given by (5.41). Consequently, y also has a Laplace transform $\bar{y}$ which is given by

$$\bar{y}(s) = \{C(sI - A)^{-1}B + D\}\bar{u}(s) \tag{5.57}$$

$$\triangleq \bar{G}(s)\bar{u}(s) \tag{5.58}$$

where $\bar{G}(s)$ defined in (5.58) is called the *transfer function matrix* of the system. This is a particular case, but an important one, of the more general result given in Section 8.

5.3 Combination of two results

In the introductory part of Section 5 the solution $x(t) = U(t)\xi(t)$ for $t > 0$ of the equation

$$\dot{\xi} = A\xi; \qquad \xi(0) = c \tag{5.59}$$

was considered. In Section 5.1 the solution of

$$\dot{x} = Ax + Bu; \qquad x(0) = 0 \tag{5.60}$$

was considered under appropriate conditions on u. Since (5.60) is linear, it follows at once that the solution to

$$\dot{x} = Ax + Bu; \qquad x(0+) = c \tag{5.61}$$

is obtained by adding the solutions of (5.59) and (5.60).

Explicitly, let u satisfy the conditions (5.55). Let x be a function satisfying

$$x(t) = 0, \quad t \leqslant 0; \qquad x(0+) = c \tag{5.62}$$

and let x be continuous for all $t > 0$ and satisfy

$$\dot{x}(t) = Ax(t) + Bu(t) \tag{5.63}$$

at each $t > 0$ for which u is continuous. Then x is uniquely defined, $x \in D_R^1$, x has a Laplace transform $\bar{x}$, and $\bar{x}$ is given by

$$\bar{x}(s) = (sI - A)^{-1}[c + B\bar{u}(s)] \tag{5.64}$$

valid for Re $s > \alpha$, where α depends on u and on A.

Equations (5.2) and (5.32) can be combined to give an expression for x,

$$x(t) = U(t)e^{At}c + \int_0^t e^{A(t-\tau)}Bu(\tau)\,d\tau \tag{5.65}$$

5.4 Comparison with operational methods

The procedure given in preceding sections for solving linear time-invariant differential equations leads to essentially the same steps as are taken in Heaviside's operational method. The latter has the slight advantage, not of much practical significance, that the condition $|u(t)| < Me^{\alpha t}$ need not be imposed. An account of the method, with a full justification, will be found in Jeffreys (1950).

Both Heaviside's method and the Laplace transform are intended to solve a given differential equation with given initial conditions, and they therefore produce a unique answer. Care is needed on this point, as there is another operational method in use for finding the general solution of the homogeneous equation (the *complementary function*). An account will be found in Lamb (1938), p. 437.

For example, given the equation

$$D^2x + 3Dx + 2x = 0 \tag{5.66}$$

the procedure is to write

$$(D^2 + 3D + 2)x = 0 \tag{5.67}$$

$$(D + 1)(D + 2)x = 0 \tag{5.68}$$

which is satisfied if either

$$(D + 1)x = 0 \tag{5.69}$$

or

$$(D + 2)x = 0 \tag{5.70}$$

Equations (5.69) and (5.70) are satisfied respectively by

$$x(t) = c_1e^{-t} \tag{5.71}$$

and

$$x_2(t) = c_2e^{-2t} \tag{5.72}$$

and so (5.67) is satisfied by

$$x(t) = c_1e^{-t} + c_2e^{-2t} \tag{5.73}$$

where c_1 and c_2 are arbitrary constants.

Notice that the equation

$$(s^2 + 3s + 2)\bar{x}(s) = 0 \tag{5.74}$$

in contrast to (5.66) gives $\bar{x} = 0$, whence $x = 0$. This is because (5.74) will arise from (5.66) only if the initial conditions are

$$x(0) = 0, \qquad Dx(0) = 0 \tag{5.75}$$

If the initial conditions are

$$x(0) = c_1 + c_2, \qquad Dx(0) = -c_1 - 2c_2 \tag{5.76}$$

to correspond with (5.73), equation (5.66) leads, by (5.26), to

$$(s^2 + 3s + 2)\bar{x}(s) = c_1(s + 2) + c_2(s + 1) \tag{5.77}$$

whence

$$\bar{x}(s) = \frac{c_1}{s + 1} + \frac{c_2}{s + 2} \tag{5.78}$$

and

$$x(t) = c_1e^{-t} + c_2e^{-2t} \tag{5.79}$$

agreeing with (5.73).

The same point arises in connection with transfer functions. Two systems respectively satisfying

$$D^2x + 3Dx + 2x = Du + u; \qquad x(0) = Dx(0) = 0 \tag{5.80}$$

and

$$Dx + 2x = u; \qquad x(0) = 0 \tag{5.81}$$

give rise to transfer functions $\bar{x}(s)/\bar{u}(s)$ which are respectively

$$\bar{g}(s) = \frac{s + 1}{(s + 1)(s + 2)} \tag{5.82}$$

and

$$\bar{g}(s) = \frac{1}{s + 2} \tag{5.83}$$

These are usually regarded as equivalent: that is, cancellation is permitted. Notice however that (5.82) implies the two initial conditions given in (5.80), whereas (5.83) implies only the single initial condition given in (5.81). Given only the transfer function $1/(s + 2)$, without knowledge of any cancellation which may have taken place, it is not possible to say how many initial conditions were imposed on the original system.

6 Sampling systems

With the increasing use of digital computers for control purposes, means are needed for analysing systems in which part of the operation is continuous in time (governed by differential equations or the like) and part is discrete (governed by difference equations). In Problem 5.3 of Chapter 5 the sample-and-hold circuit was introduced. This is one example of the devices used to convert continuous information into discrete information, which can then be converted to digital form, processed, and reconverted to analogue form for control purposes.

A more general sampling system is one which, from an input u, produces an output y defined by

$$y(t) = \begin{cases} u(r + \epsilon), & r + \epsilon < t \leqslant r + \theta + \epsilon \\ 0, & r + \theta + \epsilon < t \leqslant r + 1 + \epsilon \end{cases} \tag{6.1}$$

where

$$r = 0, 1, 2, \ldots; \quad 0 \leqslant \epsilon < 1; \quad 0 < \theta \leqslant 1 \tag{6.2}$$

and the second of equations (6.1) is disregarded if $\theta = 1$. In words, the output y is made equal to the input u immediately after the sampling time $r + \epsilon$, it is maintained at this value for time θ, and is then zero up to the next sampling time. More generally, the sampling interval is T rather than 1, but this can be accounted for by a change in the time scale.

It is easy to confirm that a system having the properties defined by (6.1) and (6.2) is a linear time-dependent system satisfying the conditions of Chapter 5, Exercise 5.2. The response at t to a unit step at t_0 is

$$H(t, t_0) = \sum_{r=0}^{\infty} U(r + \epsilon - t_0)[U(t - r - \epsilon) - U(t - r - \theta - \epsilon)] \tag{6.3}$$

where for given t_0 and t only one term (at most) of the series is non-zero. If $u \in D_R^1$ and $u(t) = 0$, $t \leqslant 0$, equation (5.24) of Chapter 5 gives

$$y(t) = \int_0^t H(t, \tau)\, du(\tau)$$

$$= -\int_0^t u(\tau)[U(t - r - \epsilon) - U(t - r - \theta - \epsilon)]\, d\left[\sum_{r=0}^{\infty} U(r + \epsilon - \tau)\right] \tag{6.4}$$

$$= \sum_{r=0}^{\infty} u(r + \epsilon)[U(t - r - \epsilon) - U(t - r - \theta - \epsilon)] \tag{6.5}$$

which agrees with (6.1).

Instead of dealing directly with the sampler defined by (6.3), it is more convenient to deal with a *staircase sampler* having step response

$$H_0(t, t_0, \epsilon) = \sum_{r=0}^{\infty} U(r + \epsilon - t_0)U(t - r - \epsilon) \tag{6.6}$$

where for subsequent developments ϵ is included explicitly in the notation. It is easily seen that

$$H(t, t_0) = H_0(t, t_0, \epsilon) - H_0(t - \theta, t_0, \epsilon) \tag{6.7}$$

Notice that

$$\begin{aligned}\lim_{\epsilon \to 0} H_0(t, 0, \epsilon) &= \sum_{r=0}^{\infty} U(t - r) \\ &= U(t) + H_0(t, 0, 0)\end{aligned} \tag{6.8}$$

Now let u be a function satisfying conditions (5.55), so that u has a Laplace transform $\bar{u}$. When u is staircase-sampled, the result $u^\dagger$ is given by

$$u^\dagger(t) = \sum_{r=0}^{\infty} u(r + \epsilon_1)U(t - r - \epsilon_1) \tag{6.9}$$

The Laplace transform of $u^\dagger$ is defined by

$$\begin{aligned}\bar{u}^\dagger(s) &= \lim_{p \to \infty} \int_0^{p+1+\epsilon_1} e^{-st} \sum_{r=0}^{p} u(r + \epsilon_1)U(t - r - \epsilon_1)\, dt \\ &= \lim_{p \to \infty} s^{-1} \sum_{r=0}^{p} u(r + \epsilon_1)[e^{-s(r+\epsilon_1)} - e^{-s(p+1+\epsilon_1)}]\end{aligned} \tag{6.10}$$

If α is the exponent occurring in (5.55), and if $\text{Re}\, s > \alpha_1 > \max\{\alpha, 0\}$

$$\begin{aligned}|u(r + \epsilon_1)e^{-s(r+\epsilon_1)}| &< Me^{\alpha(r+\epsilon_1)}e^{-\alpha_1(r+\epsilon_1)} \\ &= Me^{(\alpha-\alpha_1)\epsilon_1}[e^{\alpha-\alpha_1}]^r\end{aligned} \tag{6.11}$$

and $\sum_{r=0}^{\infty} [e^{\alpha-\alpha_1}]^r$ is absolutely convergent. Also if $\alpha \neq 0$,

$$\begin{aligned}|e^{-sp} \sum_{r=0}^{p} u(r + \epsilon_1)| &< Me^{-\alpha_1 p} \sum_{r=0}^{p} e^{\alpha(r+\epsilon_1)} \\ &= \frac{Me^{\alpha\epsilon_1}}{1 - e^{\alpha}} [e^{-\alpha_1 p} - e^{\alpha}.e^{-(\alpha_1-\alpha)p}]\end{aligned} \tag{6.12}$$

which tends to zero as $p \to \infty$. If $\alpha = 0$, a different formula is obtained but the conclusion remains true. From this it follows that the series in (6.10) is absolutely convergent and $\bar{u}^\dagger$ exists and is defined by

$$\bar{u}^\dagger(s) = s^{-1} \sum_{r=0}^{\infty} u(r + \epsilon_1)e^{-s(r+\epsilon_1)} \tag{6.13}$$

Consider now a linear time-invariant system having step response $H(t)$ and satisfying the conditions of Chapter 5, Theorem 5.1. Let $H(t)$ also satisfy

$$|H(t)| < Ne^{\beta t}, \quad t > 0 \tag{6.14}$$

and let the staircase-sampled function $u^\dagger$ be applied as input to this system. The response of the system to an input $U(t - r - \epsilon_1)$ is $H(t - r - \epsilon_1)$, and so the response to $u^\dagger$ is given by

$$y(t) = \sum_{r=0}^{\infty} u(r + \epsilon_1)H(t - r - \epsilon_1) \tag{6.15}$$

Finally, let the output y given by (6.15) be staircase-sampled at times $r + \epsilon_2$, $r = 0, 1, 2, \ldots$. These sampling times $r + \epsilon_2$ are not necessarily the same as those, $r + \epsilon_1$, used in the sampling of u. The sampled output $y^\dagger$ is given by

$$\begin{aligned} y^\dagger(t) &= \int_0^t H_0(t, \tau, \epsilon_2)\, dy(\tau) \\ &= -\int_0^t y(\tau)\, dH_0(t, \tau, \epsilon_2) \\ &= -\int_0^t \sum_{r=0}^{\infty} u(r + \epsilon_1)H(\tau - r - \epsilon_1)\, d\left[\sum_{p=0}^{\infty} U(p + \epsilon_2 - \tau)U(t - p - \epsilon_2)\right] \end{aligned} \tag{6.16}$$

For given t, each of the sums can be truncated after a finite number of terms, so that

$$y^\dagger(t) = \sum_{p=0}^{\infty}\sum_{r=0}^{\infty} u(r + \epsilon_1)H(p - r + \epsilon_2 - \epsilon_1)U(t - p - \epsilon_2) \tag{6.17}$$

which again contains a finite number of non-zero terms.

It will now be shown that $y^\dagger$ has a Laplace transform. We have, for n a positive integer,

$$\lim_{n\to\infty} \int_0^{n+1+\epsilon_2} e^{-st} \sum_{p=0}^{\infty}\sum_{r=0}^{\infty} u(r + \epsilon_1)H(p - r + \epsilon_2 - \epsilon_1)U(t - p - \epsilon_2)\, dt \tag{6.18}$$

$$= \lim_{n\to\infty} s^{-1} \sum_{p=0}^{n}\sum_{r=0}^{p} u(r + \epsilon_1)H(p - r + \epsilon_2 - \epsilon_1)[e^{-s(p+\epsilon_2)} - e^{-s(n+1+\epsilon_2)}] \tag{6.19}$$

because $H(p - r + \epsilon_2 - \epsilon_1) = 0$ for $r \geqslant p + 1$.

Also if $\operatorname{Re} s > \alpha_2 > \max\{\alpha_1, \beta\} = \max\{\alpha, \beta, 0\}$,

$$\begin{aligned} &|u(r + \epsilon_1)H(p - r + \epsilon_2 - \epsilon_1)[e^{-s(p+\epsilon_2)} - e^{-s(n+1+\epsilon_2)}]| \\ &\qquad < MNe^{\alpha(r+\epsilon_1)}e^{\beta(p-r+\epsilon_2-\epsilon_1)}[e^{-\alpha_2(p+\epsilon_2)} + e^{-\alpha_2(n+1+\epsilon_2)}] \end{aligned} \tag{6.20}$$

Summing over r,

$$\sum_{r=0}^{p} e^{(\alpha-\beta)r} = \frac{1 - e^{(\alpha-\beta)(p+1)}}{1 - e^{(\alpha-\beta)}} \tag{6.21}$$

provided that $\alpha \neq \beta$. If $\alpha = \beta$, the details which follow are slightly changed, but the conclusion remains true. Hence each of the terms in the sum over p in (6.19) is less than

$$[K_1 e^{\beta p} + K_2 e^{\alpha p}][e^{-\alpha_2 p} + e^{-\alpha_2(n+1)}] \tag{6.22}$$

where K_1 and K_2 are constants. Then

$$\begin{aligned}\lim_{n\to\infty} e^{-\alpha_2 n} \sum_{p=0}^{n} e^{\beta p} &= \lim_{n\to\infty} \frac{e^{-\alpha_2 n}[1 - e^{\beta(n+1)}]}{1 - e^{\beta}} \\ &= 0\end{aligned} \tag{6.23}$$

provided that $\beta \neq 0$: if $\beta = 0$ the conclusion remains true. Similarly,

$$\lim_{n\to\infty} e^{-\alpha_2 n} \sum_{p=0}^{n} e^{\alpha p} = 0 \tag{6.24}$$

From the fact that $\alpha_2 > \alpha$, $\alpha_2 > \beta$, it follows that the limit in (6.19) exists and can be written

$$\bar{y}^{\dagger}(s) = s^{-1} \sum_{p=0}^{\infty} \left\{ \sum_{r=0}^{p} H(p - r + \epsilon_2 - \epsilon_1) u(r + \epsilon_1) e^{-s(p+\epsilon_2)} \right\} \tag{6.25}$$

The proof shows that this series is absolutely convergent for Re $s > \alpha_2$.

6.1 The *z*-transform

The results given above are the basis for the use of the *z-transform* in the analysis of sampling systems. The input $u^{\dagger}$ to the time-invariant linear system was shown to have Laplace transform

$$\bar{u}^{\dagger}(s) = s^{-1} \sum^{\infty} u(r + \epsilon_1) e^{-s(r+\epsilon_1)} \tag{6.26}$$

the equation being valid, and the series absolutely convergent, for Re $s > \alpha_1 = \max\{\alpha, 0\}$, where α is the exponent occurring in (5.55).

Define a quantity $\bar{H}^{\dagger}(s)$ by

$$\bar{H}^{\dagger}(s) = s^{-1} \sum_{p=0}^{\infty} H(p + \epsilon_2 - \epsilon_1) e^{-s(p+\epsilon_2-\epsilon_1)} \tag{6.27}$$

Subject to (6.14), (6.27) is absolutely convergent for Re $s > \beta$. Then for Re $s > \alpha_2 = \max\{\alpha, \beta, 0\}$,

$$[s\bar{H}^{\dagger}(s)][s\bar{u}^{\dagger}(s)] = \left[\sum_{p=0}^{\infty} H(p + \epsilon_2 - \epsilon_1)e^{-s(p+\epsilon_2-\epsilon_1)}\right] \times \left[\sum_{r=0}^{\infty} u(r + \epsilon_1)e^{-s(r+\epsilon_1)}\right] \quad (6.28)$$

For Re $s > \alpha_2$ both series are absolutely convergent, and their product may be replaced by the Cauchy product (Chapter 4, Theorem 10.9).

$$\sum_{p=0}^{\infty}\sum_{r=0}^{p} H(p - r + \epsilon_2 - \epsilon_1)e^{-s(p-r+\epsilon_2-\epsilon_1)}u(r + \epsilon_1)e^{-s(r+\epsilon_1)}$$

$$= \sum_{p=0}^{\infty}\sum_{r=0}^{p} H(p - r + \epsilon_2 - \epsilon_1)u(r + \epsilon_1)e^{-s(p+\epsilon_2)} \quad (6.29)$$

But this is just the sum (6.25). It follows that for Re $s > \alpha_2$,

$$[s\bar{H}^{\dagger}(s)][s\bar{u}^{\dagger}(s)] = s\bar{y}^{\dagger}(s) \quad (6.30)$$

Now put $\epsilon_2 = 2\epsilon_1$ and let $\epsilon_1 \to 0$. By equations (6.26) and (6.27), and the corresponding equation for $\bar{y}^{\dagger}(s)$, namely

$$\bar{y}^{\dagger}(s) = s^{-1} \sum_{r=0}^{\infty} y(r + \epsilon_2)e^{-s(r+\epsilon_2)} \quad (6.31)$$

it follows that $\bar{H}^{\dagger}$, $\bar{u}^{\dagger}$, and $\bar{y}^{\dagger}$ then depend upon s only through e^{-s}. Write $e^{-s} = z^{-1}$ and introduce the notation

$$\tilde{H}(z) = s\bar{H}^{\dagger}(s) = \sum_{r=0}^{\infty} H(r+)z^{-r} \quad (6.32)$$

$$\tilde{u}(z) = s\bar{u}^{\dagger}(s) = \sum_{r=0}^{\infty} u(r+)\, z^{-r} \quad (6.33)$$

$$\tilde{y}(z) = s\bar{y}^{\dagger}(s) = \sum_{r=0}^{\infty} y(r+)\, z^{-r} \quad (6.34)$$

Then (6.30) becomes

$$\tilde{y}(z) = \tilde{H}(z)\tilde{u}(z) \quad (6.35)$$

Notice that the sampling of y was assumed to be later than that of u, that is, $\epsilon_2 > \epsilon_1$. This is the usual assumption, though often it is not explicit. If the contrary had been assumed, (6.32) would contain $H(r-)$ rather than $H(r+)$; these can differ if H is discontinuous. The series in (6.32) and (6.33) are absolutely convergent for $|z| > e^{\alpha}$, $|z| > e^{\beta}$ respectively, where α and β have the values associated with (6.26) and (6.27). Equation (6.35) has been proved for $|z| > \max\{e^{\alpha}, e^{\beta}, 1\}$. By analytic continuation (Chapter 7, Section 6) the result is true for $|z| > \rho = \max\{e^{\alpha}, e^{\beta}\}$.

The quantities defined in (6.32), (6.33), and (6.34) are known as *z-transforms*. They are sometimes written $\bar{H}(z)$, $\bar{u}(z)$, $\bar{y}(z)$ but there are two objections to this. First, it becomes impossible to use $\bar{u}$ for the function, as distinct from its value $\bar{u}(z)$ at z, because $\bar{u}$ could also mean (as it always does here) the function of s which is the Laplace transform of u. Secondly, the notation $\bar{u}(z)$, with the definition $e^{-s} = z^{-1}$, induces the error of writing $\bar{u}(\log z)$ for $\tilde{u}(z)$, where $\bar{u}$ is the Laplace transform of u; compare Exercise 6.1.

Because z-transforms such as $\tilde{u}$ are distinguished from Laplace transforms $s\bar{u}^\dagger$ only by notation, the result in Section 4 applies to them. That is to say, there is a one-to-one relationship between a staircase-sampled function $u^\dagger$, which is necessarily in D_R^1, and its z-transform $\tilde{u}$. Given a function u, there is also a unique $u^\dagger$ corresponding to it, but the converse is clearly not true. The relations between the various functions introduced so far can therefore be summarized thus:

$$\bar{u} \Leftrightarrow u \Rightarrow u^\dagger \Leftrightarrow \tilde{u}; \qquad \tilde{u}(z) = s\bar{u}^\dagger(s) \tag{6.36}$$

EXERCISE 6.1 Verify the entries in the following table, in which n is a non-negative integer. Note particularly that $\tilde{u}$ is not obtained by substituting $s = \log z$ in $\bar{u}$. Note also that, given $u^\dagger$ and u from the table, $u(t)\phi(t)$ defines another function in D_R^1 giving rise to the same $u^\dagger$, provided only that ϕ is continuous and bounded and $\phi(n) = 1$, n a non-negative integer.

$\bar{u}$	e^{-sn}/s	$(s+\alpha)^{-1}$	s^{-2}
u	$U(t-n)$	$U(t)e^{-\alpha t}$	$tU(t)$
$u^\dagger$	$\sum_{r=n}^{\infty} U(t-r)$	$\sum_{r=0}^{\infty} e^{-\alpha r}U(t-r)$	$\sum_{r=1}^{\infty} rU(t-r)$
$s\bar{u}^\dagger$	$\sum_{r=n}^{\infty} e^{-sr}$	$\sum_{r=0}^{\infty} e^{-\alpha r}e^{-sr}$	$\sum_{r=1}^{\infty} re^{-sr}$
$\tilde{u}$	$z^{-n}(1-z^{-1})^{-1}$	$(1-e^{-\alpha}z^{-1})^{-1}$	$z^{-1}(1-z^{-1})^{-2}$

If the calculations which gave rise to (6.30) are repeated with $H_0(t, t_0, \epsilon_1)$ replaced by $H_0(t - \theta_1, t_0, \epsilon_1)$ in the first staircase sampler the result is

$$[s\bar{H}^\dagger_{\theta_1}(s)][s\bar{u}^\dagger(s)] = s\bar{y}^\dagger(s) \tag{6.37}$$

where

$$s\bar{H}^\dagger_{\theta_1}(s) = \sum_{r=0}^{\infty} H(r - \theta_1 + \epsilon_2 - \epsilon_1)e^{-s(r+\epsilon_2-\epsilon_1)} \tag{6.38}$$

On using (6.7) and the linearity of the system, it follows that when the first staircase sampler is replaced by a sampler defined by (6.1), the output is defined by

$$s\bar{y}^\dagger(s) = [s\bar{H}^\dagger(s) - s\bar{H}^\dagger_{\theta_1}(s)][s\bar{u}^\dagger(s)] \tag{6.39}$$

If the second staircase sampler is replaced by a sampler defined by (6.1), with $\theta = \theta_2$, a second application of (6.7) shows that the output is defined by

$$s\bar{y}^{\dagger}(s) = [1 - e^{-s\theta_2}][s\bar{H}^{\dagger}(s) - s\bar{H}^{\dagger}_{\theta_1}(s)][s\bar{u}^{\dagger}(s)] \tag{6.40}$$

Put $\theta_1 = \theta_2 = 1$, so that the two samplers become sample-and-hold circuits. Also put $\epsilon_2 = 2\epsilon_1$ and let $\epsilon_1 \to 0$. Then all the quantities in (6.40) depend on s only through $e^{-s} = z^{-1}$, and (6.40) can be written

$$\tilde{y}(z) = (1 - z^{-1})\tilde{H}_{SH}(z)\tilde{u}(z) \tag{6.41}$$

where $\tilde{H}_{SH}$ is defined by

$$\begin{aligned}\tilde{H}_{SH}(z) &= \sum_{r=0}^{\infty} H(r+)\,z^{-r} - \sum_{r=0}^{\infty} H[(r-1)+]\,z^{-r}\\ &= \sum_{r=0}^{\infty} H(r+)(1 - z^{-1})z^{-r}\\ &= (1 - z^{-1})\tilde{H}(z)\end{aligned} \tag{6.42}$$

so that finally (6.41) can be written

$$\tilde{y}(z) = (1 - z^{-1})^2\tilde{H}(z)\tilde{u}(z) \tag{6.43}$$

Notice that the two sample-and-hold circuits used here differ from the one defined in Problem 5.3 of Chapter 5: that sampled at $t = r$, while these sample at $t = r+$, the second sampler being later than the first as the limit is taken (that is, $\epsilon_2 > \epsilon_1$).

Example 6.1 A sample-and-hold circuit is followed by an integrator and a second sample-and-hold circuit. What is the step response?

We have $u(t) = U(t)$, and by Exercise 6.1, $\tilde{u}(z) = (1 - z^{-1})^{-1}$. For an integrator, with transfer function s^{-1}, the Laplace transform of the step response is $\bar{H}(s) = s^{-2}$, so by Exercise 6.1, $\tilde{H}(z) = z^{-1}(1 - z^{-1})^{-2}$. Equation (6.43) gives

$$\tilde{y}(z) = (1 - z^{-1})^2 z^{-1}(1 - z^{-1})^{-2}(1 - z^{-1})^{-1} = z^{-1}(1 - z^{-1})^{-1} \tag{6.44}$$

and by Exercise 6.1, $y^{\dagger}(t) = \sum_{r=1}^{\infty} U(t - r)$. The waveforms are sketched in Fig. 2.

Example 6.2 A sample-and-hold circuit is followed by a system with $\bar{H}(s) = (s + 2)/s(s + 1)$ and this is followed by a staircase sampler. The input is given by $u(t) = \sin \pi t/2$. We have

$$\tilde{u}(z) = z^{-1} - z^{-3} + z^{-5} - \cdots = z^{-1}(1 + z^{-2})^{-1} \tag{6.45}$$

and

$$\bar{H}(s) = \frac{2}{s} - \frac{1}{s+1} \tag{6.46}$$

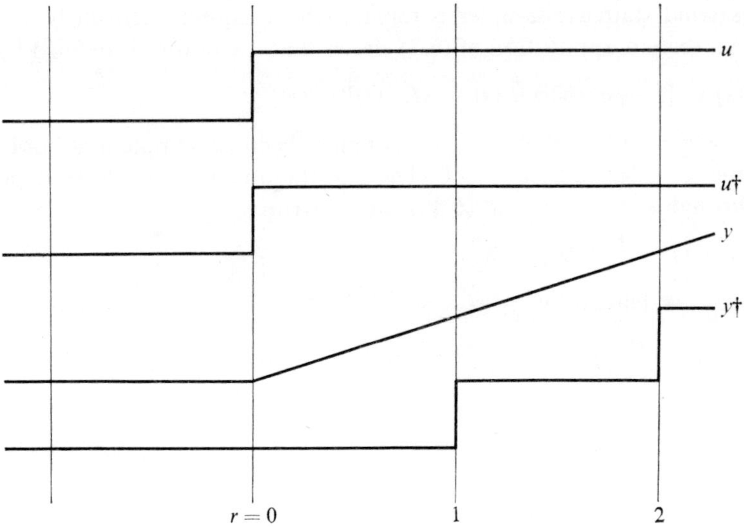

FIG. 2

whence by Exercise 6.1 and linearity

$$\tilde{H}(z) = 2(1 - z^{-1})^{-1} - (1 - e^{-1}z^{-1})^{-1} \tag{6.47}$$

Notice that only the first sampler is a sample-and-hold circuit, so from (6.37) and (6.42),

$$\begin{aligned}\tilde{y}(z) &= (1 - z^{-1})[2(1 - z^{-1})^{-1} - (1 - e^{-1}z^{-1})^{-1}]z^{-1}(1 + z^{-2})^{-1} \\ &= 2z^{-1}(1 + z^{-2})^{-1} - z^{-1}(1 - z^{-1})(1 - e^{-1}z^{-1})^{-1}(1 + z^{-2})^{-1} \quad (6.48) \\ &= 2z^{-1}(1 + z^{-2})^{-1} + \frac{e}{1 + e^2}\left\{\frac{e - 1}{1 - e^{-1}z^{-1}} - \frac{(e - 1) + (e + 1)z^{-1}}{1 + z^{-2}}\right\} \quad (6.49) \\ &= \frac{1}{1 + e^2}\left\{\frac{e(e - 1)}{1 - e^{-1}z^{-1}} + \frac{(e^2 - e + 2)z^{-1} - e(e - 1)}{1 + z^{-2}}\right\} \quad (6.50)\end{aligned}$$

where (6.49) is obtained by a partial fraction expansion of (6.48). Reference to (6.45) and to Exercise 6.1 shows that the output is

$$y^{\dagger}(t) = \frac{e(e - 1)}{1 + e^2}\left\{\sum_{r=0}^{\infty} e^{-r}U(t - r) - \sum_{r=1}^{\infty} \sin\frac{\pi r}{2}\cdot U(t - r + 1)\right\} + \frac{e^2 - e + 2}{1 + e^2}\left\{\sum_{r=1}^{\infty} \sin\frac{\pi r}{2}\cdot U(t - r)\right\} \tag{6.51}$$

EXERCISE 6.2 Examine the convergence of the series in Examples 6.1 and 6.2.

6.2 Impulse sampler

The method given in Section 6.1 is usually treated in a somewhat different way. Instead of the staircase sampler, the basic sampler is the *impulse sampler* which from an input u produces a sampled output u^* defined by

$$u^*(t) = \sum_{r=0}^{\infty} u(r+)\delta(t-r) \tag{6.52}$$

Using the approach of Chapter 5, Section 8,

$$u^*(t) = Du^{\dagger}(t) \tag{6.53}$$

Consequently, an impulse sampler can be represented (notionally) by a staircase sampler followed by a differentiator, as in Fig. 3. Section 3 shows that

$$\bar{u}^*(s) = s\bar{u}^{\dagger}(s) \tag{6.54}$$

and so under the conditions of Section 6.1

$$\tilde{u}(z) = \bar{u}^*(s) \tag{6.55}$$

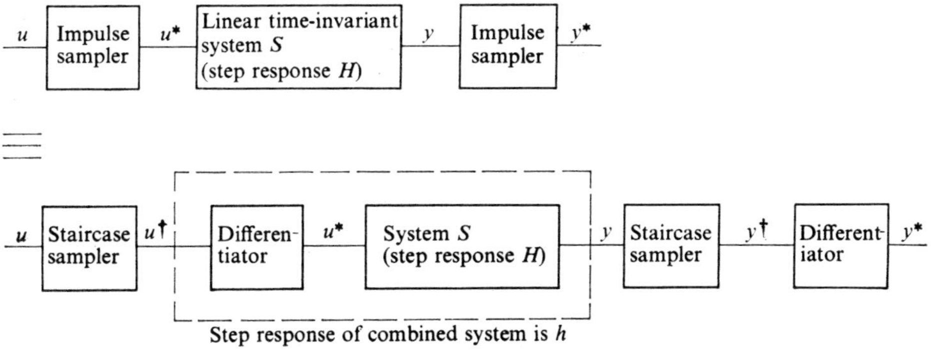

FIG. 3

Example 6.3 The table in Section 6.1 can be augmented by the following entries

$\bar{u}$	e^{-ns}/s	$(s+\alpha)^{-1}$	s^{-2}
u	$U(t-n)$	$U(t)e^{-\alpha t}$	$tU(t)$
u^*	$\sum_{r=n}^{\infty} \delta(t-r)$	$\sum_{r=0}^{\infty} e^{-\alpha r}\, \delta(t-r)$	$\sum_{r=1}^{\infty} r\, \delta(t-r)$

$\bar{u}^*$	$\sum_{r=n}^{\infty} e^{-sr}$	$\sum_{r=0}^{\infty} e^{-\alpha r} e^{-sr}$	$\sum_{r=1}^{\infty} r e^{-sr}$
$\tilde{u}$	$z^{-n}(1 - z^{-1})^{-1}$	$(1 - e^{-\alpha} z^{-1})^{-1}$	$z^{-1}(1 - z^{-1})^{-2}$

In Fig. 3, associate the first differentiator with the linear time-invariant system S. The step response of the combination, by Section 9 of Chapter 5, is the impulse response h of S. It follows from (6.30), with the same treatment of ϵ_1 and ϵ_2, that

$$\tilde{y}(z)[= \bar{y}^*(s) = s\bar{y}^\dagger(s) = s\bar{h}^\dagger(s) s\bar{u}^\dagger(s) = \bar{h}^*(s)\bar{u}^*(s)] = \tilde{h}(z)\tilde{u}(z) \qquad (6.56)$$

The function $\tilde{h}$ is called the *sampled transfer function* (or *z-transfer function*, or *discrete-time transfer function*, or *pulse transfer function*) of the linear time-invariant system having impulse response h.

Equation (6.56) can be used in the same way as (6.43) and its like. Because the behaviour of practical samplers is seldom well represented by an impulse sampler, this is often followed by a *pulse shaper*. For example, a circuit with transfer function $(1 - e^{-s})/s$ (see Section 8) converts a pulse sampler into a sample-and-hold circuit. This particular pulse shaper is called a *zero-order hold*.

EXERCISE 6.3 In the system to which (6.40) relates, let the first sampler produce an output of magnitude $1/\theta_1$, and the second an output of magnitude $1/\theta_2$. Using the approach of Section 9.1 in Chapter 5, discuss the limit as $\theta_1 \to 0$, $\theta_2 \to 0$. [Note the difficulty which arises when H is discontinuous:

$$\lim_{\theta_1 \to 0+} \{H(r + \epsilon_2 - \epsilon_1) - H(r - \theta_1 + \epsilon_2 - \epsilon_1)\}/\theta_1$$

has to be interpreted as $DH(r + \epsilon_2 - \epsilon_1)$ rather than $D_-H(r + \epsilon_2 - \epsilon_1)$. Sketch the fraction as a function of $\epsilon_2 - \epsilon_1$ near a discontinuity of H and compare equations (9.9) of Chapter 5.]

Example 6.4 The system discussed in Example 6.2 can be represented as in Fig. 4, where $(s + 2)/(s + 3)$ is $s\bar{H}(s) = \bar{h}(s)$. Since the Laplace transform of

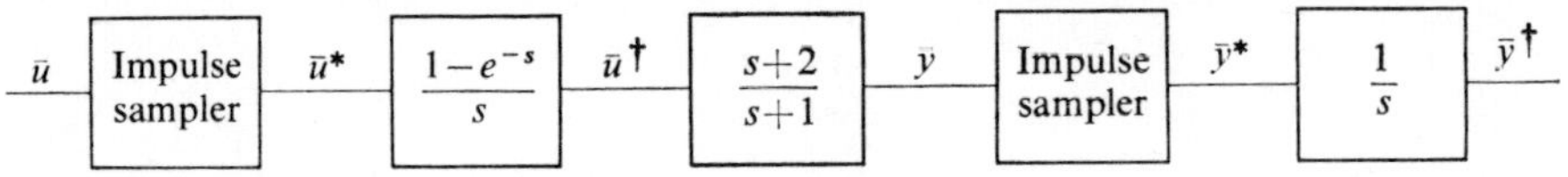

FIG. 4

$\delta(t)$ is 1, $\bar{h}(s)$ is the transfer function defined in Section 5.2. It is easy to see that the z-transform corresponding to $(1 - e^{-s})(s + 2)/s(s + 1)$ is $(1 - z^{-1})$ times the z-transform corresponding to $(s + 2)/s(s + 1)$; this is given by (6.47). Equation (6.45) gives $\tilde{u}(z)$, which by (6.54) is the same whether derived from

$\bar{u}^*(s)$ or from $s\bar{u}^\dagger(s)$. Consequently, $\tilde{y}(z)$ will again be given by (6.50) and it can be interpreted to give y^* and then integrated; or it can be interpreted directly to give $y^\dagger$. In this way the problem can be solved by the method of this section, which is the usual one.

Following the outlook of Chapter 5, Section 9, the impulse sampler will be regarded here merely as a way of paraphrasing the results in Section 6.1. To give a mathematically satisfactory account of the impulse sampler in a direct way, it would be necessary to face two kinds of difficulty. The first is the ambiguity which arises when a function is impulse-sampled at a point where it is discontinuous. The second is the fact that distributions cannot be sampled without some previous averaging.

6.3 Modified z-transform

The z-transform $\tilde{u}$ serves only to identify the sampled function $u^\dagger$ (or u^*). If it is necessary in a sampling system to find the form of signals before they are sampled, the Laplace transform may be used. Alternatively, a notional sampler may be inserted after the signal in question, and its sampling time delayed by η. On letting η run from 0 to 1, the behaviour of the unsampled function is disclosed. Such delayed sampling gives rise to the *modified z-transform.*

In Section 6, let the second staircase sampler be delayed by η. Then (6.16) becomes

$$\begin{aligned} y_\eta^\dagger(t) &= \int_0^t H_0(t, \tau - \eta, \epsilon_2)\, dy(\tau) \\ &= -\int_0^t \sum_{r=0}^{\infty} u(r + \epsilon_1) H(\tau - r - \epsilon_1) \\ &\qquad \times d\left[\sum_{p=0}^{\infty} U(p + \epsilon_2 - \tau + \eta) U(t - p - \epsilon_2)\right] \end{aligned} \tag{6.57}$$

Provided that

$$\eta + \epsilon_2 \geqslant 0 \tag{6.58}$$

so that the step in the integrator corresponding to $p = 0$ is included in the range of integration, this gives

$$y_\eta^\dagger(t) = \sum_{p=0}^{\infty} \sum_{r=0}^{\infty} u(r + \epsilon_1) H(p - r + \epsilon_2 - \epsilon_1 + \eta) U(t - p - \epsilon_2) \tag{6.59}$$

Now if

$$H(p - r + \epsilon_2 - \epsilon_1 + \eta) = 0, \quad r \geqslant p + 1 \tag{6.60}$$

the sum over r in (6.59) can be truncated at $r = p$. Equation (6.60) requires that

$$\eta + \epsilon_2 - \epsilon_1 - 1 \leqslant 0 \tag{6.61}$$

Following the same procedure as in Sections 6 and 6.1, we obtain in place of (6.30),

$$[s\bar{H}_\eta^\dagger(s)][s\bar{u}^\dagger(s)] = s\bar{y}_\eta^\dagger(s) \tag{6.62}$$

and if $\epsilon_2 = 2\epsilon_1$, $\epsilon_1 = 0+$, we may define

$$s\bar{H}_\eta^\dagger(s) = \sum_{r=0}^{\infty} H(r + \eta+)z^{-r} \triangleq \tilde{H}(z, \eta) \tag{6.63}$$

$$s\bar{y}_\eta^\dagger(s) = \sum_{r=0}^{\infty} y(r + \eta+)z^{-r} \triangleq \tilde{y}(z, \eta) \tag{6.64}$$

and the conditions (6.58) and (6.61) on η become

$$0 \leqslant \eta < 1 \tag{6.65}$$

Then (6.62) becomes

$$\tilde{y}(z, \eta) = \tilde{H}(z, \eta)\tilde{u}(z) \tag{6.66}$$

In the same way, if both samplers are impulse samplers

$$\tilde{y}(z, \eta) = \tilde{h}(z, \eta)\tilde{u}(z) \tag{6.67}$$

The notation used here is essentially the same as that in Ragazzini and Franklin (1958) if Δ is substituted for η. The modified z-transforms listed by Jury (1958) have m in place of η, and are divided by z. In both references it is the sampled output which is considered to be shifted in time, rather than the sampler. Ragazzini and Franklin advance the function by an amount Δ, while Jury retards it by an amount $1 - m$.

Example 6.5 If $u(t) = U(t)$, $h(t) = tU(t)$,

$$\tilde{h}(z, \eta) = \sum_{r=0}^{\infty} (r + \eta)U(r + \eta+)\, z^{-r}, \quad 0 \leqslant \eta < 1 \tag{6.68}$$

$$= \sum_{r=0}^{\infty} rz^{-r} + \eta \sum_{r=0}^{\infty} z^{-r} \tag{6.69}$$

$$= z^{-1}(1 - z^{-1})^{-2} + \eta(1 - z^{-1})^{-1} \tag{6.70}$$

so that (6.67) gives

$$\tilde{y}(z, \eta) = z^{-1}(1 - z^{-1})^{-3} + \eta(1 - z^{-1})^{-2} \tag{6.71}$$

$$= \sum_{r=0}^{\infty} \left[\frac{r(r + 1)}{2} + (r + 1)\eta\right] z^{-r} \tag{6.72}$$

from which

$$y(r + \eta +) = \frac{r(r + 1)}{2} + (r + 1)\eta, \quad r = 0, 1, 2, \ldots, \quad 0 \leqslant \eta < 1 \tag{6.73}$$

Since h corresponds to a double integrator and the sampled input u^* is a train of unit pulses at $r = 0, 1, 2, \ldots$, the result in (6.73) is readily confirmed. Note that $y(r + \eta +)$ in (6.73) means $\lim_{\epsilon_2 \to 0+} y(r + \eta + \epsilon_2)$.

7 Difference equations

The z-transform arises in sampled-data systems as a special case of the Laplace transform. In the solution of difference equations, the z-transform arises in a natural way from its defining series

$$\tilde{x}(z) = \sum_{r=0}^{\infty} x_r z^{-r} \tag{7.1}$$

where x_r, $r = 0, 1, 2, \ldots$, is a sequence of real numbers and z is complex. Subject to a bound

$$|x_r| < M\rho^r, \quad \rho > 0, \quad r \geqslant 0 \tag{7.2}$$

the series (7.1) is absolutely convergent for $|z| > \rho$.

Introduce the notation $x^\ddagger$ to represent the sequence $x_0, x_1, x_2, \ldots$. In other words, $x^\ddagger$ is a function having as domain S the set of non-negative integers, and having range $R \subset E_1$. Define an operator E such that $Ex^\ddagger$ is the sequence $x_1, x_2, x_3, \ldots$; $E^2x^\ddagger$ is the sequence $x_2, x_3, x_4, \ldots$; and generally for $p \geqslant 0$, $E^p x^\ddagger$ is the sequence $x_p, x_{p+1}, x_{p+2}, \ldots$. Let the z-transform $\tilde{x}$ of $x^\ddagger$ be denoted by $\tilde{x} = Z(x^\ddagger)$. Then

$$Z(E^p x^\ddagger) = \sum_{r=0}^{\infty} x_{r+p} z^{-r} \tag{7.3}$$

$$= \left\{ \sum_{r=0}^{\infty} x_r z^{-r} - (x_0 + x_1 z^{-1} + \cdots + x_{p-1} z^{-p+1}) \right\} z^p \tag{7.4}$$

$$= z^p \{ \tilde{x} - \sum_{k=0}^{p-1} x_k z^{-k} \} \tag{7.5}$$

Suppose that a homogeneous difference equation is given,

$$x_{r+n} + a_{n-1} x_{r+n-1} + \cdots + a_1 x_{r+1} + a_0 x_r = 0, \quad r = 0, 1, 2, \ldots \tag{7.6}$$

with initial conditions

$$x_{n-1} = c_{n-1}, x_{n-2} = c_{n-2}, \ldots, x_0 = c_0 \tag{7.7}$$

The solution is determined recursively: putting $r = 0$ in (7.6), x_n is determined from $c_{n-1}, c_{n-2}, \ldots, c_0$; then putting $r = 1$ in (7.6), x_{n+1} is determined from

$x_n, c_{n-1}, \ldots, c_1$; etc. A more convenient way to find the solution is to use the z-transform as follows.

Multiply (7.6) by z^{-r} and sum over r to obtain

$$\sum_{r=0}^{\infty} z^{-r}\{x_{r+n} + a_{n-1}x_{r+n-1} + \cdots + a_1x_{r+1} + a_0x_r\} = 0 \tag{7.8}$$

Provided that all the series converge, (7.8) gives

$$\sum_{r=0}^{\infty} z^{-r}x_{r+n} + a_{n-1}\sum_{r=0}^{\infty} z^{-r}x_{r+n-1} + \cdots + a_0\sum_{r=0}^{\infty} z^{-r}x_r = 0 \tag{7.9}$$

$$Z(E^n x^\ddagger) + a_{n-1}Z(E^{n-1}x^\ddagger) + \cdots + a_1Z(Ex^\ddagger) + a_0Z(x^\ddagger) = 0 \tag{7.10}$$

Each of the z-transforms can be evaluated in terms of $\tilde{x}$, z, and $c_0, c_1, \ldots, c_{p-1}$ by means of (7.5). The result is

$$\begin{aligned}
&[z^n\tilde{x}(z) - z^n c_0 - \cdots - z^2 c_{n-2} - zc_{n-1}]\\
&+ a_{n-1}[z^{n-1}\tilde{x}(z) - z^{n-1}c_0 - \cdots - zc_{n-2}]\\
&+ \cdots\\
&+ a_1[z\tilde{x}(z) - zc_0]\\
&+ a_0\tilde{x}(z) = 0
\end{aligned} \tag{7.11}$$

which may be compared with equation (5.26). On solving for $\tilde{x}(z)$, putting the result in partial fractions, and consulting a table of z-transforms, the required sequence $x^\ddagger$ satisfying (7.6) and (7.7) can be found.

To complete the justification of this method of solving (7.6) and (7.7) it is necessary to show first that the series in (7.9) converge for large enough $|z|$, and secondly that the sequence $x^\ddagger$ obtained from the table is the only one having the given transform $\tilde{x}$. The first point gives little difficulty. Let

$$c = \max\{|c_0|, |c_1|, \ldots, |c_{n-1}|\} = \max\{|x_0|, |x_1|, \ldots, |x_{n-1}|\} \tag{7.12}$$

$$a = \max\{|a_0|, |a_1|, \ldots, |a_{n-1}|, 1\} \tag{7.13}$$

Then determining $x_n, x_{n+1}, \ldots$ recursively,

$$\begin{aligned}
|x_n| &= |a_{n-1}c_{n-1} + a_{n-2}c_{n-2} + \cdots + a_0c_0|\\
&\leqslant nac
\end{aligned} \tag{7.14}$$

and because $nac \geqslant c$,

$$nac \geqslant \max\{|x_1|, |x_2|, \ldots, |x_n|\} \tag{7.15}$$

so that

$$\begin{aligned}
|x_{n+1}| &= |a_{n-1}x_n + a_{n-2}x_{n-1} + \cdots + a_0x_1|\\
&\leqslant na(nac)
\end{aligned} \tag{7.16}$$

and

$$(na)^2c \geqslant \max\{|x_2|, |x_3|, \ldots, |x_{n+1}|\} \tag{7.17}$$

Proceeding in this way, we find

$$|x_{n+r}| \leqslant (na)^r nac \tag{7.18}$$

which is the same as (7.2) and shows that the series in (7.1) is absolutely convergent for $|z| > na$ when the z_r, $r = 0, 1, 2, \ldots$ are determined by (7.6) and (7.7). The convergence of the other series in (7.9) follows from (7.10) and (7.5).

To establish the second point let there be two sequences, $x^\ddagger$, $y^\ddagger$, such that

$$\tilde{x}(z) = \sum_{r=0}^{\infty} x_r z^{-r} = \sum_{r=0}^{\infty} y_r z^{-r} = \tilde{y}(z) \tag{7.19}$$

where both series are absolutely convergent for $|z| > \beta > 0$. Let $q^\ddagger = x^\ddagger - y^\ddagger$, so that

$$\tilde{q}(z) = \sum_{r=0}^{\infty} q_r z^{-r} = \sum_{r=0}^{\infty} (x_r - y_r) z^{-r} = 0 \tag{7.20}$$

where the series is absolutely convergent for $|z| > \beta$. By Exercise 4.1 of Chapter 5 it follows that every coefficient $x_r - y_r$ in the last series is zero, and so $y^\ddagger = x^\ddagger$.

The non-homogeneous difference equation can be dealt with similarly. No new point arises, so the discussion is omitted.

Example 7.1 If a sequence $x^\ddagger$ is defined by

$$x_r = a^r, \quad r = 0, 1, 2, \ldots \tag{7.21}$$

then $\tilde{x}$ is defined by

$$\tilde{x}(z) = \sum_{r=0}^{\infty} a^r z^{-r} = (1 - az^{-1})^{-1} \tag{7.22}$$

absolutely convergent for $|z| > |a|$. (Compare Example 6.3.)

Example 7.2 Let the sequence $x^\ddagger$ be defined by the difference equation

$$x_{r+3} - 6x_{r+2} + 11x_{r+1} - 6x_r = 0, \quad r = 0, 1, 2, \ldots \tag{7.23}$$

with

$$x_2 = 1, \quad x_1 = x_0 = 0 \tag{7.24}$$

Then (7.11) gives

$$z^3\tilde{x}(z) - z - 6z^2\tilde{x}(z) + 11z\tilde{x}(z) - 6\tilde{x}(z) = 0 \tag{7.25}$$

$$\tilde{x}(z) = z/(z^3 - 6z^2 + 11z - 6) \tag{7.26}$$

$$= z^{-2}/(1 - 6z^{-1} + 11z^{-2} - 6z^{-3}) \tag{7.27}$$

$$= \frac{1}{2(1 - z^{-1})} - \frac{1}{1 - 2z^{-1}} + \frac{1}{2(1 - 3z^{-1})} \tag{7.28}$$

and by Example 7.1 the sequence $x^{\ddagger}$ is defined by

$$x_r = \tfrac{1}{2} - 2^r + \tfrac{1}{2}3^r, \quad r = 0, 1, 2, \ldots \tag{7.29}$$

7.1 Comment on notation

It will have been noticed that although a distinction in notation has been made between $x^{\dagger}$, x^*, and $x^{\ddagger}$, no corresponding distinction has been made between the z-transforms: all of $x^{\dagger}$, x^*, and $x^{\ddagger}$ have their z-transform written $\tilde{x}$. The justification for this is that $\tilde{x}$ depends only on a set of real numbers which suffice to define the parent function. For a given $\tilde{x}$ this set of numbers will be the same, whether they are derived from $x^{\dagger}$, x^*, or $x^{\ddagger}$. In all cases

$$\tilde{x}(z) = \sum_{r=0}^{\infty} x_r z^{-r} \tag{7.30}$$

provided that x_r is interpreted as the rth member of the sequence $x^{\ddagger}$; or the strength of the rth impulse of x^*; or the height of the rth incremental step in $x^{\dagger}$.

8 Transfer function matrix resumed

When an input $u \in D_R^1$ is applied to a linear time-invariant system with step response $H(t)$, the output y, subject to the conditions of Chapter 5, Section 5, is given by

$$y(t) = \int_0^t H(t - \tau)\, du(\tau) \tag{8.1}$$

If, in addition,

$$|u(t)| < Me^{\alpha t}, \qquad |H(t)| < M_1 e^{\alpha_1 t}, \quad t \geqslant 0 \tag{8.2}$$

u and H will have Laplace transforms $\bar{u}$, $\bar{H}$ respectively.

It will now be shown that y has a Laplace transform given by

$$\bar{y}(s) = s\bar{H}(s)\bar{u}(s) \tag{8.3}$$

provided that H satisfies the stronger condition that, for all t, its variation on $[0, t]$ is less than $Ne^{\beta t}$ for some fixed N and $\beta \geqslant 0$. This condition implies the less stringent condition in (8.2), as can be seen by expressing the function H, of bounded variation, as the difference of two monotonic functions. On the other hand, consider $H \in D_R^1$ defined by

$$H(t) = U(t) \operatorname{sign}\,[\sin \pi e^{t^2}] \tag{8.4}$$

where

$$\operatorname{sign} x = \begin{cases} 1, & x > 0 \\ 0, & x = 0 \\ -1, & x < 0 \end{cases} \tag{8.5}$$

The function defined by (8.4) satisfies $|H(t)| < 2$, but its variation on $[0, t]$, where $\exp t^2$ is a positive integer, is $2 \exp t^2$ and there do not exist N and β such that $2 \exp t^2 < N \exp \beta t$, all t.

The defining limit for $\bar{y}$ is

$$\bar{y}(s) = \lim_{a \to \infty} \int_0^a e^{-st} y(t)\, dt \tag{8.6}$$

where $y \in D_R^1$ by Chapter 5, Theorem 10.3, so that the integral exists. Also

$$\begin{aligned} |y(t)| &= \left| \int_0^t u(\tau)\, dH(t - \tau) \right| \\ &< Me^{\alpha t} Ne^{\beta t} \end{aligned} \tag{8.7}$$

from which it follows that the limit in (8.6) exists if $\operatorname{Re} s > \alpha + \beta$.

Integrating (8.1) by parts and inserting the result in (8.6),

$$\bar{y}(s) = -\lim_{a \to \infty} \int_0^a e^{-st} \left\{ \int_0^t u(\tau)\, dH(t - \tau) \right\} dt \tag{8.8}$$

In the inner integral $0 \leqslant t \leqslant a$, and because $H(t - \tau) = 0$, $\tau \geqslant t$, the value of the inner integral is unchanged if its upper limit is changed from t to a. So

$$\bar{y}(s) = -\lim_{a \to \infty} \int_0^a e^{-st} \left\{ \int_0^a u(\tau)\, dH(t - \tau) \right\} dt \tag{8.9}$$

The first stage of the proof is to show that

$$\int_0^a e^{-st} \left\{ \int_0^a u(\tau)\, dH(t - \tau) \right\} dt = -\int_0^a u(\tau) \left\{ \int_0^a e^{-st}\, dH(t - \tau) \right\} d\tau \tag{8.10}$$

where on the left-hand side the variable of integration in the inner integral is τ and on the right-hand side it is t. In the proof, results which were established for real variables in Chapter 5 will be freely used for complex variables. All that is involved in this is the separate application of the earlier results to real parts and to imaginary parts: products of complex variables do not occur.

Since $H \in D_R^1$, write

$$H(x) = \sum_{i=1}^{m} H_i U(x - q_i) + \sum_{i=1}^{l} (x - r_i)\eta_i U(x - r_i) + \phi(x) \tag{8.11}$$

where the $q_i \in [0, a]$ and $r_i \in [0, a]$ are such that ϕ has continuous first derivative on an open interval containing $[0, a]$. On substituting (8.11) into (8.10) and using linearity, three terms are obtained on each side: these will be shown to be equal in pairs.

First,

$$\int_0^a e^{-st}\left\{\int_0^a u(\tau)\, dU(t - \tau - q_i)\right\} dt = -\int_0^a e^{-st}u(t - q_i)\, dt \tag{8.12}$$

while

$$-\int_0^a u(\tau)\left\{\int_0^a e^{-st}\, dU(t - \tau - q_i)\right\} d\tau = -\int_0^{a-q_i} u(\tau)e^{-s(\tau+q_i)}\, d\tau \tag{8.13}$$

$$= -\int_0^a u(t - q_i)e^{-st}\, dt \tag{8.14}$$

Here the change in the upper limit of integration in (8.13) occurs because the inner integral is zero when $\tau + q_i \geqslant a$. The change of variable $t = \tau + q_i$ has been introduced in (8.14), and the lower limit of integration has been changed, making use of the fact that $u(t - q_i) = 0,\ t \leqslant q_i$. From (8.12) and (8.14)

$$\int_0^a e^{-st}\left\{\int_0^a u(\tau)\, d\left[\sum_{i=1}^m H_i U(t - \tau - q_i)\right]\right\} dt$$

$$= -\int_0^a u(\tau)\left\{\int_0^a e^{-st}\, d\left[\sum_{i=1}^m H_i U(t - \tau - q_i)\right]\right\} d\tau \tag{8.15}$$

Secondly,

$$\int_0^a e^{-st}\left\{\int_0^a u(\tau)\, d[(t - \tau - q_i)U(t - \tau - q_i)]\right\} dt$$

$$= -\int_0^a e^{-st}\left\{\int_0^{t-q_i} u(\tau)\, d\tau\right\} dt \tag{8.16}$$

$$= -\int_0^a e^{-st}f(t - q_i)\, dt \tag{8.17}$$

where

$$f(t) = \int_0^t u(\tau)\, d\tau \tag{8.18}$$

Note that u can be written

$$u(t) = \sum_{i=1}^n u_i U(t - p_i) + v(t) \tag{8.19}$$

where $v(t)$ is continuous, and on using Chapter 5, Theorem 10.1, it follows that $f \in D_R^1$ and is continuous and $D_- f = u$. Also

$$-\int_0^a u(\tau)\left\{\int_0^a e^{-st}\, d[(t - \tau - q_i)U(t - \tau - q_i)]\right\} d\tau$$

$$= -\int_0^{a-q_i} u(\tau)\left\{\int_{\tau+q_i}^a e^{-st}\, dt\right\} d\tau \tag{8.20}$$

because the inner integral on the left-hand side is zero if $\tau + q_i \geqslant a$. Now integrate (8.20) by parts, using equation (7.25) and Theorem 10.1 of Chapter 5, to obtain

$$-\left[f(\tau)\int_{\tau+q_i}^{a} e^{-st}\,dt\right]_0^{a-q_i} - \int_0^{a-q_i} f(\tau)e^{-s(\tau+q_i)}\,d\tau = -\int_0^a e^{-st} f(t-q_i)\,dt \tag{8.21}$$

From (8.17) and (8.21) it follows that

$$\int_0^a e^{-st}\left\{\int_0^a u(\tau)\,d\left[\sum_{i=1}^{l}(t-\tau-q_i)\eta_i U(t-\tau-q_i)\right]\right\}dt$$
$$= -\int_0^a u(\tau)\left\{\int_0^a e^{-st}\,d\left[\sum_{i=1}^{l}(t-\tau-q_i)\eta_i U(t-\tau-q_i)\right]\right\}d\tau \tag{8.22}$$

Thirdly, because ϕ has continuous first derivative,

$$\int_0^a e^{-st}\left\{\int_0^a u(\tau)\,d\phi(t-\tau)\right\}dt = -\int_0^a e^{-st}\left\{\int_0^a u(\tau)\,D\phi(t-\tau)\,d\tau\right\}dt \tag{8.23}$$
$$= -\int_0^a\left\{\int_0^a D\phi(t-\tau)\,df(\tau)\right\}dg(t) \tag{8.24}$$

where (8.23) uses Theorem 7.1, Chapter 5, and (8.24) uses the same theorem twice, with f given by (8.18) and g by

$$g(t) = \int_0^t e^{-s\tau}\,d\tau \tag{8.25}$$

In (8.24), $D\phi$ is continuous, and therefore uniformly continuous on $[-a, a]$; f has bounded variation on $[0, a]$ because it is in D_R^1; and by direct evaluation, the real and imaginary parts of g have bounded variation on $[0, a]$. The functions f, Re g, and Im g can each be expressed as the difference between two non-decreasing functions, giving eight separate contributions to the integral (8.24). It suffices to consider one of these, so we may assume without loss of generality that both f and g in (8.24) are real and non-decreasing.

Choose δ such that $|x_2 - x_1| < 2\delta$ implies $|D\phi(x_2) - D\phi(x_1)| < \epsilon$. Choose an integer $N > 1/\delta$ and divide the range of integration of τ and of t in (8.24) into N equal parts. The integral in (8.24), by equation (3.2) of Chapter 5, can be written as a sum of N^2 integrals of the form

$$I_{ij} = \int_{b_i}^{b_{i+1}}\left\{\int_{c_j}^{c_{j+1}} D\phi(t-\tau)\,df(\tau)\right\}dg(t) \tag{8.26}$$

Apply the mean-value theorem of Section 4.1, Chapter 5, twice to give

$$I_{ij} = \int_{b_i}^{b_{i+1}}\left\{D\phi(t-\tau_{ij})[f(c_{j+1}) - f(c_j)]\right\}dg(t)$$
$$= D\phi(t_{1_i} - \tau_{1j})[f(c_{j+1}) - f(c_j)][g(b_{i+1}) - g(b_i)] \tag{8.27}$$

where $b_i \leqslant t_{1i} \leqslant b_{i+1}$, $c_j \leqslant \tau_{1j} \leqslant c_{j+1}$.

In the same way, the integral

$$\int_0^a u(\tau)\left\{\int_0^a e^{-st}\, d\phi(t-\tau)\right\} d\tau = \int_0^a \left\{\int_0^a D\phi(t-\tau)\, dg(t)\right\} df(\tau) \tag{8.28}$$

can be expressed as a sum of N^2 integrals

$$\begin{aligned} K_{ij} &= \int_{c_j}^{c_{j+1}} \left\{\int_{b_i}^{b_{i+1}} D\phi(t-\tau)\, dg(t)\right\} df(\tau) \\ &= D\phi(t_{2i}-\tau_{2j})[f(c_{j+1})-f(c_j)][g(b_{i+1})-g(b_i)] \end{aligned} \tag{8.29}$$

where $b_i \leqslant t_{2i} \leqslant b_{i+1}$, $c_j \leqslant \tau_{2j} \leqslant c_{j+1}$. Because

$$|(t_{1i}-\tau_{1j})-(t_{2i}-\tau_{2j})| < 2\delta, \quad \text{all } i, j \tag{8.30}$$

it follows that

$$|D\phi(t_{1i}-\tau_{1j})-D\phi(t_{2i}-\tau_{2j})| < \epsilon, \quad \text{all } i, j \tag{8.31}$$

and because g and f have bounded variation on $[0, a]$,

$$\sum_{i=0}^{N-1} |g(b_{i+1})-g(b_i)| < M_2 \tag{8.32}$$

$$\sum_{i=0}^{N-1} |f(c_{j+1})-f(c_j)| < M_3 \tag{8.33}$$

Therefore equations (8.23) onwards give

$$\begin{aligned} \int_0^a e^{-st}\left\{\int_0^a u(\tau)\, d\phi(t-\tau)\right\} dt - \int_0^a u(\tau)\left\{\int_0^t e^{-st}\, d\phi(t-\tau)\right\} d\tau \\ = \sum_{i=0}^{N-1}\sum_{j=0}^{N-1} (I_{ij}-K_{ij}) \quad (8.34) \\ < \epsilon M_2 M_3 \quad (8.35) \end{aligned}$$

Because ϵ is arbitrary, the left-hand side of (8.34) must be zero.

It has now been established that (8.10) is true. Make a change of variable $t - \tau = w$ on the right-hand side to give

$$\begin{aligned} \int_0^a e^{-st} y(t)\, dt &= \int_0^a u(\tau)\left\{\int_0^{a-\tau} e^{-s(w+\tau)}\, dH(w)\right\} d\tau \\ &= \int_0^a u(\tau)e^{-s\tau}\left\{\left(\int_0^a - \int_{a-\tau}^a\right) e^{-sw}\, dH(w)\right\} d\tau \end{aligned} \tag{8.36}$$

Then for $\operatorname{Re} s > \gamma > \alpha + \beta$

$$\left|\int_{a-\tau}^a e^{-sw}\, dH(w)\right| < e^{-\gamma(a-\tau)} N[e^{\beta a} - e^{\beta(a-\tau)}] \tag{8.37}$$

and

$$\left|\int_0^a u(\tau)e^{-s\tau}\left\{\int_{a-\tau}^a e^{-sw}\,dH(w)\right\}d\tau\right|$$

$$< \int_0^a Me^{\alpha\tau}e^{-\gamma\tau}e^{-\gamma(a-\tau)}N[e^{\beta a} - e^{\beta(a-\tau)}]\,d\tau \tag{8.38}$$

$$= MN\left\{\frac{e^{-(\gamma-\alpha-\beta)a} - e^{-(\gamma-\beta)a}}{\alpha} - \frac{e^{-(\gamma-\alpha)a} - e^{-(\gamma-\beta)a}}{\alpha-\beta}\right\} \tag{8.39}$$

provided that $\alpha \neq 0$, $\alpha - \beta \neq 0$. If $\alpha = 0$ or $\alpha = \beta$, different formulae result but the conclusions drawn below remain true.

Finally, integrate by parts in (8.36) to give, because $H(0) = 0$,

$$\int_0^a e^{-sw}\,dH(w) = e^{-sa}H(a) + \int_0^a H(w)se^{-sw}\,dw \tag{8.40}$$

and for $\operatorname{Re} s > \gamma > \alpha + \beta$

$$|e^{-sa}H(a)| < e^{-\gamma a}Ne^{\beta a} = Ne^{-(\gamma-\beta)a} \tag{8.41}$$

Then use (8.39) and (8.41) in (8.36) and let $a \to \infty$. The result is

$$\int_0^\infty y(t)e^{-st}\,dt = \int_0^\infty u(t)e^{-st}\,dt\,.\,s\int_0^\infty H(t)e^{-st}\,dt \tag{8.42}$$

or

$$\bar{y}(s) = s\bar{H}(s)\bar{u}(s) \tag{8.43}$$

All the Laplace transforms exist for $\operatorname{Re} s > \alpha + \beta$. Collecting all the conditions, the result just proved can be stated as a theorem.

Theorem 8.1 Let a system have the following properties:

(a) It is linear.
(b) It is non-anticipative.
(c) If the input u satisfies $u(t) = 0$, $t \leqslant 0$, and $|u(t)| < M_1(T)$, $0 < t \leqslant T$, then the output y satisfies $|y(t)| < M_2(M_1, T)$, $0 < t \leqslant T$.
(d) If $H(t)$ is the response to $U(t)$, then $H(t - \tau)$ is the response to $U(t - \tau)$, $\tau > 0$.
(e) The response $H(t)$ to a unit step $U(t)$ is in D_R^1 and has variation on $[0, T]$ less than $Ne^{\beta T}$, for some fixed N and β and for all T.

Let the input u to the system be in D_R^1, let $u(t) = 0$, $t \leqslant 0$, and let $|u(t)| < Me^{\alpha t}$ for some fixed M and α, and for all $t > 0$. Then u, H, and y have Laplace transforms $\bar{u}$, $\bar{H}$, and $\bar{y}$ with exponential convergence respectively for $\operatorname{Re} s > \alpha$, $\operatorname{Re} s > \beta$, and $\operatorname{Re} s > \alpha + \beta$, and $\bar{y}$ is given by

$$\bar{y}(s) = s\bar{H}(s)\bar{u}(s) \tag{8.44}$$

Example 8.1 The step response of a system is

$$H(t) = e^{-(t-1)}U(t-1) \tag{8.45}$$

which has Laplace transform

$$\bar{H}(s) = e^{-s}/(s+1) \tag{8.46}$$

The input is

$$u(t) = U(t) \sin \omega t \tag{8.47}$$

having

$$\bar{u}(s) = \omega/(s^2 + \omega^2) \tag{8.48}$$

The output has Laplace transform

$$\bar{y}(s) = \frac{\omega s e^{-s}}{(s+1)(s^2+\omega^2)} = \frac{\omega e^{-s}}{1+\omega^2}\left\{\frac{-1}{s+1} + \frac{s+\omega^2}{s^2+\omega^2}\right\} \tag{8.49}$$

whence

$$y(t) = \frac{\omega}{1+\omega^2} U(t-1)\{-e^{-(t-1)} + \omega \sin \omega(t-1) + \cos \omega(t-1)\} \tag{8.50}$$

It will be noticed that (8.45) could not arise from a system of the type considered in Section 5.2.

EXERCISE 8.1 Obtain (8.50) from (8.1).

When a system has l inputs and m outputs, the above results generalize in a natural way: u, $\bar{u}$ are then l-vectors, y, $\bar{y}$ are m-vectors, and H, $\bar{H}$ are $m \times l$ matrices. The matrix

$$\bar{G}(s) \triangleq s\bar{H}(s) \tag{8.51}$$

is called the *transfer function matrix* of the system. Equation (8.44) becomes, with this notation,

$$\bar{y}(s) = \bar{G}(s)\bar{u}(s) \tag{8.52}$$

A particular case of (8.52) was given in (5.58) for a system having

$$H(t) = \{CA^{-1}[e^{At} - I_n]B + D\}U(t) \tag{8.53}$$

$$s\bar{H}(s) = s\int_0^\infty e^{-st}\{CA^{-1}[e^{At} - I_n]B + D\}\,dt \tag{8.54}$$

$$= C(sI_n - A)^{-1}B + D \tag{8.55}$$

8.1 Relation with the impulse response

In Section 9 of Chapter 5, the impulse response h was defined by

$$h(t) = DH(t) \tag{8.56}$$

the derivative being taken in the distribution sense, or the equivalent. By equation (3.12), this gives

$$\bar{h}(s) = s\bar{H}(s) \tag{8.57}$$

Consequently, if h is generalized to an $m \times l$ matrix, (8.51) shows that its Laplace transform $\bar{h}$ is simply $\bar{G}$, the transfer function matrix.

The impulse response h may contain delta functions: that is, it may in effect be unbounded on a finite interval. If so, however, the delta functions must be such that condition (e) of Theorem 8.1 is satisfied, otherwise $\bar{y}$ may not exist.

Example 8.2 Let h be given by

$$h(t) = D\{U(t) \operatorname{sign} [\sin \pi e^{t^2}]\} \tag{8.58}$$

where sign x is defined by (8.5). Let the input be

$$u(t) = U(t) \cos \pi e^{t^2} \tag{8.59}$$

Then

$$y(t) = \int_0^t u(\tau)h(t - \tau)\, d\tau = \delta(t) + \sum_{i=1}^{\infty} \{\delta(p_i - t) + \delta(t - p_i)\} \tag{8.60}$$

where the p_i have those values which make $\exp p_i^2$ a positive integer. But $\bar{y}$ does not exist in any strip $\operatorname{Re} s > \alpha$ because if $\operatorname{Re} s = \sigma$ and $q = \exp p_r^2$

$$\sum_{i=1}^{q} \exp(-\sigma p_i) > \exp p_r^2 . \exp(-p_r\sigma) = \exp p_r(p_r - \sigma) \tag{8.61}$$

so that the series does not converge however large σ may be.

8.2 Note on time-dependent systems

Subject to appropriate conditions, linear time-invariant systems have a transfer function matrix $\bar{G}(s) = s\bar{H}(s)$. The transfer function of two such systems in cascade is $\bar{G}_2\bar{G}_1$ if $\bar{G}_1$ operates first, or $\bar{G}_1\bar{G}_2$ if $\bar{G}_2$ operates first. In each case it is assumed that the matrices are conformable so that the appropriate product exists. In general $\bar{G}_2\bar{G}_1 \neq \bar{G}_1\bar{G}_2$, but if each of the systems has only one input and only one output, then the order in which the systems operate does not affect their combined transfer function.

These properties do not, in general, carry over to time-dependent linear systems. Such systems do not generally have transfer functions; nor can the order in which they operate be changed, even if each has only one input and one output, without changing their combined effect.

Example 8.3 If the input to a staircase sampler is $U(t)$, Section 6.1 shows that the input and output have Laplace transforms respectively

$$\bar{u}_1(s) = s^{-1}, \quad \bar{y}_1(s) = \bar{u}_1^{\dagger}(s) = s^{-1}(1 - e^{-s})^{-1} \tag{8.62}$$

If the input is $tU(t)$, these are replaced by

$$\bar{u}_2(s) = s^{-2}, \quad \bar{y}_2(s) = \bar{u}_2^{\dagger}(s) = s^{-1}e^{-s}(1 - e^{-s})^{-2} \tag{8.63}$$

But the two ratios $\bar{y}_1(s)/\bar{u}_1(s)$ and $\bar{y}_2(s)/\bar{u}_2(s)$ are not equal: a transfer function cannot be defined for the sampler, which is a time-dependent system.

Example 8.4 The step response of a sample-and-hold circuit followed by an integrator is $tU(t)$. The step response of an integrator followed by a sample-and-hold circuit is $\sum_{r=1}^{\infty} U(t - r)$. The order in which the two systems operate affects their combined result.

There is an apparent exception to this general rule in the z-transfer function. For example, let the staircase-sampled input be $u^{\dagger}$, with Laplace transform $\bar{u}^{\dagger}(s)$. Let this be applied to a linear time-invariant system followed by a staircase sampler (which is time-dependent), and let the Laplace transform of the sampled output $y^{\dagger}$ be $\bar{y}^{\dagger}$. Then equation (6.30) shows that the ratio $\bar{y}^{\dagger}(s)/\bar{u}^{\dagger}(s)$ is the same for all $u^{\dagger}$ satisfying the appropriate conditions.

The reason for this apparent exception is that the class of input functions $u^{\dagger}$ is restricted to those which are constant except at $t = 0, 1, 2, \ldots$. The only time shifts permitted are therefore shifts by an integral number of sampling periods. Under such shifts it is easy to confirm that the linear time-invariant system followed by its staircase sampler obeys the criterion for time-invariance. That is to say, if the response to $U(t)$ is $\mathscr{H}(t)$, then the response to $U(t - n)$, n a positive integer, is $\mathscr{H}(t - n)$.

EXERCISE 8.2 Prove the statement in the preceding sentence.

Example 8.5 The response $\mathscr{H}(t)$ to an input $U(t)$ of an integrator followed by a staircase sampler is $\mathscr{H}(t) = \sum_{r=0}^{\infty} rU(t - r)$. The response to an input $U(t - n)$ is $\sum_{r=n}^{\infty} (r - n)U(t - r) = \sum_{r=0}^{\infty} rU(t - r - n) = \mathscr{H}(t - n)$, where n is a positive integer.

Subject to the same restriction on the class of inputs, blocks consisting each of a linear time-invariant system followed by a staircase sampler can be connected in cascade in any order without changing their total effect. The same is true when each block consists of a linear time-invariant system followed by a sample-and-hold circuit, or by an impulse sampler. Blocks with staircase samplers can be interchanged with blocks having sample-and-hold circuits, but it is better not to interchange these types with blocks having impulse samplers: if this were done a mixed type of formula would have to be used instead of (6.35) or (6.56).

Example 8.6 Two blocks B_1, B_2 consist respectively of an integrator followed by a sample-and-hold circuit, and of a linear time-invariant system having step response e^{-t} and followed by a sample-and-hold circuit. Here $\bar{H}_1(s) = s^{-2}$, $\bar{H}_2(s) = s^{-1}(s+1)^{-1}$ and on using partial fractions and the table in Section 6.1, $\tilde{H}_1(z) = z^{-1}(1-z^{-1})^{-2}$, $\tilde{H}_2(z) = (1-z^{-1})^{-1} - (1-e^{-1}z^{-1})^{-1}$. The cascade connection of B_1 followed by B_2, and the cascade connection of B_2 followed by B_1, both give the same output when operating on a staircase-sampled input. If the z-transform of the input is $\tilde{u}(z)$, the z-transform of the output in either case is

$$\tilde{y}(z) = (1-z^{-1})^2\{z^{-1}(1-z^{-1})^{-2}[(1-z^{-1})^{-1} - (1-e^{-1}z^{-1})^{-1}]\}\tilde{u}(z) \tag{8.64}$$

Here the first factor $(1-z^{-1})^2$ arises from the use of two sample-and-hold circuits instead of two staircase samplers: compare (6.41) and (6.42).

Example 8.7 The cascade connection of the two linear time-invariant systems in Example 8.6 gives

$$\bar{H}_1(s)\bar{H}_2(s) = s^{-3}(s+1)^{-1} = s^{-1} - s^{-2} + s^{-3} - (s+1)^{-1} \tag{8.65}$$

If the two linear time-invariant systems operate on an input $u^{\dagger}$ and are followed by a sample-and-hold circuit, the output $y^{\dagger}$ has z-transform

$$\tilde{y}(z) = (1-z^{-1})\{(1-z^{-1})^{-1} - z^{-1}(1-z^{-1})^{-2} + \tfrac{1}{2}z^{-1}(1+z^{-1})(1-z^{-1})^{-3} - (1-e^{-1}z^{-1})^{-1}\}\tilde{u}(z) \tag{8.66}$$

Notice that this is not the same as (8.64): many errors in the use of z-transforms come from failing to remember this distinction.

9 Inverse Laplace transform

So far, it has been assumed that if a Laplace transform $\bar{x}$ has been obtained, the corresponding x will be found by means of a table. This is the most usual practice with dynamical systems. For completeness, however, it will now be shown how x can be obtained from $\bar{x}$ by an integral formula closely related to the defining integral for $\bar{x}$ in terms of x. This procedure becomes important chiefly when the Laplace transform is used for solving partial differential equations, which are outside the scope of this book; consequently the result will be stated without proof.

Theorem 9.1 Let $f \in D_R^1$ satisfy $f(t) = 0$, $t \leqslant 0$ and $|f(t)| < Me^{\alpha t}$, $t > 0$. Let the Laplace transform of f, valid for $\operatorname{Re} s > \alpha$, be $\bar{f}$. Then if $\alpha_1 > \alpha$, $f(t) = \phi(t-)$, where

$$\phi(t) = \frac{1}{2\pi i}\int_{\alpha_1 - i\infty}^{\alpha_1 + i\infty} e^{st}\bar{f}(s)\,ds \tag{9.1}$$

and the integral is to be interpreted as the Cauchy principal value. More explicitly

$$f(t) = \lim_{\epsilon\to 0+}\left\{\lim_{\gamma\to+\infty}\frac{1}{2\pi i}\int_{\alpha_1 - i\gamma}^{\alpha_1 + i\gamma} e^{s(t-\epsilon)}\bar{f}(s)\,ds\right\} \tag{9.2}$$

Example 9.1 If f is the unit step $U(t)$, then $\bar{f}$ is s^{-1}, valid for $\operatorname{Re} s > 0$. Equation (9.1) gives

$$\phi(t) = \frac{1}{2\pi i}\int_{\alpha_1 - i\infty}^{\alpha_1 + i\infty} s^{-1}e^{st}\,dt, \quad \alpha_1 > 0 \tag{9.3}$$

which, by Chapter 7, Exercise 5.2, leads to

$$\phi(t) = \begin{cases} 0, & t < 0 \\ \frac{1}{2}, & t = 0 \\ 1, & t > 0 \end{cases} \tag{9.4}$$

Then $\phi(t-) = \phi(t)$ except at $t = 0$, where $\phi(t-) = 0$. Consequently (9.1) gives the result

$$f(t) = \phi(t-) = U(t) \tag{9.5}$$

which is correct.

Many accounts of the subject define the unit step by (9.4), and make similar changes at every discontinuity of a function. This avoids the introduction of the intermediate function ϕ, and is the most convenient course when the Laplace transform alone is under consideration. For our applications, however, this procedure agrees less well with the phenomena which are to be represented, and it is also less convenient when dealing with Stieltjes convolution.

Proofs of Theorem 9.1 will be found in many textbooks; for example, Apostol (1957) and Titchmarsh (1937). The main point of difficulty is to justify the change in order of two improper integrations:

$$\phi(t) = \frac{1}{2\pi i}\int_{\alpha_1 - i\infty}^{\alpha_1 + i\infty}\left\{\int_0^\infty e^{s(t-\tau)}f(\tau)\,d\tau\right\}ds$$
$$= \frac{1}{2\pi i}\int_0^\infty\left\{\int_{\alpha_1 - i\infty}^{\alpha_1 + i\infty} e^{s(t-\tau)}f(\tau)\,ds\right\}d\tau \tag{9.6}$$

The methods used are similar to those which have been demonstrated previously (compare Section 8, for example) but with extra difficulties arising from the limiting processes which are implied by the improper nature of the integrals. The proof would therefore be excessively tedious without a preliminary development of theorems which we have avoided.

Further reading

Further mathematical development of the subject for reference purposes will be found in Apostol (1957), Titchmarsh (1937), and Widder (1946). As a technique for the solution of problems, with applications in many fields of physics and engineering, the Laplace transform is considered by Carslaw and Jaeger (1949), Van der Pol and Bremmer (1964), and Doetsch (1961). Tables of Laplace transforms are given by Doetsch (1961) and by Van der Pol and Bremmer (1964); a particularly extensive list of Fourier transforms will be found in Campbell and Foster (1931).

Applications of the z-transform to control systems will be found in Ragazzini and Franklin (1958) and in Jury (1958). These (particularly the latter) contain tables of z-transforms and modified z-transforms. The notation used for modified z-transforms should be checked before using the tables (see Section 6.3).

References

Apostol, T. M., 1957, *Mathematical analysis*, Addison-Wesley, Reading, Mass.

Bromwich, T. J. I'A., 1942, *An introduction to the theory of infinite series*, Macmillan, London.

Campbell, G. A. and R. M. Foster, 1931, *Fourier integrals for practical applications*, Bell System Technical Monograph B-584, Van Nostrand, Princeton, N.J.

Carslaw, H. S. and J. C. Jaeger, 1949, *Operational methods in applied mathematics*, Oxford University Press, London.

Doetsch, G., 1961, *Guide to the applications of Laplace transforms*, Van Nostrand, Princeton, N.J.

Jeffreys, H. and B. S. Jeffreys, 1950, *Methods of mathematical physics*, Cambridge University Press, London.

Jury, E. I., 1958, *Sampled-data control systems*, John Wiley, New York.

Lamb, H., 1938, *Infinitesimal calculus*, Cambridge University Press, London.

Ragazzini, J. R. and G. F. Franklin, 1958, *Sampled-data control systems*, McGraw-Hill, New York.

Titchmarsh, E. C., 1937, *Introduction to the theory of Fourier integrals*, Oxford University Press, London.

Van der Pol, B. and H. Bremmer, 1964, *Operational calculus*, Cambridge University Press, London.

Widder, D. V., 1946, *The Laplace transform*, Princeton University Press, Princeton, N.J.

Zadeh, L. A. and C. A. Desoer, 1963, *Linear system theory*, McGraw-Hill, New York.

Problems

1 A system has a single input u and a single output y related by

$$a_0 y + a_1\, Dy + \cdots + a_n\, D^n y = b_0 u + b_1\, Du + \cdots + b_{n-1}\, D^{n-1} u$$

and the polynomials $a_0 + a_1 s + \cdots + a_n s^n$ and $b_0 + b_1 s + \cdots + b_{n-1} s^{n-1}$ have no root in common. Let $u = 0$ and let the initial conditions be $y(0) = c_0$, $Dy(0) = c_1, \ldots, D^{n-1}y(0) = c_{n-1}$, where the c_i are real constants. Show that the solutions y of the differential equation for all possible c_i span an n-dimensional function space. [There is a one-to-one relation between solutions and initial conditions, by Chapter 6, Section 2, and this preserves sums and scalar multiples: hence proceed as in Chapter 3, Example 1.7.]

2 Show that the functions $s^r/(a_0 + a_1 s + \cdots + a_n s^n)$, $r = 0, 1, \ldots, n - 1$, span an n-dimensional function space. [The Laplace transformation sets up a one-to-one relation between these functions and the solutions considered in Problem 1, and this relation preserves sums and scalar multiples.]

3 Let the transfer function of the system in Problem 1 be $\bar{g}(s) = \bar{y}(s)/\bar{u}(s)$. Write $s^r\bar{g}(s) = \phi_r(s) + \psi_r(s)$, $r = 0, 1, 2, \ldots, n - 1$ where the ψ_r are polynomials and $\phi_r(\infty) = 0$. Show that the functions ϕ_r, $r = 0, 1, 2, \ldots, n - 1$, define an n-dimensional function space. [Show that if the functions ϕ_r were linearly dependent, the polynomials $a_0 + a_1 s + \cdots + a_n s^n$ and $b_0 + b_1 s + \cdots + b_{n-1} s^{n-1}$ would have a common factor. Then express the ϕ_r in terms of the functions in Problem 2.]

4 A system satisfies the differential equations

$$\dot{x} = Ax + bu$$
$$y = c^T x$$

where

$$A = \begin{pmatrix} -2 & 2 \\ 1 & -3 \end{pmatrix}, \qquad b = \begin{pmatrix} 0 \\ 1 \end{pmatrix}, \qquad c = \begin{pmatrix} 1 \\ 0 \end{pmatrix}$$

Find the transfer function $\bar{y}(s)/\bar{u}(s)$ and the response $y(t)$ (from zero initial conditions) to an input $u(t) = U(t) \cos 100\pi t$.

5 If in Problem 4, b is replaced by $\begin{pmatrix} 2 \\ 1 \end{pmatrix}$, show that the resulting transfer function can arise from a simpler system.

6 Without using the z-transform, find the response of the system in Problem 4 to an input

$$u(t) = \sum_{r=0}^{\infty} (-1)^r \{U[t - 0{\cdot}01r] - U[t - 0{\cdot}01(r + 1)]\}$$

7 The system in Problem 4 is preceded and followed by a sample-and-hold circuit. By z-transform methods find the response (from zero initial conditions) to an input $U(t) \cos 100\pi t$ when (a) the samplers both operate at the instants $t = 0, 0{\cdot}01, 0{\cdot}02, \ldots$, and (b) when the first sampler operates at the instants $t = 0, 0{\cdot}01, 0{\cdot}02, \ldots$, and the second sampler at the instants $t = 0{\cdot}005, 0{\cdot}015, 0{\cdot}025, \ldots$. Compare with the result of Problem 6.

8 Verify that the transfer function of the system in Problem 4 is given by the formula

$$g(s) = \begin{vmatrix} sI - A & b \\ -c^T & 0 \end{vmatrix} \div |sI - A|$$

9 Solve the difference equation

$$x_{r+2} + 5x_{r+1} + 4x_r = 0; \quad x_1 = 1, x_0 = -1$$

by using the z-transform and also by direct evaluation. Show that the solution can arise from a simpler difference equation.

10 A discrete-time system obeys the equations

$$x_{r+1} = Ax_r + bu_r$$
$$y_r = c^T x_r$$

where

$$A = \begin{pmatrix} -2 & 2 \\ 1 & -3 \end{pmatrix}, \quad b = \begin{pmatrix} 0 \\ 1 \end{pmatrix}, \quad c = \begin{pmatrix} 1 \\ 0 \end{pmatrix}$$

Find the response y_r, $r = 0, 1, 2, \ldots$, to an input $u_r = (-1)^r$, $r = 0, 1, 2, \ldots$, when the initial condition is $x_0 = \begin{pmatrix} 0 \\ 0 \end{pmatrix}$.

Author index

Subject index

Section numbers are shown in parentheses